APPLICATIONS OF THE NEWER TECHNIQUES OF ANALYSIS

PROGRESS IN
ANALYTICAL CHEMISTRY
Based upon the Eastern Analytical Symposia

Series Editors:
Ivor L. Simmons
M&T Chemicals, Inc., Rahway, New Jersey

and Galen W. Ewing
Seton Hall University, South Orange, New Jersey

PROGRESS IN ANALYTICAL CHEMISTRY
VOLUME 6

APPLICATIONS OF THE NEWER TECHNIQUES OF ANALYSIS

Edited by
Ivor L. Simmons
M&T Chemicals, Inc.
Rahway, New Jersey

and

Galen W. Ewing
Seton Hall University
South Orange, New Jersey

PLENUM PRESS • NEW YORK-LONDON

Library of Congress Cataloging in Publication Data

Eastern Analytical Symposium, Atlantic City, 1972.
  Applications of the newer techniques of analysis.

  (Progress in analytical chemistry, v. 6)
  Includes bibliographical references.
  1. Spectrum analysis—Congresses. I. Simmons, Ivor L., ed. II. Ewing,
Galen Wood, 1914–   ed. III. Title.
QD95.E28   1972   543'.085   73-14877
ISBN 978-1-4684-3320-3   ISBN 978-1-4684-3318-0 (eBook)
DOI 10.1007/978-1-4684-3318-0

PREFACE

This volume presents a group of selected papers given at the November, 1972, meeting of the Eastern Analytical Symposium. As has always been the intent of the series, "Progress in Analytical Chemistry," the papers are written by authorities who are also active workers in their fields. Included are applications of Raman Spectroscopy, X-Ray Diffraction, Emission Spectroscopy, Nuclear Magnetic Resonance, Liquid Chromatography, Thin-Layer Chromatography, Pyrolysis Gas Chromatography, Mass Spectrometry, Gas Chromatography and the powerful marriage of the last two.

In modern analysis there is a constant search for applications of existing instrumentation in new ways to produce results hitherto unobtainable. In every case both the versatility of the researcher and the flexibility of the method used are amply illustrated.

Thanks are due to the authors of the papers for their efforts in coping with the many problems of producing a manuscript in finished form enabling rapid publication.

Ivor L. Simmons

Galen W. Ewing

CONTENTS

RAMAN SPECTROSCOPY FOR THE STUDY OF NEMATIC LIQUID CRYSTALS

Bernard J. Bulkin
Hunter College, College of the City University
of New York, 695 Park Avenue, New York,
New York 10021

Liquid crystals, primarily as a consequence of their potential electro-optic applications, have attracted considerable scientific interest in the past decade. This interest has extended to a large number of theoretical studies of the structure and properties of liquid crystals, as well as to a variety of spectroscopic techniques which nicely complement the theoretical work.

A rather late entry into the spectroscopic studies has been vibrational spectroscopy. While it is true that Maier and Englert did some vibrational spectroscopy several years ago (1), and a few other early works were reported (2), most infrared and Raman work has been published since 1969. In this paper we will review the published Raman spectroscopic studies, with emphasis on the work from our laboratory, but with comments and discussion of the work of others as well. It is inappropriate to enter into any detailed discussion of the structure and properties of liquid crystals herein, except as applied to the particular experiments which are described. For such discussion, the reader is referred to any of a number of standard texts and reviews (3).

Liquid crystals are often divided into a number of subcategories. In this paper we confine ourselves to a discussion of thermotropic, nematic materials. Figure 1 shows a schematic representation of the nematic phase. The molecules have a very anisotropic (cigar-like) shape, and are

1

Figure 1.   Schematic representation of the nematic phase.

shown to be aligned with their long axes parallel. In a nematic phase, the correlation length for the long axes may
extend over several hundred angstroms. This may be increased
considerably by the application of electric or magnetic
fields, or through the use of any of a number of rubbing or
chemical alignment techniques.

One of the most studied series of nematic liquid crystals are the alkoxy azoxy benzene homologs, whose structure
is shown below. These molecules show the characteristic anisotropic shape, as well as the dipolar groups along the
molecule which provide strong intermolecular forces along
certain directions. The polarizable end groups are also
important.

It is interesting to compare the nematic structure
shown in Figure 1 with the crystal packing diagram shown in
Figure 2 for the methyl homolog of the above compounds (this)
compound is known as para-azoxy anisole or PAA). As can be
seen from Fig. 2, the molecules already have their long axes
parallel, and the aligned dipoles clearly provide considerable stabilization energy for the crystals. If one can induce some disorder in the crystal along the long axes, a
nematic phase will be present.

Bulkin, Grunbaum and Santoro (4) discussed the changes
which took place in the infrared spectrum at the crystal-
nematic (c-n) phase transition in the context of such an

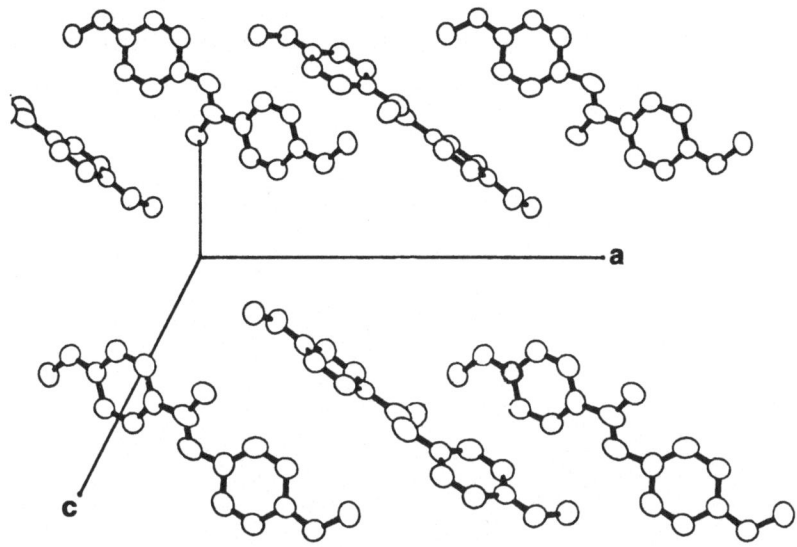

Figure 2.   Para-azoxyanisole packing diagram (after
Krigbaum et al., Acta Cryst. B26, 97 (1970).)

order-disorder phenomenon. Certain infrared bands, attri-
buted to sum and difference modes between lattice vibrations
and internal vibrations, disappeared at the phase transition.
Further, it was found that this disappearance was not sudden
but occured gradually over a period of several degrees in
the crystalline phase. Thus, viewed from the crystalline
side, the c-n transition appeared to be other than a simple
first order phase transition.

If such changes taking place in the infrared spectrum
were indeed due to changes in the lattice vibrations, then
more direct confirmation of their existence should come
from a study of these vibrations. Such was the aim of the
study by Bulkin and Prochaska (5a). Bulkin and Grunbaum (5b)
had previously reported the lattice vibration region spectrum
of PAA in the crystalline phase, and Amer, Shen and Rosen had
also reported the spectrum in the crystalline, nematic and
isotropic phases (9). Unfortunately, the latter spectrum
suffered from a particularly high background in the low fre-

quency region and some bands were not observed. A later
paper of Amer and Shen (10) has improved these conditions

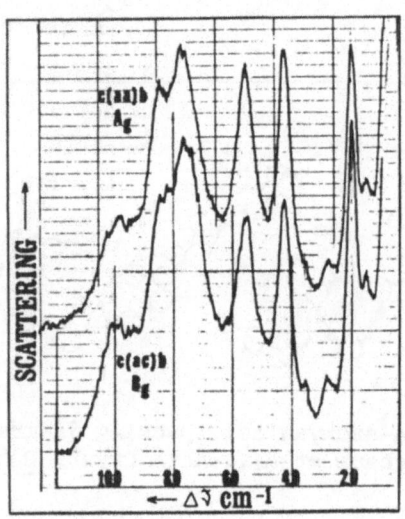

Figure 3. Raman spectra of the lattice vibration region
          of a PAA single crystal at -90°. The spectra
          show two different positions of the polarization
          analyzer, isolating polarizability components
          of A_g and B_g symmetry. (From ref. 5a).

     The results of ref. 5a are shown in Figure 3. PAA is
a monoclinic crystal containing 4 molecules per unit cell
in a C_{2h}^5 factor group. Twelve Raman active lattice vibrations are predicted, and these are seen at low temperatures
(-90) in the single crystal. In the polycrystalline sample,
at room temperature, composite bands are seen with several
modes overlapped. This spectrum is shown in Figure 4.

     When the temperature of the polycrystalline sample is
raised to 90°, the relative intensities of all bands change,
and the frequencies shift slightly, all to lower frequency
by about 1-2 cm^{-1}.

     As one comes closer to the c-n phase transition, these

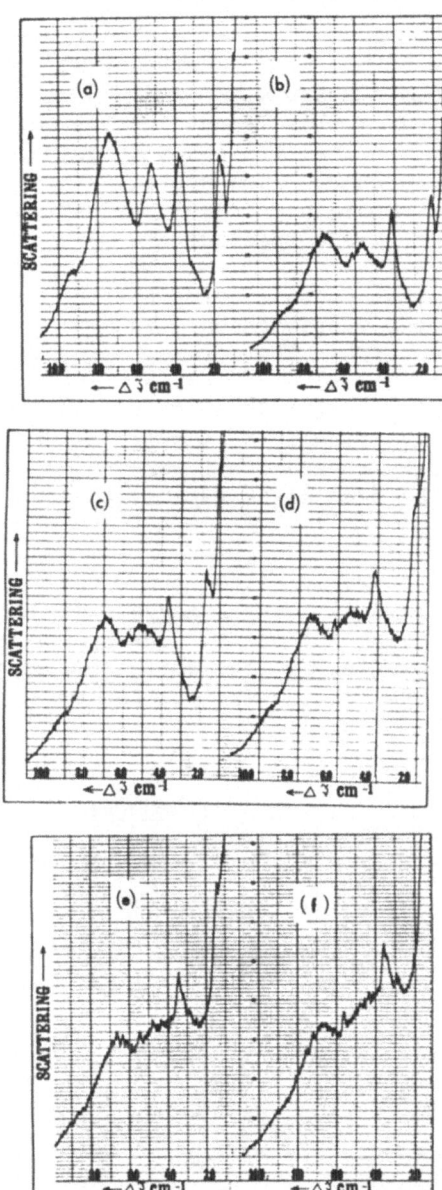

Figure 4.   Temperature dependence of the Raman spectrum of
polycrystalline PAA as the crystal-nematic phase
transition is approached.   (a) 25°; (b) 90°; (c)
116°; (d) -(f)   0.3° temperature increments a-
bove (c) in the crystalline phase.

changes become more dramatic. At 116° (Fig. 4c), about two
degrees below the actual c-n transition, we see that the
band which had been a sharp maximum at 55 cm^{-1} is now extremely broad. Clearly this is not due to the temperature
broadening of the band, as the lower frequency mode at 37
cm^{-1} is still quite sharp. The subsequent spectra, Fig. 4d-f,
show that as the temperature is raised just up to the transition, the changes in the spectrum are more pronounced.

It is clear from these spectra that as the temperature
increases, a lattice vibration band is moving very rapidly
to lower frquency shift. The movement of this intense band
first manifests itself as a change in relative intensities,
then as a broadening of the 55 cm^{-1} band as it begins to
overlap it strongly, finally as a shift of the maximum in
scattering intensity toward the lower frequency region of
the spectrum as the band moves rapidly towards zero frequency.

As the temperature is raised above the c-n transition,
the width of the Rayleigh wing changes suddenly, as does
the overall scattering intensity. The spectrum in the nematic phase is shown in Fig. 5a. While similar to that reported by other workers (9), in that two broad maxima are

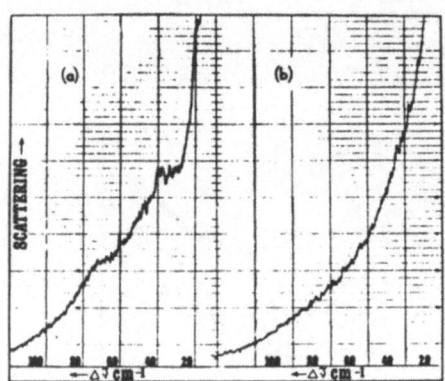

Figure 5.   (a) Raman spectrum of nematic PAA, 120°.   (b)
            Raman spectrum of isotropic PAA, 140°.

observed in this region, we find that a relatively sharp feature also exists near 37 cm^{-1}. Further, our spectra indicate that the broad bands are centered near 70 and 45 cm^{-1}, rather than at 40 and 52 cm^{-1} as reported in (1).

In the isotropic liquid, Fig. 5b, the bands seen in the nematic phase have disappeared, or are hidden under the Rayleigh scattering. This phase transition has been discussed previously (9,10) and will not be treated further here. We believe that a broad band due to a pseudo-lattice mode may be present in the isotropic liquid. However, our experimental conditions are not yet adequate for its observation.

Although the intensity of Raman bands is a complicated function of temperature and frequency (complicated because of instrumental factors) it is clear that the intensity changes found in this frequency range at temperatures below the c-n transition are other than what one would expect.

Two possible explanations should be considered for the change in relative intensity observed in Fig. 4. First, such changes are similar to changes in intensity of infrared bands at comparable temperatures (4). The infrared bands occurred well into the mid-infrared region, however, and were assigned to combination modes between lattice vibrations and internal vibrations. One expects that if a direct decoupling of molecules from certain lattice forces takes place well below the observed transition temperature, then a change in the relative intensity of the lattice vibrations will occur. Such a result is consistent with the qualitative model of the c-n transition, namely, that it is a phase transition in which some of the lattice forces are maintained while others are lost.

While we believe that the explanation for the relative intensity changes described above is operative, there is evidence that these changes are also caused by a shift in frequency of at least one of the lattice vibrations. As the temperature of a crystal is changed, all lattice vibrations tend to change frequency, generally to lower frequency shifts as temperature is increased. In certain ferroelectrics, however, it has been found that particular lattice vibrations move to lower Raman frequency shifts very rapidly as a solid-solid phase transition is approached. Such lattice frequencies are known as 'soft modes'. In a crystal with twelve

strongly overlapped lattice vibration bands, a rapid shift
in frequency of one or more of the bands will appear as a
relative intensity change. Thus, to cause the changes observed between Fig 4a and 4b, the existence of a soft mode
in the broad, highest frequency band of lattice vibrations
is postulated.

The possibility of a soft mode gains further support
from the spectra as the c-n transition is approached. The
spectra of Figs. 4c-f are best understood by assuming that
almost all of the lattice vibrations are staying very nearly
constant in frequency and bandwidth, while one mode, originally located near 68 cm^{-1}, shifts rapidly towards $\Delta\nu = 0$.
To our knowledge, this is the first observation of a soft
mode in a phase transition involving an organic crystal.

Kobayashi (18), considering the infrared study of
Bulkin et al (4) developed a statistical mechanical treatment
of the crystal - nematic phase transition. This work, which
appeared at about the same time as the Raman spectroscopic
studies just presented, predicted the existence of a soft
mode at this transition. Ford (19) also has considered this
problem from a theoretical point of view. He calculated the
order parameter, S, as a function of temperature for PAA
starting at about 30° below the c-n phase transition, and
extending about 20° above the transition to the isotropic
phase. His results showed that the order parameter begins
to decrease well below the c-n transition, changes slowly
in the nematic phase, then drops off to a zero gradually a-
bove the n-l transition.

The nature of the c-n transition was also remarked on
by Billard, Delhaye and Merlin (11) in their study of the
liquid crystal MBBA, whose structure is shown below:

Their spectra showed an abnormal broadening of the lattice
vibration bands in the Raman spectrum of the crystalline

material. This spectrum was also reported by Borer, Mitra and Brown (14).

In the nematic phase (Fig. 5a), a spectrum which is in accord with expectations for an ordered fluid is found. For a crystal with strong intermolecular interactions, the nearest neighbors have the overwhelming influence on the lattice vibration frequencies. These also determine the low frequency, pseudo-lattice modes of the liquid crystal. However, because of the short time scale, and some variation in environment across the sample, these higher temperature modes are considerably broader than those in the crystal.

It is also clear that certain of the modes which were present in the crystal should be absent in the nematic phase. In particular, it is known that the molecules have considerable motional freedom along the long axes, whereas motion in directions normal to these axes is severely restricted. Little is known about rotational freedom, but we postulate that this is also very nearly unhindered about the long axes of the molecules in the nematic phase.

In Fig. 5a there are two broad bands, both of which are centered very near the position of maxima in the crystal (70 and 40 cm^{-1}). These maxima are undoubtedly due to the translational and rotational modes whose frequencies are primarily determined by the strong interaction between the azoxy and methoxy groups on neighboring molecules. It is this dipolar interaction which is largely responsible, along with the anisotropic molecular shape, for the existence of the nematic phase.

Recently, Bulkin and Lok (6) have reported the far infrared spectrum of crystal, nematic, isotropic and solution phase PAA and nematic, isotropic and solution phase MBBA. The results were different from those seen in the Raman spectrum in the following ways. First, the most intense lattice vibrations found in the far infrared spectrum are at higher frequencies than those seen in the Raman spectrum. Thus strong lattice modes near 125 cm^{-1} are found. Second, in the nematic phase a broad band, very intense, with maximum absorption near 100 cm^{-1} is found for both cases. In the case of PAA, a weaker band is also seen near 55 cm^{-1}. Again, this intermolecular mode is a different one from that observed in the Raman spectra. Finally, the intermolecular modes weaken but do not disappear in the isotropic liquid,

although they do vanish in dilute CCl_4 solution. These results indicate that the modes observed in the far infrared
reflect the nearest neighbor, short range interactions,
while those in the Raman arise from the longer range order.
The far infrared bands of the isotropic phases of the nema-
togenic materials are similar to those observed for many
polar liquids in this same spectral region. More information
on their assignment is found in ref. 6.

Schnur et al. (7,8) have made a careful study of the
internal modes of the alkoxyazoxy benzenes. Some of these
had been **reported** previously (10,15). The most important
observation of these studies was a change in the integrated
intensity of the so-called accordion mode of the alkyl side
chain as one went through the various phase transitions.
The intensity of this band decreased rapidly in the solid
phase, was approximately constant in the nematic phase until
just before the n-l transition, then decreased further in
the isotropic liquid. Indeed, it is interesting to note the
close resemblance between Schnur's curve (ref. 7, Fig. 2)
and the curve in Ford's paper discussed above.

Priestly and Pershan (16) prepared highly aligned samples of a nematic phase. This was an interesting material
in that it had a -CN group along the long axis of the molecule (actually at an angle of approx. 10°). If one measures
the depolarization ratio of this band for different
alignments of the nematic phase, the order parameter, S, can
be calculated. To do this, 180° scattering has to be used
rather than the more customary 90° orientation usually used
for Raman measurements. The authors have shown that because
the Raman effect has its origin in a second rank tensor, the
method yields $\langle \cos^4 \Theta \rangle$ as well as $\langle \cos^2 \Theta \rangle$ usually obtained
in measurements of S. It is not clear what the utility of
knowing $\langle \cos^4 \Theta \rangle$ is, however.

A problem with such measurements, discussed further
below, is that the sample must be highly aligned, virtually
a nematic singly crystal, to make these measurements. Otherwise, multiple scattering and refractive index discontinui-
ties will depolarize the incoming and scattered light. Such
was the difficulty in a recent publication of Wang (17). To
take this effect into account, Priestly and Pershan made
measurements as a function of thickness and extrapolated to
a thin film value. The problem with this method of correction
is that wall effects might influence the degree of alignment

in thin films. Despite all the limitations discussed above, these experiments are extremely interesting in providing the first measurements of S from vibrational spectroscopy.

REFERENCES

1. W. Maier and G. Englert, Z. Elektrochem., 62, 1020 (1958) and references therein

2. A.S. L'vova, U.A. Chikov, U.A. Zulov, and M.M. Sushchinskii, Opt. Spektr., 23, 168 (1967)

3. G. Gray, Molecular Structure and the Properties of Liquid Crystals, Academic Press (New York), 1962.

4. B.J. Bulkin, D. Grunbaum and A.V. Santoro, J. Chem. Phys., 51, 1602 (1969).

5. a. B.J. Bulkin and F.T. Prochaska, J. Chem. Phys. 54, 635 (1971).

   b. B.J. Bulkin, and D. Grunbaum, in Liquid Crystals and Ordered Fluids, edited by R. Porter and J. Johnson (Plenum, New York, 1970), p. 303.

6. B.J. Bulkin and W.B. Lok, J. Phys. Chem., 77, 326 (1973).

7. J.M. Schnur, Phys. Nev. Lett., 29, 1141 (1972).

8. J.M. Schnur, M. Hass, W.L. Adair, 41A, 326 (1972).

9. N.M. Amer, Y.R. Shen, and H. Rosen, Phys. Rev. Lett., 27, 718 (1970).

10. N.M. Amer, and Y.R. Shen, J. Chem. Phys., 56, 2654 (1972).

11. J. Billard, M. Delhaye, J-C. Merlin, and G. Vergoten, Compt. Rend., 273B, 1105 (1971).

12. B.J. Bulkin and K. Krishnan, J. Amer. Chem. Soc., 93, 5998 (1971).

13. B.J. Bulkin, J. Lephardt, and K. Krishnan, Mol. Cryst. Liq. Cryst., in press.

14. W.J. Borer, S.S. Mitra, and C.W. Brown, Phys. Rev. Lett.,
    27, 379 (1971).

15. A.S. Zhadnova, L.F. Morozova, G.V. Peregudov, and M.M.
    Sushchinskii, Opt. Spektr. 24, 209 (1968).

16. R. Priestly and S.P. Pershan, presented at the IV th
    International Liquid Crystal Conference, Kent, Ohio
    (1972).

17. C.H. Wang, J. Amer. Chem. Soc., 94, 8605 (1972).

18. K.K. Kobayashi, Molec. Cryst. Liq., Cryst., 13, 137
    (1971).

19. W.G.F. Ford, J. Chem. Phys., 56, 6270 (1972).

ACKNOWLEDGMENT

It is a pleasure to acknowledge the contributions to this work made by several co-workers, particularly F. Prochaska, W.B. Lok, D. Grunbaum, K. Krishnan and J.O. Lephardt. The continued support of the American Cancer Society and the U.S. Army Research Office-Durham is also acknowledged.

RECENT DEVELOPMENTS IN ARC EMISSION SPECTROSCOPY: INTRODUCTORY REMARKS

Ramon M. Barnes

Chemistry Dept. Univ. of Massachusetts

Amherst, Mass. 01002

This symposium on Recent Developments in Arc Emission Spectroscopy includes three papers describing three new developments in techniques and methodology of arc spectrochemistry. Why, in the first place, is a symposium on arc emission spectroscopy necessary? Figure 1 provides some idea of the reason. Plotted in the figure are the number of publications cited in spectrochemical analysis each year for the past 52 years. The curve on the left is taken from Meggers and Scribner's Index to Literature on Spectrochemical Analysis between 1920 and 1955. The middle curve comes from 16 volumes of Spectrochemical Abstracts for the years 1933 through 1970. And the values to the right are from the annual reviews in Emission Spectrometry, published biennially in Analytical Chemistry covering the period 1946 through 1972. Although each source may provide a different absolute index, the trends are clearly apparent. Since 1920 a steady growth in the field of spectrochemical analysis, except for the time during World War II, has firmly established emission techniques as essential in modern analysis. Unfortunately, the active spectrochemist does not find time to read everything new appearing in the literature, and for that reason this symposium will focus his attention on three of the key developments in arc spectrochemistry in the recent past.

What have been some of the historical developments in arc spectroscopy leading to the work presented in this symposium? An arc can be considered a self-sustaining electrical discharge

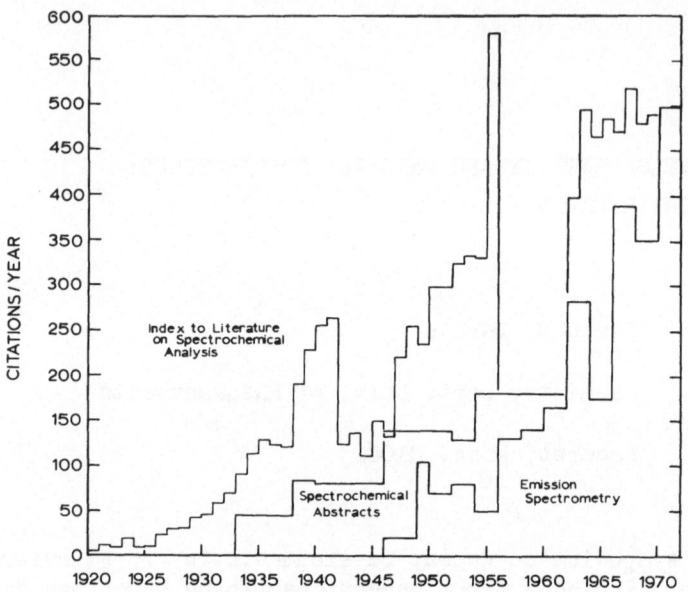

Figure 1. Publications in Spectrochemical Analysis

between electrodes in a gas or vapor with a low voltage drop
across the electrode gap except in the regions adjacent to the
electrodes. In spectrochemical applications the discharge
conditions remain continuous during relatively long durations
ranging from milliseconds to minutes. The name probably resulted from the shape of the luminous discharge caused by the
buoyant forces due to the hot gases produced near the discharge volume which forced the column to <u>arch</u> or <u>arc</u> (1).

Although the general use of arcs for spectrochemistry did
not become popular for approximately 50 years following the
commercial distribution of electricity in 1880-90, pioneers
relied upon laboratory sources, notably batteries and later
dynamos, for major discoveries in spectroscopy during the 1800s.
Until a theory of electrical discharges through gas was developed in the early 1900s, these initial observations lacked a
sound foundation based upon understanding the arc processes.

During most of the 1800s, chemical batteries served as
the primary source of electrical energy, and the cells developed by Volta, Bunsen, Grove, and Daniell were commonly used

for the generation of arcs. Volta described the first battery in 1792, and by 1802 Sir Humphry Davy had succeeded in generating brilliant discharges between charcoal or metal electrodes with a Voltaic pile (2). Davy applied arcs primarily to study the electrical decomposition of materials. In 1835 Wheatstone described the prism spectra obtained from spark generators and a Voltaic battery between two charcoal electrodes (3). From these observations he suggested that metallic substances could be more readily distinguished by spectral than chemical methods. Daniell also produced brilliant arcs between charcoal electrodes in 1839 (4). In developing a zinc-carbon battery in the early 1840s, Bunsen observed the prism spectra of arcs between two carbon or metal electrodes powered by the battery (5). He made some of the first photometric intensity measurements on these arcs. Foucault investigated the arc in the mid 1840s (6) and reported in 1849 a comparison of the solar spectrum with the spectra of arcs between carbon or metal electrodes (7). Robiquet extended these arc studies to other metals in 1859 (8). Bunsen and Kirchhoff's description of the spectra from flames and sparks in 1859-1861 established the science of spectrochemical analysis (9). Beginning in 1872 Lockyer presented a series of papers which clearly demonstrated the quantitative ability of spectrochemical analysis using arc and spark sources. In 1874 he used a horizontal arc, powered by 30 Grove cells. He photographed the spectrum to determine the radial distribution of lines in the arc on which he and Roberts based their analyses (10). In 1879 Lockyer improved the arc performance by substituting a Siemens dynamo-electric machine, driven by a gas engine (11). He studied the distribution of lines along the axis of a vertical arc and observed reversal of lines (12). In 1873 Secchi reported the spectrum of iron and other metals in an arc powered by a 50-cell Bunsen battery (13). Of the three electrode arrangements used, he anticipated the development of the globule and Pfund arcs when he used beads of iron placed in the positive carbon electrode. Liveing and Dewar, in their study of line reversal of metallic elements, developed a graphite arc in the form of a crucible operated by 25 Grove cells in 1878 (14).

During the final decades of the nineteenth century, the arc spectrum of numerous elements were observed and cataloged as the result of successful efforts by Rowland to rule concave gratings and construct a concave grating spectrograph with linear dispersion. Spectrum analysis remained the major spectroscopic application of the arc until the 1930s. In contrast, spectrochemical developments occurring in the 50 years following Lockyer's analysis almost exclusively employed spark or flame sources.

In 1908 Pfund described an arc which was positionally stabilized by the formation of a metal bead as the cathode (15). The Pfund iron arc was adopted as a secondary wavelength standard. In 1934 Milbourn introduced the globule arc, which used a graphite electrode to support the metal bead, for the spectrochemical analysis of copper base metals (16). The copper globule arc is the basis for modern methods of the analysis of high purity copper (17). Molten metal electrodes in arc spectrochemistry were brought to a sophisticated level in 1958 by Fassel and co-workers (18) for the determination of gases in metals. Inspired by the success of this technique, Gordon in 1964 began the development of a metal cathode for high precision arc analyses (19) described in this symposium by Franklin, Gordon, and Hambidge.

The design of new arc sources came in the mid 1930s with the Pfeilsticker's ignited arc (20) and Duffendack's high-voltage ac arc (21). The perfection of arc techniques for spectrochemical analysis during the 1930s was illustrated by the development of the cathode layer technique by Mannkopff and Peters (22), Strock (23) and Scott and Mitchell (24). Gerlach and Schweitzer described their modified internal standard method in 1931 (25), and Slavin popularized the total energy method and emphasized the importance of fractional volatilization and the difficulties of using internal standards in arc analysis in 1938 (26).

Scribner and Mullin's development of the carrier distillation method during the Second World War (27) stimulated study of chemical reactions in the electrode cavity and provided a sensitive new technique for refractory materials. Universal or generalized arc methods became popular in the early 1950s through the methods of Harvey (28), Jaycox (29), Addink (30) and others. One of the most successful empirical approaches to a generalized arc method has been the application of a common matrix technique, in which the sample and standards were diluted by a common material (31).

Stallwood developed a simple concentric air jet in 1954 in an attempt to provide a generalized arc method (32). Although Stallwood's method worked, the real impact of his investigations has resulted from the concept of controlling the

discharge, and the electrode and sample vaporization with a
flowing gas stream (33). The Stallwood jet also helped to
initiate interest in gas sheaths and controlled atmospheres
in arc methods. These interests were further stimulated in
1959 by the description of the plasma arc jet (34). The
plasma jet probably started many spectrochemists thinking
about improved methods to obtain arc spectra.

One of these approaches taken by Jones and Dahlquist
at Applied Research Laboratories in 1967 was to separate the
arc sampling and excitation processes for solid and molten
metals (35). The recent applications of this approach at ARL
and the National Bureau of Standards are presented in this
symposium by Golightly and Jones. *

The use of discharges at reduced pressures and in noble
gas atmospheres date to studies by Angström in 1852 (36),
Plücker and Hittorf between 1858 and 1865 (37), and Pollok in
1912 (38). Paschen introduced the hollow cathode discharge
in 1916 (39), which Schüler (40) and Sawyer (41) studied and
further developed. In 1947 McNally et al. used the hollow
cathode for quantitative analysis (42), and during the 1960s
Walsch et al. used them for sources and detectors in absorption and fluorescence spectroscopy (43). The glow discharge
has been remarkably well characterized and studied, and the
sampling or sputtering of metal electrodes, especially the
cathode, has found extensive use in hollow cathode lamps. A
new configuration glow discharge in which a flat metal sample
was used as the cathode was described in 1967 by Grimm (44),
and since then a number of groups have undertaken detailed
studies on the applications and properties of the discharge
lamp. Included among these have been Boumans (45), and Dogan
et al. (46). The recent investigations at Dortmund are described in this symposium by Laqua and colleagues. *

This brief historical introduction has emphasized the
growth of arc methods in the history of spectroscopy. However, only within the past decade have arc methods been developed based upon a fundamental understanding of the energy
and material transport processes related to the arc. The arc
holds many more potentials, and through developments of the
type described in this symposium some of these potentials
may be realized.

---

*The editors regret that conditions beyond their con-
trol have prevented the publication of these papers.

ACKNOWLEDGEMENTS

The support of the National Science Foundation and the American Chemical Society, administors of the Petroleum Research Fund, through NSF Grant GP25909 and PRF Grant 2055G3 is gratefully acknowledged.

REFERENCES

(1) H. Davy, On the Magnetic Phenomena Produced by Electricity, Phil. Trans., 7-19 (1821).

(2) H. Davy, An Account of Some Experiments on Galvanic Electricity, J. Roy. Inst.,1,165-7 (1802).
H. Davy, Elements of Chemical Philosophy, Vol. I, p.152 (1812).
H. Ayrtron, The Electric Arc, "The Electrician" Printing and Publishing Co., London, 1902.

(3) C. Wheatstone, On the Prismatic Decomposition of Electric, Voltaic,and Electro-Magnetic Sparks, Chem. News, 3, 198-201 (1861).
C. Wheatstone, On the Prismatic Decomposition of Electric Light, Ann. Phys., 36, 148-9 (1835).

(4) J.F. Daniell, Fifth Letter on Voltaic Combinations, with Some Accounts of the Effects of a Large Constant Battery, Phil. Trans., 89-96 (1839).

(5) R.W. Bunsen, Ljuset af hydro-elektriska staplars urladdning, Ofver. Kongl. Vetens. Akad. Förhandl. Stockholm, 1, 144-6 (1844).
R.W. Bunsen, The Application of Carbon in the Voltaic Battery, Ann. Phys., 54, 417-30 (1841).
R.W. Bunsen, The Preparation of a Grove Series with Platinum Replacing Carbon, Ann. Phys., 55, 265-76 (1842).
J. Reiset, A New Construction of the Voltaic Pile, Ann. Chim. Phys. [3] 7, 355-8 (1843).
J. Reiset, New Documents on the Bunsen Pile, Ann. Chim. Phys. [3] 8, 28-36 (1843).
J. Berzelius, Jahresberichte Fortschr. Chem. Min., 25, 20 (1846).

(6) M. Fizeau and L. Foucault, Research on the Intensity of Light Emitted by Carbon on the Experience of Davy, Ann. Chim. Phys.,[3] 11, 370-83 (1844).

(7) L. Foucault, Note on the Light of the Voltaic Arc, Ann. Chim. Phys. [3] 58, 476-8 (1860).

(8) E. Robiquet, Research on the Lines of the Solar Spectrum and of the Different Electrical Spectra, Comp. Rend. 49, 606-10 (1959).

(9) G. Kirchhoff and R. Bunsen, Chemical Analysis Through
     Spectral Observations, Ann. Phys.,110, 161-189 (1860).
(10) J.N. Lockyer and W. C. Roberts, On the Quantitative Ana-
     lysis of Certain Alloys by Means of the Spectroscope,
     Phil. Trans., 164, 495-9 (1874).
(11) J.N. Lockyer, Note on Some Spectral Phenomena Observed
     in the Arc Produced by a Siemens' Machine, Proc. Roy.
     Soc.,28, 452-8 (1879).
(12) J.N. Lockyer, Note on Some Phenomena Attending the Re-
     versal of Lines, Proc. Roy. Soc., 28, 428-32 (1879).
(13) A. Secchi, Spectra of Iron and Some Other Metals in the
     Voltaic Arc, Comp. Rend.,77, 173-7 (1873).
(14) G.D. Liveing and J. Dewar, On the Reversal of the Lines
     of Metallic Vapours, Proc. Roy.Soc.,28, 353-8, 367-72 (1879).
(15) A.H. Pfund, Metallic Arcs for Spectroscopic Investigations,
     Astrophys. J.,27, 297-8 (1908).
(16) M. Milbourn, The Spectrographic Detection and Estimation
     of Minute Quantities of Impurities in Copper, J. Inst.
     Metals, 55, 275-82 (1934).
     M. Milbourn, The Spectrographic Analysis of Copper and
     Copper-Rich Alloys by the Arc Method, J. Inst. Metals,69,
     441-63 (1943).
(17) W.E. Publicover, Spectrochemical Analysis of Oxygen-Free
     Electrolytically Pure Copper by a Globule Arc Procedure,
     Anal. Chem.,37, 1680-4 (1965).
(18) V.A. Fassel and R.W. Tabeling, The Spectrographic Deter-
     mination of Oxygen in Titanium and Titanium Alloys, Anal.
     Chem.,30, 179-82 (1958).
(19) W.A. Gordon and G.B. Chapman, Quantitative Direct-Current
     Arc Analysis of Random Compositions of Micrograms Residues
     in Siver Chloride Common Matrix, Spectrochim. Acta, 25B,
     123-37 (1970).
(20) K. Pfeilsticker, Intermittent Arc with H.F. Excitation, Z.
     Electrochem., 43, 717-21 (1937).
(21) O.S. Duffendack and K.B. Thompson, Developments in the
     Quantitative Analysis of Solutions by Spectroscopic Means,
     Proc. Am.Soc. Testing Materials, 36,II, 301-9 (1936).
(22) R. Mannkopff and C. Peters, Quantitative Spectrum Analysis
     with the Aid of the Negative Glow (Cathode Layer) in the
     Arc, Z. Phys.,70, 444-53 (1931).
(23) L. Strock, Spectrum Analysis with the Carbon Arc Cathode
     Layer ("Glimmschict"), Hilger, London, 1936.
     L. Strock, Quantitative dc Arc Spectrochemical Analysis:
     Development in the Decade 1930-1940, Appl. Spectrosc.,
     23, 309-16 (1969).

(24) R.L. Mitchell, The Spectrometric Analysis of Soils,Plants, and Related Materials, Commonwealth Agricultural Bureau, Farnham Royal, England, 1964.
R.O. Scott, The Spectrographic Determination of Trace Elements in the Cathode Layer Arc by the Variable Internal Standard Method, J. Soc. Chem.Ind., 65, 291-7 (1946).

(25) W. Gerlach and E. Schweitzer, Foundations and Methods of Chemical Analysis by the Emission Spectrum, Hilger, London, 1931.

(26) M. Slavin, Quantitative Analysis Based on Spectral Energy Ind. Eng. Chem.,Anal.Ed.,10, 407-11 (1938).

(27) B.F. Scribner and H.R. Mullin, Carrier Distillation Method for Spectrographic Analysis and Its Application to the Analysis of Uranium-Base Materials, J.Res. Natl. Bur. Stand., 37, 379-89 (1946).

(28) C.E. Harvey, A Method of Semiquantitative Spectrographic Analysis, Applied Research Laboratories, Glendale,Calif., 1947.

(29) E.K. Jaycox, A Spectrochemical Procedure of General Applicability, Anal. Chem.,27, 347-50 (1955).

(30) N.W.H. Addink, J.A.M. Dikhoff,C. Schipper, A.Witmer, and T.Groot, Spectrochemical Analysis by Means of the D.C. Carbon Arc,Spectrochim.Acta,7, 45-59 (1955).
N.W.H. Addink, DC Arc Analysis, MacMillan, London, 1971.

(31) A.J. Mitteldorf and D.O. Landon, Semiquantitative Spectrochemical Analysis Using Spex Standards, Appl. Spectrosc., 10, 12-14 (1956).

(32) B.J.Stallwood, Air-Cooled Electrodes for the Spectrochemical Analysis of Powders, J. Opt.Soc.Am., 44, 171-6 (1954).

(33) P.W.J.M. Boumans and F.J.M.J. Maessen, Influences of the Physical and Chemical Processes in the Electrode Cavity and the Gaseous Atmosphere in the Arc on the Emission Characteristics of the D.C. Arc for Spectrochemical Analysis, Spectrochim. Acta, 24B, 585-610; 611-28 (1969).

(34) M.Margoshes and B.F. Scribner, The Plasma Jet as a Spectroscopic Source, Spectrochim. Acta, 15, 138-45 (1959).
V.V. Korolev and E.E. Vainshtein, Application of the Plasma Generator as a Source of Excitation in Spectral Analysis, J.Anal. Chem. USSR,14, 731-35 (1959).

(35) J.L. Jones, R.L. Dahlquist, and R.E. Hoyt, A Spectroscopic Source with Improved Analytical Properties and Remote Sampling Capabilities, Appl. Spectrosc.,25, 628-35 (1971).

(36) A.J. Ångström, Optical Investigations, Svensk.Akad.Handl., 327-342 (1852); Ann. Phys.,94, 141-65 (1855); Phil. Mag. (4),9, 327-42 (1855).

(37) J. Plücker and J.W. Hittorf, On the Spectra of Ignited
     Gases and Vapours, with Especial Regard to the Different
     Spectra of the Same Elementary Gaseous Substances, Phil.
     Trans., 155, 1-29 (1865).
(38) H. Pollok, On the Vacuum Tube Spectra of the Vapours of
     Some Metals and Metallic Chlorides, Sci. Proc.Roy. Dub-
     lin Soc.,13, 202-218; 253-68 (1912).
(39) F. Paschen, Bohrs Helium Lines, Ann. Phys.,(4), 50,
     901-40 (1916).
(40) H. Schüler, On a New Light Source and Its Possible Appli-
     cations, Z.Phys.,35, 323-37 (1926).
     H. Schüler and A. Michel, About Two New Hollow-Cathode
     Discharge Tubes, Spectrochim. Acta, 5, 322-6 (1952).
(41) R.A. Sawyer, Excitation Processes in the Hollow Cathode
     Discharge, Phys. Rev., 36, 44-50 (1930).
(42) J.R. McNally Jr., G.R. Harrison, and E. Rowe, A Hollow
     Cathode Source Applicable to Spectrographic Analysis for
     the Halogens and Gases, J. Opt.Soc.Am.,37, 93-8 (1947).
(43) W.G. Jones and A. Walsh, Hollow Cathode Discharges. The
     Construction and Characteristics of Sealed-Off Tubes for
     Use as Spectroscopic Light Sources, Spectrochim.Acta,
     16, 249-54 (1960).
     J.V. Sullivan and A. Walsh, High Intensity Hollow Cathode
     Lamps, Spectrochim.Acta, 21, 721-6 (1965).
     J.V. Sullivan and A. Walsh, The Application of Resonance
     Lamps as Monochromators in Atomic Absorption Spectroscopy,
     Spectrochim. Acta, 22, 1843-52 (1966).
(44) W. Grimm, Glow Discharge Lamp for Routine Spectral Analy-
     sis, Naturwissensch., 54, 586 (1967).
     W. Grimm, A New Glow Discharge Lamp for Optical Emission
     Spectral Analysis, Spectrochim. Acta, 23B, 443-54 (1968).
(45) P.W.J.M. Boumans, Studies of Sputtering in a Glow Dis-
     charge for Spectrochemical Analysis, Anal. Chem., 44,
     1219-28 (1972).
(46) M. Dogan, K. Laqua, and H. Massman, Spectrochemical Ana-
     lysis with a Glow Discharge Lamp as the Light Source,
     Spectrochim. Acta, 26B, 631-49 (1971); 27B, 65-88 (1972).

ADVANCES IN THE STATIC-ARGON, DC-ARC SPECTROCHEMICAL SOURCE

W.A. Gordon[1], K.M. Hambidge[2] and M.L. Franklin[2]

[1]N.A.S.A.,Lewis Research Center, Cleveland, Ohio

[2]University of Colorado Medical Center, Denver, Colo.

The advanced dc-arc spectrochemical source discussed here
was developed by William A. Gordon at N.A.S.A., Lewis Research
Center, Cleveland, Ohio. It consists of a dc current arc in a
static-argon gas chamber. Figure 1 is a photograph of a typical
spectrochemical analysis system. This unit is in operation at
the University of Colorado Medical Center in Denver, Colorado,
under the direction of Michael Hambidge, M.D. It employs a
standard Jarrell-Ash 1.5 meter Paschen-Runge mount direct reading
spectrometer. The spectrometer offers a reciprocal linear dispersion of 5.5 Å/millimeter along the focal curve yielding a
spectral band pass of 0.4 Å for typical exit slit widths. Photo-
multipliers are arranged behind the exit slits along the focal
curve located in the rear of the spectrograph and each of the
currents from these photomultiplier tubes is integrated by
storing a charge on separate capacitors. After analysis each
capacitor is discharged through an electronic measuring circuit
and the signal is recorded. This is the same model spectrometer
which is employed at N.A.S.A., Lewis Research Center with the
static-argon arc at that location. The inert gas chamber itself
is located in front of the entrance slit on the optical bar of
the spectrometer. Some of the auxiliary equipment necessary to
operate the system is visible on the right side of the photograph.
A standard spectrochemical dc-arc supply source is located below
the arc chamber. A manifold system containing gas flow valves
and a pressure gauge is visible. The gas flow system which controls the argon pressure that is backfilled into the arc chamber
is also connected to a vacuum pump located on the floor.

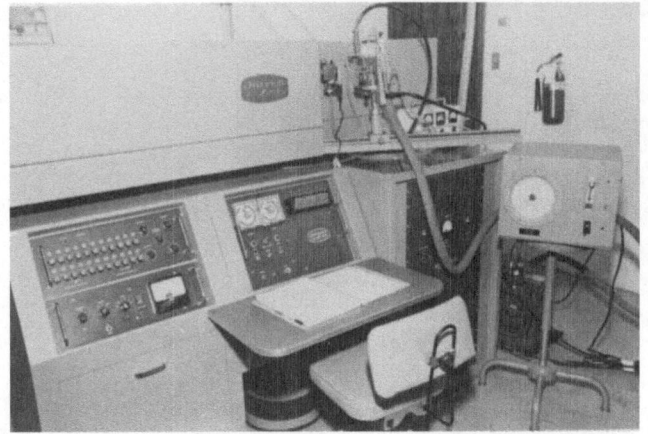

Fig. 1 - Spectrochemical analysis system

Fig. 2 - Controlled atmosphere static-argon excitation chamber

Figure 2 is a close-up view of the arc chamber (1,2). The chamber is made of fused silica quartz and is 6-½ inches in diameter and 4-½ inches in height. Each end of the quartz cylinder is sealed by a rubber gasket against a stainless steel plate. These plate assemblies are water cooled and the inner surfaces have been silver plated. The plating reduces the amount of elemental contamination for trace elements analyzed with this source. The top stainless steel plate can be elevated by means of the clamp assembly shown on the left side of the chamber for loading or unloading sample electrodes. The upper electrode - the cathode - is secured in an electrode mount with a set screw. This mount has a bellows seal on the inside which allows vertical adjustment of the upper electrode. The arc gap is maintained at 12mm. Secured in the electrode holder is a 1/8 inch diameter pyrolytic graphite rod electrode. A small hole has been drilled in the end of this electrode and a 0.05 inch diameter tantalum wire is inserted to protrude 6mm. The spherical tantalum electrode is then formed or grown at a specific current by arcing to the tantalum wire. There are 10 anode electrodes located in the turret. Each of these sample electrodes is a porous carbon material with a point forming a 90° angle at the tip. Silver chloride spectrographic buffer is precipitated in the electrode tip itself by sequential doping with silver nitrate and hydrochloric acid. Two mm above the sample electrodes is a tantalum plate cover supported by an insulating boron nitride post. This cover prevents deposition of any vaporized material from the electrodes being analyzed onto any of the other sample electrodes contained in the chamber. The last item is an anti-fogging tube located on top of the electrode cover plate. The tube is made of quartz and has a small open aperture in the end of the quartz tube nearest the electrode to be arced. Any sample material vaporized from the electrode must first drift through the small entrance aperture, then drift down a long straight line path to the other end of the chamber before it can be deposited on the quartz area which transmits the atomic emission radiation. The result is that this area remains clear during the analysis of 10 electrodes. The entire chamber is tilted toward the spectrometer at an angle of approximately 5 degrees from the vertical, and a section of the arc approximately 5mm high and 5mm wide, beginning 1mm above the sample electrode tip, is focused into the spectrograph by the entrance optics.

A considerable amount of information has been abstracted from N.A.S.A. technical notes which Bill Gordon has published over the past years. These publications sketch in the development of this chamber to its present form and discuss many of the spectrochemical properties that this arc manifests. The two areas of application to which this arc has been put are as

follows: first, at N.A.S.A. in the analysis of high temperature
alloys; second, at the University of Colorado Medical Center for
the analysis of trace metals in biological materials.

One method of arc stabilization is the use of rapidly
flowing inert gases around the arc column. Other methods which
are useful for arc stabilization include the use of intense
magnetic fields to stabilize the arc column or the rotation of
the electrodes during arcing. Optical defocusing methods are
often used for reducing the effect of non-uniform illumination
of the spectrograph slit due to the arc wander. A recent stabilization method reported in 1967 is the use of metal salts contained
in hollow graphite cathodes (3). Using a graphite cathode in
which a tiny hole has been drilled, and packed with a metal salt,
resulted in the positional stabilization of the dc-arc column.

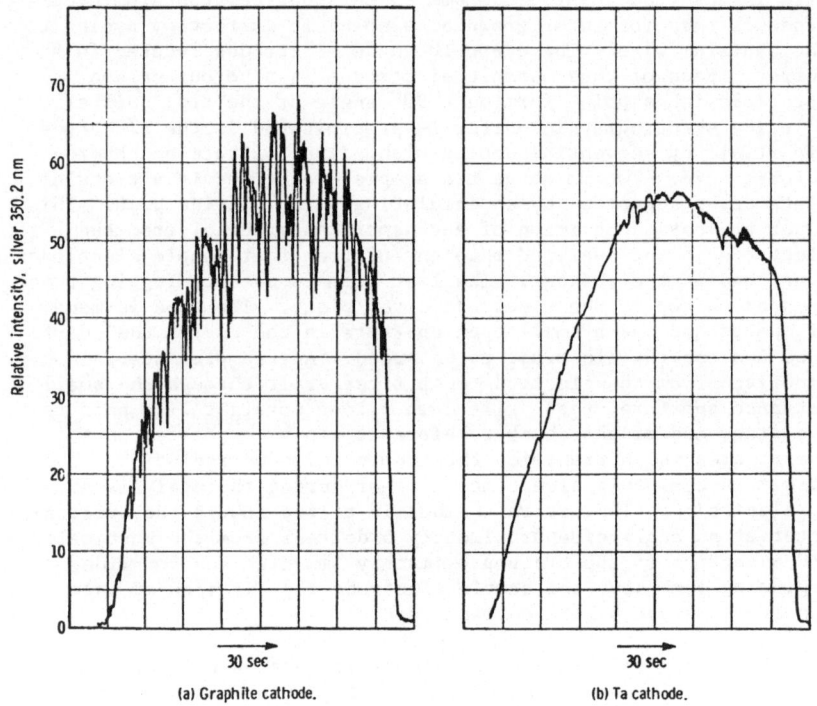

(a) Graphite cathode.   (b) Ta cathode.

- Vaporization profiles of silver. Argon arc, 30 amperes.

Fig. 3 - (from Ref. #4)

The elimination of arc wander in the static-argon arc was achieved about that same time at N.A.S.A. by using a tantalum sphere attached to the end of a graphite cathode in place of the conventional graphite electrode (4). Figure 3 illustrates the results of improvement in positional stability achieved when a Ta spherical cathode is substituted for the usual graphite counter electrode. Note the decrease in signal fluctuations for the silver in the spectrochemical buffer. Metal cathodes cannot be used in air because of rapid destruction by oxidation, but they can be used in inert atmospheres. Tantalum spherical electrodes in use last from between 5-200 arcings. Recently, a spherical tantalum counter electrode has been used with a flowing gas system at atmospheric pressure. When used in spectrochemical sources such as the Stallwood jet (5), or the plasma jet reported by Scriver and Margoshes (6), it reduces the amount of arc wander. One such system is in routine use at the Colorado State University to measure trace metals collected from the atmosphere in porous carbon electrodes. Scogerboe and Seeley report an increase in precision of a factor of 1.5 to 3 when using the spherical tantalum electrode, and that electrode lifetime is greater than 100 determinations (7).

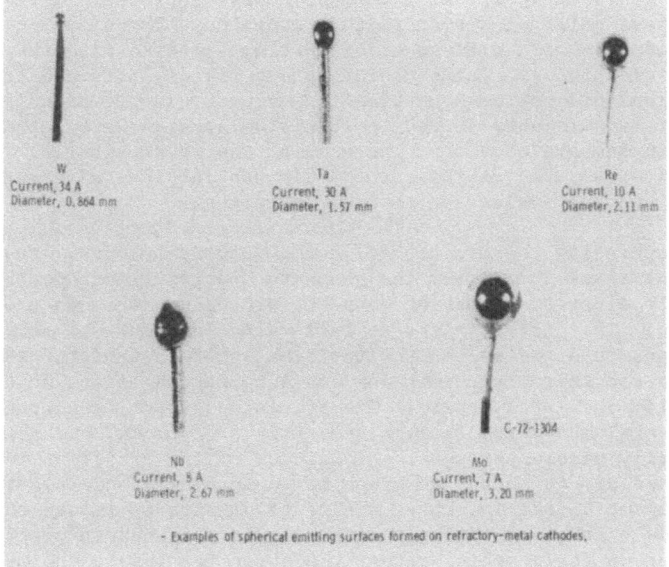

- Examples of spherical emitting surfaces formed on refractory-metal cathodes.

Fig. 4 - (from Ref. #8)

Figure 4 pictures spherical electrodes formed from several different metals, such as tungsten, tantalum, rhenium, niobium and molybdenum (8). In the case of rhenium, niobium and molybdenum large cathode spheres are formed at low current. This gave rise to another form of arc instability because the spheres tended to move quickly during the arcing. Low periodic axle rotations with similar motions of the arc column were observed in these molten cathodes. The instability was caused by the combined effects of large sphere volumes and increased metal fluidity of the spheres. Optical pyrometer measurements indicated that there was a large temperature gradient in these spheres. The temperatures were significantly higher on the tip of the spheres nearest the arc column than they were at the point of attachment to the metal wire inbedded in the graphite electrode.

Tungsten produced a very small spherical metal electrode, and the cathodes did not give maximum arc stability. The arc columns were not always symmetrical but showed the instability characteristic of small emitting areas. The small tungsten spherical electrode was very similar to the small cathode spot of the graphite cathode.

The tantalum spherical cathodes, however, produced molten spheres having larger diameters than tungsten. The emission of the cathode was more diffuse with tantalum compared with the tungsten electrodes. The tantalum sphere rigidly attached to the wire and did not move physically during arcing. Optical pyrometer measurements showed the tantalum spheres were rather uniform in temperature over the area of the sphere itself. The temperature measured at the cathode tip was in close agreement with the handbook value for the melting point.

In operating the arc the following sequence occurs. The initiating spark jumps from the graphite counter electrode to the sample electrode breaking down the arc column through the argon gas. The current rises to full value in about 0.1 sec., and in less than one second the tantalum sphere has heated so that electron thermionic emission can support the entire 30 amp dc-arc from the metal sphere. The arc cathode spot then transfers to the tantalum sphere, forms a symmetric arc column, and remains positionally stable through the remainder of the arc lifetime. It is necessary to shape the graphite support electrode tip to a sharp point before inserting the tantalum wire to reduce the likelihood of the cathode spot attaching to the support electrode and moving up the graphite rather than transferring to the metal sphere tip. The pyrolytic graphite, manufactured by Poco Carbon, Decatur, Texas, is especially good material for this support electrode.

In arcs it is generally accepted that the electron emission is controlled by the thermal energy at the cathode. The Richardson-Dushman equation describes the energy balance using the work function of various materials, the cathode temperature, and the current density. The cathode spot observed with graphite cathodes is believed to be caused by the variability of the work function of graphite. The work functions are critically dependent upon the crystalline structure and the smoothness of the graphite surface. The graphite cathode continuously undergoes changes in the degree of graphitization in the arc. In contrast to graphite the spherical metal surfaces formed from metal wires are highly uniform, so uniform that they have mirror surfaces. The area of electron emission is well defined by the surface of the sphere. Consequently, the emission from these spherical metal cathodes does not move over the cathode surface. The size of the metal cathodes is determined by the amount of current. An interesting outgrowth of this work has been the determination of work functions for various metals by applying the Richardson-Dushman equation and accurately measuring sphere temperature and the diameter of the sphere formed. The procedure for work function determination is experimentally advantageous because surface condition of the specimen is not critical, high vacuum is not required, and the anode-to-cathode spacing between the arc electrodes is not critical. The experimental procedure involved striking an arc to a metal wire cathode to form a melted ball having an emitting area defined by its diameter.

The Richardson equation predicts a linear relationship between the log of the emitting current density divided by the cathode temperature squared and the elementary electronic charge divided by Boltzmann's constant times the cathode temperature. From such a plot the slope gives the negative value of the work function, and the intercept is log of the effective emission constant. A least squares and a quadratic fit over a range of 300° Kelvin around the melting temperature of tantalum had equal standard deviations for quadratic and linear fits, thus indicating a linear fit is acceptable. The effective work functions are listed in the following table showing a comparison between accepted values and the work function determination by dc current arc method.

Metal	Melting Point	Work Functions (V) By Arc Method	Recommended Value
Mo	2883	4.3	4.3
Nb	2741	4.0	3.99
Re	3453	5.1	5.0
Ta	3269	4.13*	4.12
w	3683	4.5	4.5

*Standard deviation for five consecutive determinations 0.013V. (Table from Ref. #8)

A measurement of the current, the effective diameter of the sphere with a micrometer, and the temperature measurement with an optical pyrometer are all that are necessary for the determination of a metal work function.

There are several benefits derived from arc excitation in gases other than air. Flowing gas systems of argon-nitrogen mixtures, pure argon and helium have been reported. Since all the flowing gas systems operate at atmospheric or slightly positive pressure, a more versatile dc-arc inert gas chamber was designed in which a static pressure of inert gas could be varied from about 1 torr to 1000 torr (1). In trying to determine the gas pressure producing maximum sensitivity, the line-to-background ratio was determined and used.

Several publications report only the absolute intensity of the emission line. When a sample is arced in pure argon at pressures above atmospheric, the positive column of the dc-arc becomes constricted and therefore brighter. More light, both line emission and background, is transmitted to the detector. This does not, however, improve the ability to measure the concentration of analyte present (provided sufficient light is reaching the detector to give an accurately measured signal). Thus, the line-to-background ratio should be used to determine conditions which lead to sensitivity improvements.

The free-running arc between graphite electrodes in a static-argon gas at atmospheric pressure is about as unstable as the arc in the air. An anode flame as long as 5 cm develops at a 30 amp current and revolves around the vertical electrode axis. The discharge in argon tends to become more stable when metal salts are vaporized from the anode into the arc column. If, however, the pressure in the chamber is reduced below atmospheric, the positive column increases in volume and the

discharge becomes more stable. Reducing the argon pressure
reduces the photometric brightness of the arc as verified by
measuring the spectral background as a function of argon pressure
at three widely separated wavelengths.

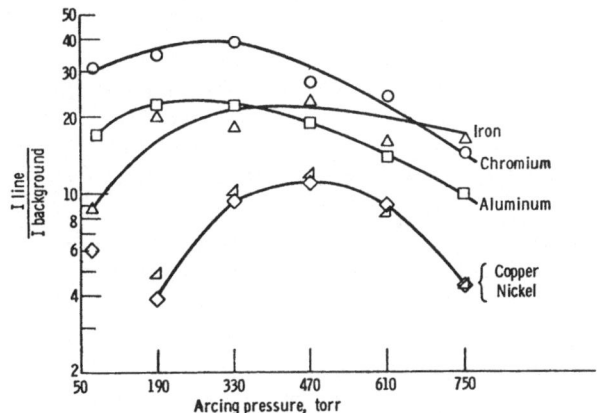

- Effect of arcing pressure on line-to-background intensity ratios
of typical elements when arced in presence of silver chloride. Weight of
elements, 5×10^{-8} gram.

Fig. 5 - (from Ref. #1)

Figure 5 is a plot of the ratio of the line-to-background
vs. the pressure. For many of the elements the line-to-background
ratio increases as the pressure decreases over the range from
700-100 torr. If the pressure is lowered further, the result is
a glow discharge extending over a large area of the chamber and
surrounding a large number of the metal objects in the metal arc
chamber. Using a silver chloride spectroscopic buffer, the
peak sensitivity for chromium is at 300 torr, and the optimum
pressure for nickel analysis is at approximately 500 torr.

If the intensity of argon spectrochemical emission from the
static-argon arc with no spectroscopic buffer is recorded,
intense lines are only produced near the cathode. This "cathode-
layer-effect" can be enhanced by decreasing the supporting
atmosphere pressure. This is the technique used to determine

halogens, where the atomic lines require higher energies for excitation. However, for the determination of the common metal elements using a spectroscopic buffer of AgCl, no cathode layer effect was observed for neutral or singly-ionized spectra. The cathode layer effect was not useful for the determination of these elements, and masking the intense emission from the tantalum sphere and the cathode fall region decreased the background and improved the line-to-background ratio. Moving the observation closer and closer to the anode improved the analytical sensitivity. In summary, the simple expedience of reducing the arc pressure and properly masking the cathode area produced enhancement in line-to-background ratios.

The effect of the silver chloride precipitated in the porous carbon anodes is to act as a spectroscopic buffer: the presence of the silver atoms in the arc column holds the arc plasma temperature constant. This buffering effect increases the neutral atom emission in the arc column. In trace element work, using blood serum, in which there is approximately a million times more sodium than trace metals, the addition of silver chloride buffer is used to increase the line-to-background ratios by a factor of two to three.

It was determined that the temperature of the carbon electrode containing the sample was dependent only on the arc current and not on the pressure inside the chamber itself. This was done by using an optical pyrometer to measure the surface temperature of the anode electrode. Reducing the pressure and therefore the current density in the arc column itself can be done without disturbing the temperature of the electrode furnace effect, which boils atoms into the arc column itself. The excitation of the atomic spectra from the vaporized sample occurs in the arc column but the rate of sample introduction is determined by the heat generated at the graphite anode containing the specimen. The anode temperature and thus the rate of sample vaporization are primarily determined by the arc current, the anode material, the anode geometry, sample form, and the composition and pressure of the atmosphere. Variations in any of these factors between analyses can affect the reproducibility of sample vaporization.

In practice these factors are kept constant for each analytical procedure. High uniformity of electrode-geometry and material density is maintained by suppliers of spectrographic electrodes. The use of uniform samples with respect to weight, particle size and density is also a common practice. Electronic control over the arc current has been used to ensure a constant current level regardless of changes in the vapor composition in the arc column. In spite of these precautions, relative standard deviations of from 10-50% for repetitive arcings often occur. Because of these analytical errors, the direct arcing procedure is seldom used for quantitative analysis.

Improvement of this situation has been achieved by using the internal standard technique. This technique compensates for experimental variations by taking the ratio of the analytical line intensity to the line intensity of an element that is added to all samples in known concentration. The major limitation in the internal standard technique is caused by the difficulty of achieving good compensation for many elements with a single internal standard.

Achieving very reproducible sample vaporization would eliminate the need for internal standards. One method of attempting this is to use electronics to precisely control the spectral light emission from an arc. Such a servo-controlled arc system has been developed at N.A.S.A., Lewis (9). With this method, fluctuations in sample vaporization can be reduced to as low as 1%. Atomic emission output control of the dc-arc was carried out using two thyratrons to control both half-cycles of a fullwave rectified 60 cycle, 30 amp arc source. The optical intensity from the dc-arc was collected by a direct reading spectrometer. A photomultiplier aligned along the focal curve intercepted the atomic emission from the silver spectroscopic buffer. The signal from the photomultiplier tube was connected to an electrometer amplifier and compared with a reference signal from a curve following programmer. Any differences in the two signals resulted in a control signal which was directed to the thyratron tubes. Thus, monitoring the intensity of the silver line emission allowed the closed loop controller to provide effective control of time-dependent light intensity emitted from silver chloride vaporized in the dc-arc. Compared with samples vaporized at constant current, both the instantaneous and the time-integrated intensities were more repeatable when the control system was used. Figure 6 shows a typical comparison of 5 repetitive vaporizations at constant current and at controlled current. The instantaneous fluctuations of the silver line intensity were reduced when controlled light intensities were used. If perfect control of the major element intensity could be maintained, a built-in internal standard in the form of a constant background would result.

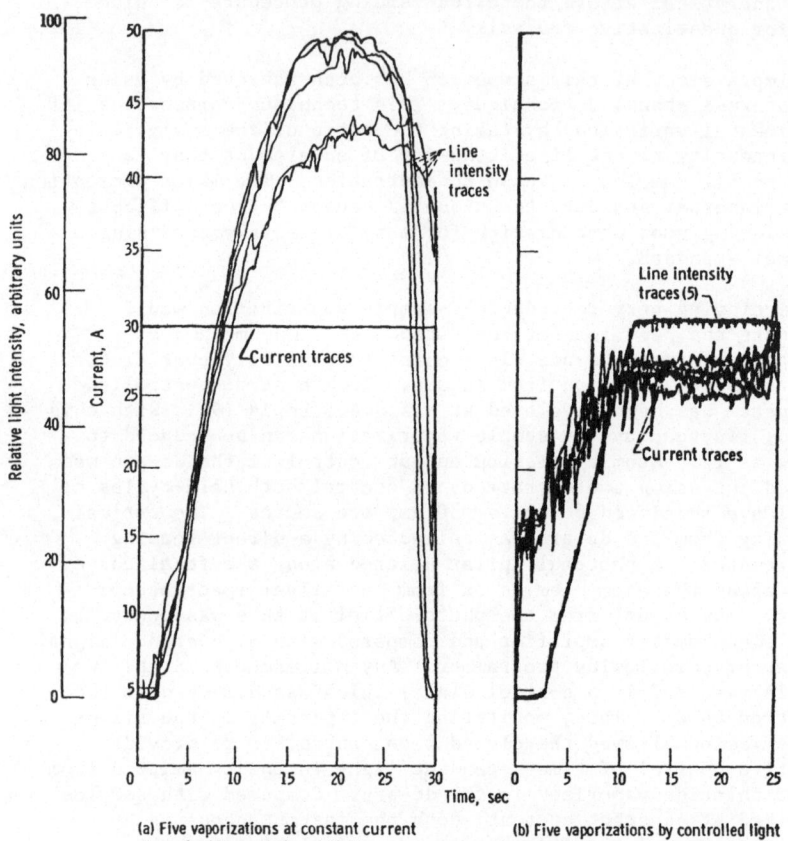

(a) Five vaporizations at constant current
(vaporized to completion at 30 A).

(b) Five vaporizations by controlled light
intensity (arc terminated after 25 sec).

- Comparison of repeatability of silver light emission with time. Arc atmosphere,
argon; pressure, 345 torr; cathode, Ta-tipped graphite; sample, 4.0 milligrams of AgCl in
anode cavity; spectral line controlled, Ag, 3502 Å.

Fig. 6 - (from Ref. #9)

The ability to vaporize the analyte elements reproducibly
into the dc-arc by controlling and programming the current
allows the spectrometer to be calibrated for absolute amounts of
each of these elements. A spectrochemical procedure was recently
described for quantitatively determining random combinations
of 20 programmed elements over the range of 10 nanograms to
100 micrograms absolute amount (10). The advantage of this pro-

cedure derive from the refinements of the static-argon arc source. Because of good detectability of analyte elements in the presence of larger quantities of silver chloride than formerly possible, dilution ratios of 1:400 are feasible with silver chloride serving as a common matrix. The use of a tantalum-tipped cathode for eliminating errors caused by arc wander also improves results. Finally, a method for feedback controlling either arc light intensities or arc current is a necessity for reproducible vaporization. The most important procedural advance was reduction of matrix effects through use of large relative amounts of spectroscopic buffer. (The term "matrix effect" is used here to denote any mechanism which either increases or decreases atomic emission from a given amount of material in the sample.) Standard solutions for each of the analyte elements were prepared and determined by the dc-arc method described. All of the element channels of the direct-reading spectrometer are recorded for each standard. The standard element channel provides data for the analytical curve for that element, while the responses from the element channels, other than the one being calibrated, provide the necessary information for making automatic corrections for direct spectral line interferences. The complete calibration data obtained is used to construct a 20-by-20 matrix array of response curves. Twenty element-calibration curves and 380 interference-response curves are calculated from these data. Both the element calibration data and the interference data are stored in a digital computer.

A test of the accuracy expected from such a procedure was run using a 10 microgram residue of a NBS standard stainless steel alloy. The calculation of percentage composition is done in three fashions. The first method is based on absolute quantities of analyte elements and is accurate if the mass balance is 100% and if matrix effects are negligible. The results were $\pm$ 9% relative standard deviation (RSD) computed from the average of the absolute standard deviation differences for metal elements present in greater than 1% by weight. Method two assumes that the mass of non-programmed elements in the sample is negligible. This method can provide compensation for matrix effects caused by non-quantitative recovery of sample elements from the electrodes. The accuracy in this case was $\pm$ 9% RSD. The third method provides compensation for matrix effects of all types but is less efficient than the other two methods, because it requires the preparation and arcing of a synthetic standard. This procedure also provides compensation for a system drift if the standards are arced with the samples. The accuracy here was $\pm$ 4.4% RSD. Since the result of the last method (obtained with the use of comparison standards) is best, it is apparent that the other two techniques are not correcting for all the matrix effects. (Nine months elapsed between calibration and the NBS standard determination.) This

technique does, however, correct for matrix effects strictly due
to interelement spectral interferences.

An optical drift correction system has been designed and is
currently being used to improve this analysis procedure (11).
An incandescent bulb, operated at a low constant current is
used as an optical irradiance standard. The lamp is mounted
inside the spectrometer near the grating, and the light from
this standard lamp is used to correct for drift in the PM
detectors, PM supply voltage, and electronic readout. It is
felt that this system will aid especially in long term drift
correction.

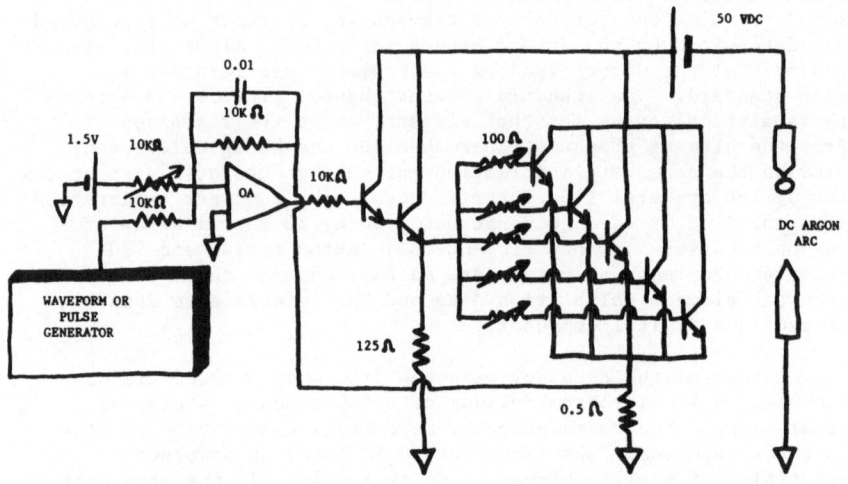

SCHEMATIC DIAGRAM OF THE SOLID STATE, DC ARGON ARC, CURRENT CONTROLLER

Fig. 7 - (from Ref. #12)

Present methods of dc-arc power supplies used in analytical
chemistry do not control the current in a rigorous fashion. The
usual control units consist of a large inductor which limits the
maximum current from a half or full-wave rectified supplied
voltage yielding a pulsating dc current with as much as 40%
peak-to-peak ripple. The employment of an accurately controlled
current source would enable the standardization of excitation
source waveform variables. In addition, waveforms could be
shaped to separate the sample heating and introduction rate
from the excitation conditions in the analytical plasma. This
could be accomplished by pulsing the arc between high and low

currents. A solid-state controlled current arc source has been recently designed to allow this (12). The unit uses operational feedback to duplicate any uni-directional current waveform from dc to 100 KHz. A circuit diagram is shown in Figure 7, illustrating how this is accomplished. The diagram consists of a pulse generator which is capable of producing square wave pulses, asymmetric square wave pulses, or triangular waves. This information is fed into an operational amplifier which has the capability of $\pm$ 50 volts output. The output from the operational amplifier powers a Darrington circuit which provides sufficient current to drive the high-power transistors at the output. The output transistors are configured as a source follower. A 1/2 ohm resistor is a portion of the source load, while the arc voltage drop itself is the remainder. Since the voltage across a constant resistor is fed back to the operational amplifier, the current is controlled. The Joule heat produced by the current flow in the sample carbon electrode determines the vaporization rate of sample material into the arc column. Because all the transistors controlling the current are enclosed in a feedback loop, the open loop gain of the operational amplifier of 100,000 compensates for any gain change in the rest of the system. If the $\pm$ 50 volt supply to the operational amplifier is reasonably ripple-free, ripple fluctuations in the 20 amp supply for the arc itself are controlled, and virtually pure dc flows through the arc itself. Preliminary work done at N.A.S.A. indicates that pulsing the arc increased the Joule heat at the sample electrode, and more sample was vaporized into the dc-arc column. Rapidly decreasing the arc current to only a few amperes decreased the background intensity of the dc-arc but still allowed excitation of the atomic species in the arc column. It remains to be seen if pulsing the arc can significantly increase sensitivity by affecting a change in the line-to-background ratio. It does allow dynamic changes in the arc which change the instantaneous excitation conditions. The system is also useful for programming the current from low initial values to high final values. In addition, feedback control using major element line channels or spectroscopic buffer channels can be more closely maintained with a system having a frequency response of 100 KHz, compared to systems where the frequency response is typically 120 Hz.

One major area of application of the Gordon dc-arc has been in the determination of trace elements in biological samples. The static-argon arc has resulted in a very stable arc source which has been used for three years. Data have been recorded at the University of Colorado Medical Center resulting in the analysis of 2000 hair samples and 3000 blood plasma samples for simultaneous determination of Cr, Mn, Zn, Fe, Mg and Cu. The improvement in arc stability achieved by the introduction of a tantalum sphere at the tip of a graphite cathode electrode operated in an inert argon gas has resulted in increased sensitivity and repro- ducibility of analysis, when compared with the dc-arc in air analysis of blood plasma reported in 1971 (12).

The biological sample preparation uses low-temperature oxygen plasma ashing to eliminate organic material. This treat- ment has resulted in a more stable arc, together with improved accuracy and reproducibility of analyses. Recovery of chromium with RF-induced oxygen plasma ashing is above 97%, and the ashed oxides are yielded in a small volume sample which is ideal for subsequent processing.

A major problem in the determination of trace metals in biological samples by atomic emission spectroscopy results from the fluctuation of spectral background. Since biological samples typically contain a large amount of alkaline metal elements such as sodium or potassium, the spectral background continuum is rather high when compared with water standards. The presence of high relative concentrations of sodium in the sample affects the excitation condition and causes a slope reduction by a factor of three in the standard curves from 1 to 5 ng Cr. This matrix effect starts when the sodium/chromium ratio reaches 10,000:1.

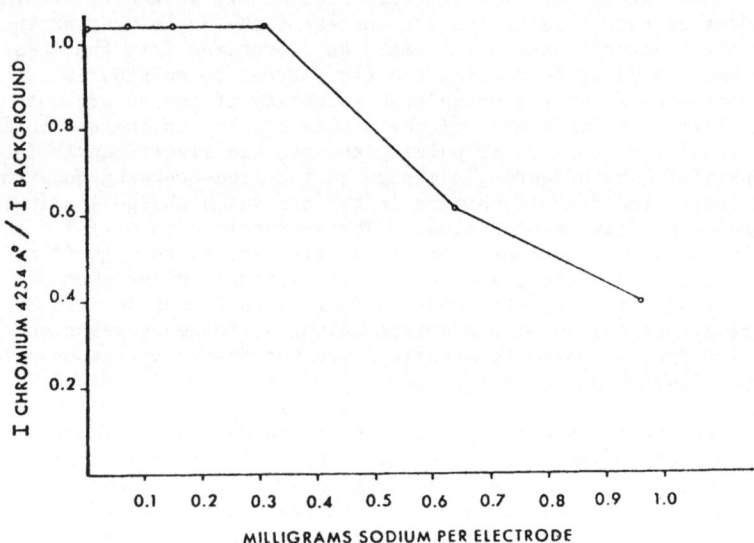

Fig. 8 - Effect of increasing increments of sodium on line: background ratio for 3 nanograms of chromium.

Figure 8 plots the effect of increasing increments of sodium on chromium line-to-background ratios for a constant quantity of 3 nanograms of chromium. For quantities larger than 0.3 mg of sodium per electrode, the chromium line-to-background ratio decreases very rapidly.

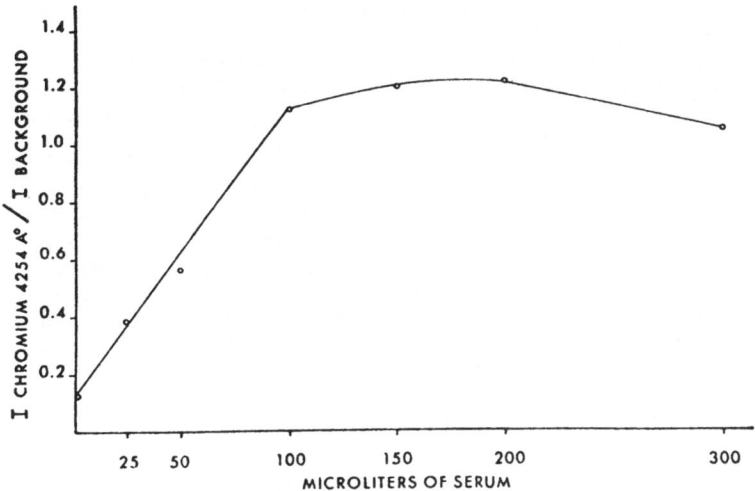

Fig. 9 - Variation in chromium line:background ratio with increasing increments of serum-ash sample.

Figure 9 illustrates the change in line-to-background ratio for increasing increments of serum ash. The data indicate that when the sample size is increased to the equivalent of more than about 150 μl serum, there is no further increase in the line-to-background ratio.

The electrode-to-electrode excitation conditions seem to fluctuate in spite of holding all parameters such as excitation current, pressure of argon, spectroscopic buffer and the like constant; and multiple electrode averaging is necessary to attain reasonable precision.

In order to reduce the effect of background fluctuation and low line to background ratio due to the quantities of alkaline elements present, a background correction method is necessary. Figure 10 illustrates the system developed at the University of Colorado Medical Center which is an extension of a system

described by Leys in 1969 (14). The background corrector
drawing shows light entering the entrance slit and passing
through the entrance refractor quartz plate. It is then
focused onto the diffraction grating and directed through the
exit slits onto the photocathode of a photomultiplier. The
quartz refractor plate is connected physically to a rotary
solenoid. Through the use of suitable electronics, the rotary
solenoid can be made to step from monitoring the atomic emission
line plus the background to a position approximately 1 Å away
where background alone is monitored. A mercury-wetted reed
relay separates the line-plus-background signal from the background signal and integrates both signals on separate capacitors.
Plotting the ratio of line to background minus one vs. concentration for various standard additions yields the initial sample
concentration at the 1-ordinate intercept. For a refractor
plate movement of $30°$, the background was measured at a distance
of approximately 1 Å longer wavelength than the emission line
position. It is necessary that there be no spectral interference
at this position. This was checked for the cases of manganese
and chromium in blood serum by recording the background signal
at small refractor plate angle increments on both sides of the
spectral lines.

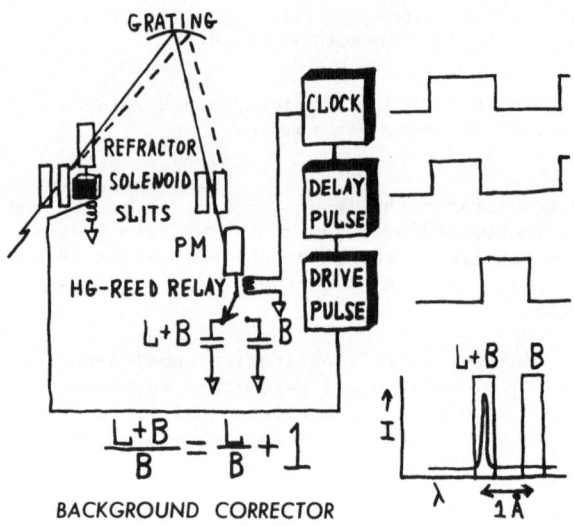

BACKGROUND CORRECTOR

$$\frac{L+B}{B} = \frac{L}{B} + 1$$

Fig. 10 - (from Ref. #14).

The increase in precision by a factor of three over background correction using two separate photomultiplier channels separated by about 50 Å from the emission line due to physical slit requirements, results from several advantages offered by this particular system. First, the system yields more stable operation because it only employs one photomultiplier as a detector. The same photocathode area of the single photomultiplier detector is used; thus, no compensation for varying sensitivity is required. Long term drift in photomultiplier sensitivity does not affect the measured signal ratio. Because both the spectral background and line-plus-background information are recorded on the same sample, separate samples are not required to determine the spectral blank value. This conserves sample, because no additional samples are required to get an off-profile average signal.

Easier sample calculations are possible from this single ratio compared to those made with data from two separate photomultiplier channels where sensitivities have to be compensated and normalized. Finally, a more representative background signal, indicated by the increase in precision, means that sample-to-sample fluctuations are compensated better by monitoring the background position just adjacent to the atomic emission signal (1 Å separation) than by measuring non-adjacent background.

Michael Hambidge has applied the static-argon dc-arc to the analysis of biological samples for studies of human trace mineral nutrition including, in particular, the development of techniques for the detection of human deficiencies of nutritionally essential trace elements.

The primary aim of these studies has been to develop means of detecting individuals with chromium deficiency (2,15,16,17). Chromium is one of several elements that have recently been recognized as essential micronutrients, and this element is now known to be necessary for normal glucose tolerance (18). The concentration of chromium in body fluids such as blood plasma and urine are only a few parts per billion; accurate measurement of these concentrations has presented a major analytical challenge, for which the static-argon dc-arc has proved invaluable.

Simultaneously, with these chromium studies and because of the potential for multielement analysis with emission spectrochemical techniques, data has been acquired on other elements of biological importance. One important result of including other elements in these studies has been the first documentation of zinc deficiency in otherwise normal young growing children in the United States (19). These children, who were detected because of their remarkably low hair zinc concentrations, have an impairment of taste acuity which can be corrected with dietary zinc supplementation. Typically, they also have a poor appetite and relatively poor growth, and it seems likely that zinc deficiency has prevented realization of their full growth potential. Preliminary studies suggest that at least a marginal zinc deficiency may be a relatively common problem in young growing children in this country.

REFERENCES

(All N.A.S.A. documents are available through Clearinghouse for Federal Scientific and Technical Information, Springfield, Virginia 22151. The figures from all references are used by permission.)

1. Gordon, W.A.: N.A.S.A., T.N. D-2598. Use of temperature buffered argon arc in spectrochemical trace analysis. (1965)
2. Hambidge, K.M.: Use of static-argon atmosphere in emission spectrochemical determination of chromium in biological materials. Anal. Chem., 43:103 (1971).
3. Mellichamp, J.W.: Cored cathodes for stabilization of the dc-arc. Appl. Spectros., 21:23 (1967).
4. Gordon, W.A.: N.A.S.A., T.N. D-4236. Stabilization of dc-arcs in static-argon atmospheres for use in spectrochemical analysis (1967).
5. Stallwood, B.J.: Air cooled electrodes for the spectrochemical analysis of powders. J. Opt. Soc. Am., 44:171 (1945).
6. Margoshes, M., Scribner, B.F.: The plasma jet as a spectroscopic source. Spectrochim. Acta., 15:138 (1959).
7. Seeley, J., Ph.D.: Thesis, Dept. of Chem., Colorado State University, Fort Collins, Colorado (1973). (Dr. R. Scogerboe, Director)
8. Gordon, W.A., Chapman, G.B.: N.A.S.A., T.N. D-6888. Determination of work functions near melting points of refractory metals by using a direct-current arc (1972).
9. Gordon, W.A.: N.A.S.A., T.N. D-4769. A servocontroller for programming sample vaporization in direct current arc spectrochemical analysis (1968).
10. Gordon, W.A., Chapman, G.B.: Quantitative direct-current arc analysis of random compositions of microgram residues in silver chloride common matrix. Spectrochim. Acta., 25:123 (1969).
11. Franklin, M.L., Gordon, W.A.: N.A.S.A., T.N. D-6313. Use of incandescent lamps to measure optical and detector drifts in photoelectric spectrometers (1971).
12. Franklin, M.L., Gordon, W.A.: N.A.S.A., T.N. Operational feedback arc controller for atomic spectrochemical studies (in preparation).

13. Niedermeier, W., Griggs, J.H., Johnson, R.S.: Emission spectrometric determination of trace metals in biological fluids. Appl. Spectros., 25:53 (1971).

14. Franklin, M.L., Hambidge, K.M.: Abstracts Rocky Mountain Regional Meeting, Am. Chem. Soc., Colorado State University, Fort Collins, Colorado, June 30, 1972. #2: Application of a photoelectric system for monitoring atomic emission line and spectral background intensities used for multi-element biological trace metal analysis.

15. Hambidge, K.M., Baum. J.D.: Hair chromium concentrations of human newborn and changes during infancy. Am. J. Clin. Nutr., 25:376 (1972).

16. Hambidge, K.M., Franklin, M.L., Jacobs, M.A.: Changes in hair chromium concentrations with increasing distances from hair roots. Am. J. Clin. Nutr., 25:380 (1972).

17. Hambidge, K.M., Franklin, M.L., Jacobs, M.A.: Hair chromium concentration: effects of sample washing and external environment. Am. J. Clin. Nutr., 25:389 (1972).

18. Hambidge, K.M.: Chromium nutrition in the mother and growing child. In: Newer Trace Elements in Nutrition, edited by W. Mertz and W.E. Cornatzer, N.Y.; Marcel Dekker, 1971, chapter 9.

19. Hambidge, K.M., Hambidge, C., Jacobs, M.A., Baum, J.D.: Low levels of zinc in hair, anorexia, poor growth and hypogeusia in children. Pediat. Res., 6:868 (1972).

20. Present address for M.L. Franklin: Dept. of Chemistry, University of Missouri, Columbia, Missouri 65201.

21. Supported by Public Health Service Grant 1-RO1-AM-12432 from the National Institute of Arthritis and Metabolic Diseases. Salary paid in part by grant RR-69 from the General Clinical Research Center's Program of the Division of Research Resources, National Institutes of Health, Bethesda, Maryland.

THE USE OF CALCULATED X-RAY POWDER PATTERNS IN THE INTERPRETATION OF QUANTITATIVE ANALYSIS

C. Clark*, D.K. Smith and G.G. Johnson, Jr.**

Dept. of Geosciences, Pennsylvania State Univ.

University Park, Pennsylvania 16802

ABSTRACT

The calculated powder diffraction patterns of well characterized substances are being computed for inclusion in the Powder Diffraction File. The intensities reported in these patterns include relative integrated and peak height intensities which are simulated using computer programs allowing for various aberrations, geometric conditions, and resolutions. A scaling factor is reported with these patterns which allows for the conversion of these intensities to an absolute scale for application in quantitative analysis. By absolute scaling all patterns or by using a basis of $I/I_{\alpha-Al_2O_3}$ quantitative interpretation of experimental patterns is permitted.

INTRODUCTION

Perhaps the most important change in the history of crystallography has stemmed from the development and avail-

---

* Formerly with the Department of Geosciences; now at the Materials Research Laboratory, Pennsylvania State University, University Park, Pennsylvania 16802.

** Present Address: Materials Research Laboratory, Pennsylvania State University, University Park, Pennsylvania 16802.

ability of ultra-high-speed computers. Because of the past emphasis on single-crystal studies, elaborate computer programs have been written to optimize the procedures of checking proposed models. As features of single crystals become better understood, more advanced studies involving poly-crystalline samples become conceivable; however, similar programs for polycrystalline conditions are less common, and most tend to study individual features of a poly-crystalline pattern.

The calculation of the set of integrated intensity values from a specific atomic arrangement and unit cell is not difficult. The fundamental equation is:

$$I(hk\ell)_{rel.} = \alpha p \; LPAT \left| F(hk\ell) \right|^2 g$$

where $I(hk\ell)_{rel.}$ is the diffracted intensity on a relative scale usually with $I_{max}=100$, α is a scale factor used to place the data on the relative scale, p is the multiplicity factor, L is the Lorentz factor, P is the polarization factor, A is the absorption factor, and $F(hk\ell)$ is the structure factor. The factor g is a function of many specialized features unique to powder patterns such as preferred orientation, strain, particle size and shape and experimental conditions, and is not always easy to determine.

Three programs have been written which evaluate the integrated intensity for conditions where g = 1. The programs by Smith (1963, 1967), Clark, Smith and Johnson (1973), Jahanbagloo and Zoltai (1966), Jahanbagloo, Vand and Johnson (1968), Jietschko and Parthe (1965) and Yvon, Jeitechko and Parthe (1969) are identical in all essential features. The comments in this discussion are generally applicable to all these programs. The sequence of steps in the generation of a set of integrated intensities is straightforward. The necessary input data include the unit cell constants, the space group symmetry equivalent positions, the atom position and thermal parameters, their atomic scattering factors, the linear absorption coefficient and x-ray wavelengths to be used. Additional useful data include the rules of extinction and the Miller indices of the strongest reflection. Using the space group data (which defines the Laue group) and the unit cell parameters, the full set of nonredundant Miller indices and corresponding d-spacings is generated. Reflections not allowed by space group extinctions are eliminated to minimize the calculations. Integrated intensities for each hkℓ are then generated by calculating the temperature-modified structure factor, applying the appropriate multiplicity, Lorentz-polarization and absorption factors. Finally, a scale factor is introduced to reduce the calculated intensities to the desired relative scale. If the input structure data are of high quality, usually requiring an R-factor less than 0.10, the integrated intensities match very closely with experimental data. The calculation of the structure factors for powder intensities is the same as for single crystals and yields no special problems. An important input variable in most programs is an atom multiplier which is used for distributing fractional atoms throughout the structure on special symmetry positions or for partial, mixed or statistical site occupancy. This multiplier is especially useful in powder work for order-disorder and solid-solution studies.

ABSOLUTE INTEGRATED INTENSITY

For most purposes involving studies with powder patterns, only relative values of the intensities are necessary; however, for some purposes, such as in quantitative comparisons between samples absolute intensities are more useful. For such purposes, calculated absolute intensities provide useful checks on the calibration samples and prove useful as standards.

For the standard diffractometer, the expression for the absolute intensity reflected from a thick, flat, single-phase sample of theoretical density is

$$I(hk\ell) = [I_o A \left(\frac{e^2}{mc^2} \right)^2 \frac{\lambda^2}{32R}] [\frac{1}{2\mu v^2}] [p \frac{1+\cos^2 2\theta}{\sin^2\theta\cos\theta} F(hk\ell)^2$$

$$e \frac{2B \sin^2\theta}{\lambda^2} A(\theta)g]$$

Where I_o = intensity of incident beam/unit area
      A = cross-sectional area of incident beam
      e = electronic charge
      m = electronic mass

```
c        = velocity of light
λ        = x-ray wavelength
R        = distance to detector slit
μ        = linear absorption coefficient
v        = unit cell volume
p        = multiplicity factor
θ        = Bragg angle
F(hkℓ)   = structure factor
B        = temperature parameter
A(θ)     = absorption factor (angularly dependent portion)
g        = experimental condition function
```

This expression may be dissected into three parts as indicated by the square brackets. The first term is constant for a given diffractometer system, the second term is a function of the sample only, and the third term is a function of the sample and the diffraction angle. For relative intensities only the third term need be evaluated. To place the relative intensities on an absolute scale is simply a matter of multiplying by the appropriate constants. Where experimental work is to be done on the same or a similar diffraction system, the first term is common and may be ignored or applied by the user.

Thus the calculated intensities may be placed on a relative absolute scale (i.e., ignoring the first term) by multiplying all relative intensities by $v = 1/(2\alpha\mu V^2)$ where α is the scale factor used to place the intensities on the chosen relative scale.

PROFILE SIMULATION AND PEAK INTENSITIES

The most direct intensity value obtainable from the calculation is the integrated value. These numbers are quantitative and are directly comparable with the total energy per unit time obtained from the sample as it satisfies each Bragg condition i.e., the area under a diffractometer profile. For most quantitative work, therefore, it is desirable to measure these areas, thus eliminating instrumental observations and other effects which cause the line shape to deviate from an easily interpreted analytic profile. However, in many studies not requiring the rigor of quantitative measurements much time could be saved if peak intensities could be calculated which were more meaningful with respect to experimental patterns. This comparison between observed and calculated patterns could be made readily without the need for area measurement.

The diffractometer profile is a convolution of the x-ray spectral distribution with the sample and instrumental aberrations. The spectral distribution is Cauchy-like in that the tails decay as $I/[1 + K^2(x - x_o)^2]$. The peak region is more nearly Gaussian, but no one analytical function has been found to express the total profile. Thus, both the Cauchy and Gaussian functions have been used to analyze line profiles under various conditions. Figure 1 shows the relative shapes of the two curves.

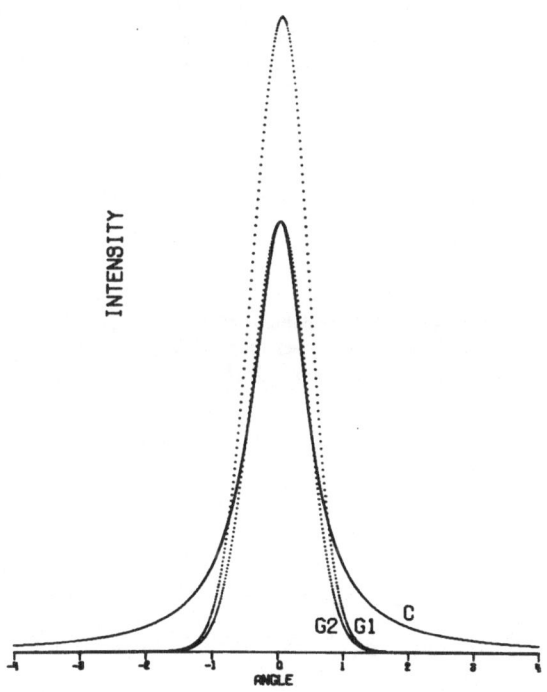

Fig. 1.   Comparison of the Cauchy and Guassian Function
           Showing Relative Shape of the Two Curves.

A very close approximation to an experimental diffraction pattern can be obtained by assuming the Cauchy shape for
both the α_1 and α_2 components of each diffraction profile.
The appropriate half width, ω, is used along with the integrated intensity, and each profile is calculated at intervals
of $\theta(2\theta)$. By centering each profile at the appropriate, θ_0,
the composite trace is achieved by summing all contributions
at each θ value along the trace. This resulting single
valued function, $I(\theta)$, represents the theoretical intensity
distribution as a function of diffraction angle and should
be directly comparable with experimental patterns.

Assuming the Cauchy profile is acceptable, the only unknown in the calculation is the half-width, ω. The angular
dependence of the half-width is a complex function of many
of the geometric aberrations of the x-ray optical system.
Although some of the aberrations are known, it is proved
more practical at this stage to use empirical functions
measured from experimental traces for specific diffracto-
meters. Figure 2 shows some of the points so determined.

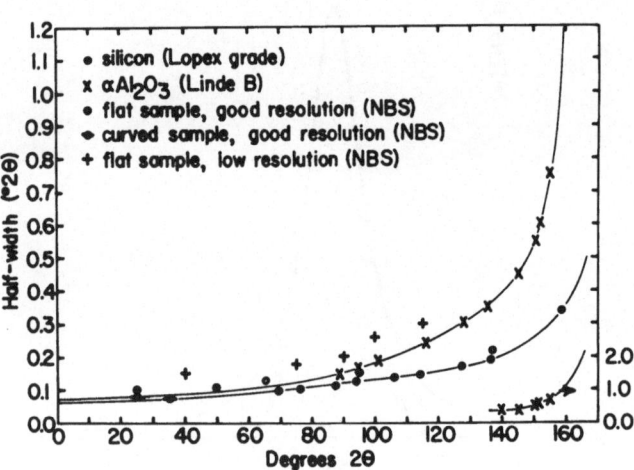

Fig. 2.   Resolution as a Function of 2-Theta.

The two lines through the Si and $\alpha\text{-Al}_2\text{O}_3$ points represent
the functions used by the computer to produce the patterns
presented in this article. The Si function is representative
of a very well crystallized sample; the $\alpha\text{-Al}_2\text{O}_3$ is Linde B
material with a particle size of 0.03, and thus, the patterns
affected by crystallite size effects. A desired half-width
at a specific 2θ angle is given to the computer, and the
appropriate function is determined by an interpolation or
extrapolation proportionally between Si and $\alpha\text{-Al}_2\text{O}_3$ curves
over the full 2θ range.

The patterns produced by the computer are esthetically
pleasing. The experienced experimentalist will be immediately skeptical that anything so perfect could possibly be
of any value. Figures 3a, b and c illustrate a comparison
between the calculated pattern and NBS standard intensity
patterns. The patterns show a close fit in the low-order
lines, with a slightly poorer agreement in the back reflection region. This deviation is in part due to the sensitivity of the half-width dependence on diffractometer
characteristics in the high angle region, to the inaccu-
racies still remaining in the structure and to the variation
in sample preparation. More detailed discussion giving
several examples of calculated versus experimental traces
has been presented by Smith (1968) and Borg and Smith (1969).

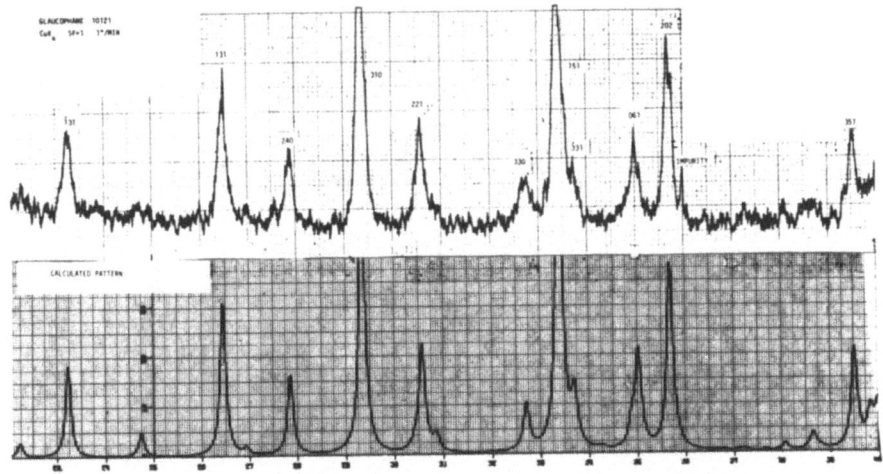

Fig. 3a. Glaucophane Pattern, Experimental
          Top; Calculated Bottom.

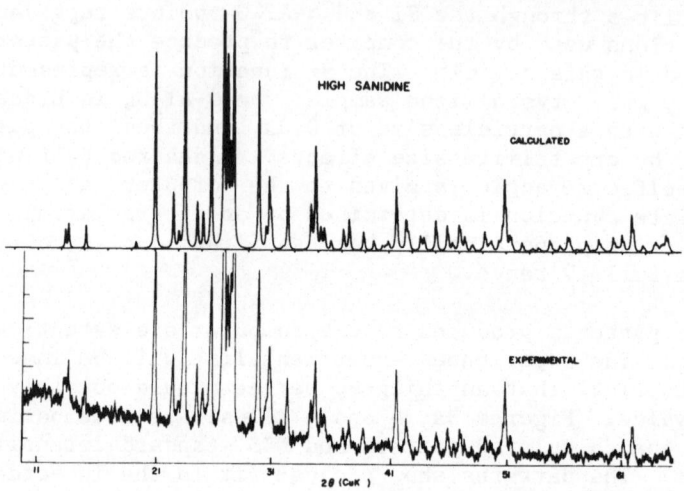

Fig. 3b. High Sanidine Pattern, Calculated
Top, Experimental Bottom.

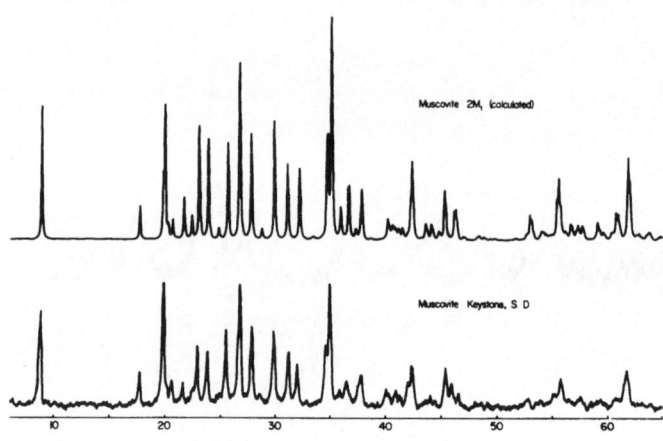

Fig. 3c. Muscovite (2m$_1$) Calculated
Top, Experimental Bottom.

These approximate traces are useful for semiquantitative studies but caution must be used when interpretations are based on them. Until we have a more accurate knowledge of the true spectral profile, any simulation of the diffractometer results will be only a fair approximation. The most readily obtainable unit of data from the diffractometer, the peak intensity, is the most difficult to synthesize with any degree of confidence because of the extreme sensitivity of the peak to the other parameters involved. The measured diffractometer standards presently being incorporated in the Powder Diffraction File are based on peak intensity measurements. The use of peak intensities is not objectionable because the main aim of the Powder Diffraction File is to aid in identification. However, for quantitative purposes peak intensities are unreliable as indicated, and the Powder Diffraction File is now including calculated integrated intensities as well as peak height intensities.

Applications for calculated patterns include the determination of powder data for material whose structure has been determined but whose powder pattern has not been recorded. Other needs include standards for samples whose preferred orientation is difficult to eliminate, for series of isostructural compounds, for members of isostructural solid solutions, for disordered states, for checks on experimental data, for calculation of expected patterns of postulated structures in pressure temperature studies, and for testing form factor ionization.

In order to illustrate the potential value of the calculated patterns in isomorphous series, the solid solution of olivines between forsterite and fayalite have been studied by Jahanbagloo, Vand and Johnson (1968). The unit cell data of 16 analyzed olivines show the existence of linear relationships between the lattice constants and the Fa-content in olivines. The following linear equations, obtained in this work, with a least-squares technique, can be used to determine the Fa% from the unit cell parameters in unanalyzed olivines.

$$Fa\% = -7188.27 + 1511.77 \text{ a} \pm 5.8\%$$
$$Fa\% = -3317.44 + 325.53 \text{ b} \pm 3.8\%$$
$$Fa\% = -4877.01 + 815.40 \text{ c} \pm 3.7\%$$
$$Fa\% = -1525.96 + 5.265 \text{ v} \pm 3.6\%$$

The lattice constants versus composition can also be obtained from the following relations:

$$a = 4.7556 + .0642 \text{ Fa\%} \pm .004\text{Å}$$
$$b = 10.1922 + .3035 \text{ Fa\%} \pm .001\text{Å}$$
$$c = 5.9817 + .1212 \text{ Fa\%} \pm .004\text{Å}$$
$$v = 289.89 + 18.780 \text{ Fa\%} \pm .69\text{Å}^3$$

The x-ray powder patterns of six members of the olivine series between forsterite and fayalite (with intervals of 20 percent Fa) have been calculated. These patterns can be used as standards for the x-ray powder patterns of olivines. The lattice constants for these patterns were obtained from equations and their atomic coordinates and temperature factors were estimated by interpolation and extrapolation of two members of the olivine series: forsterite ($Fo_{90}Fa_{10}$) and hortonolite ($Fo_{47}Fa_{53}$) refined by Gibbs (1964). The variation of calculated intensities with composition for several of the lattice planes have been illustrated in Figure 4. Such graphs and equations can be used as standards for the determination of Fa% in unanalyzed olivines.

The relative abundance of the component minerals in a mixture can be calculated from their x-ray powder patterns provided that all the constituent minerals are identified and their calculated intensities, on an absolute scale, are available. The calculated absolute intensities per unit volume of a substance can be obtained by multiplying its "Intensity Normalizing Factor" by the relative calculated intensities reported in its calculated powder patterns.

If the observed relative integrated intensities of the diffraction lines of each component in a mixture are divided by the corresponding absolute intensities of the calculated patterns and are corrected for the absorption effect of the mixture, they will be proportional to the quantities of the components present. The theory of these calculations are described by Klug and Alexander (1954). In the technique developed in this work, first an approximate composition is calculated by neglecting the absorption effect, and then both the composition and the absorption coefficient of the mixture are refined in consecutive cycles until convergence is obtained. The reliability of this quantitative analysis is a function of the quality and resolution of the observed pattern and increases as a greater number of lines becomes available for each component. The validity of this technique has been tested on five artificial mixtures and the result has been shown in Table I.

Table I. The Analysis of Five Artificial Mixtures with the Aid of Calculated Powder Patterns.

Examples	Minerals	Prepared (wt. %)	Analyzed (wt. %)
Mixture	Quartz	28.2	27.2
No. 1	Fluorite	55.3	58.5
	Calcite	16.5	14.3
Mixture	Halite	50.0	48.7
No. 2	Fluorite	30.0	28.8
	Rutile	20.0	22.5
Mixture	Niter	90.0	92.5
No. 3	Sylvite	10.0	7.5
Mixture	Quartz	65.0	69.2
No. 4	Pyrite	35.0	30.8
Mixture	Calcite	65.0	68.2
No. 5	Hematite	35.0	31.8

This technique is comparable to the use of an interval standard as an experimental reference intensity as established by members of the Joint Committee on Powder Diffraction Standards (1972). Alpha Al_2O_3 (synthetic corundum) was chosen as the internal standard to be mixed with the material. This was chosen because it is available commercially as Linde "A" in approximately one micron particle size, because of its purity and chemical stability, and because of its freedom from orientation due to shape in sample preparation. A mixture of sample and standard, 1:1 by weight is used. It is necessary to run only that portion of the pattern that includes the strongest line of the sample and the Al_2O_3 (see Figure 5). For corundum the strongest line has $hk\ell = 113$, d = 2.085Å ($2\theta = 43.36°$ for CuKα.) The direct ratio of the peak height of the two lines is reported as I/I_{cor}. In a few instances the strongest line of one material may fall on lines of the other. In this case the second strongest line is measured, and based on previous

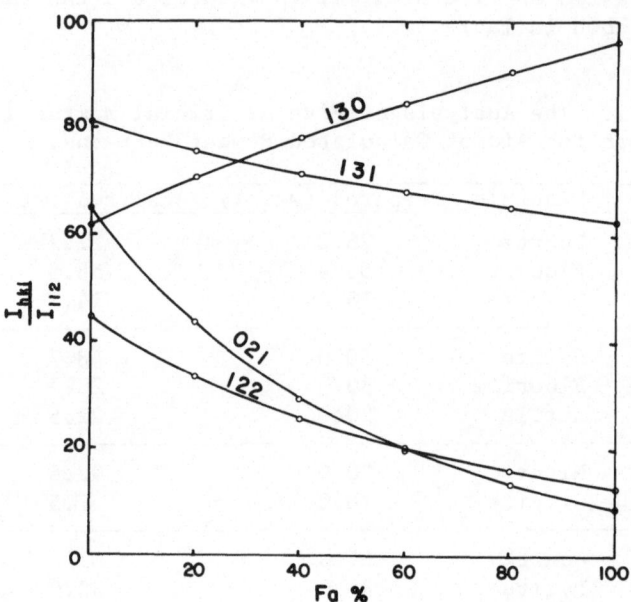

Fig. 4.   Variation of Intensities of Some of the
Lattice Planes Versus Composition in Olivines.

knowledge of the relative peak heights a correction is made
thus enabling one to reconstruct the value of the strongest
lines.   The value of I/I_{cor} is reported only to one decimal
place.   It should be noted that I/I_{cor} applies to peak
heights while the calculated scale factor applies to integrated intensities.   An I/I_{cor} could be calculated for each
pattern.

Quantitative phase analysis by x-ray powder diffraction
methods has had varying degrees of success depending on the
nature of the material being studied and the amount of effort
expended to eliminate common experimental problems such as
preferred orientation.   Calculated patterns may be used in
several ways to facilitate quantitative studies.   In the
case of solid solutions, where the crystal structures are
accurately known as a function of composition, calculated
intensities could yield determinative curves which could be
applied to experimental data.   In phase mixtures, the cal-

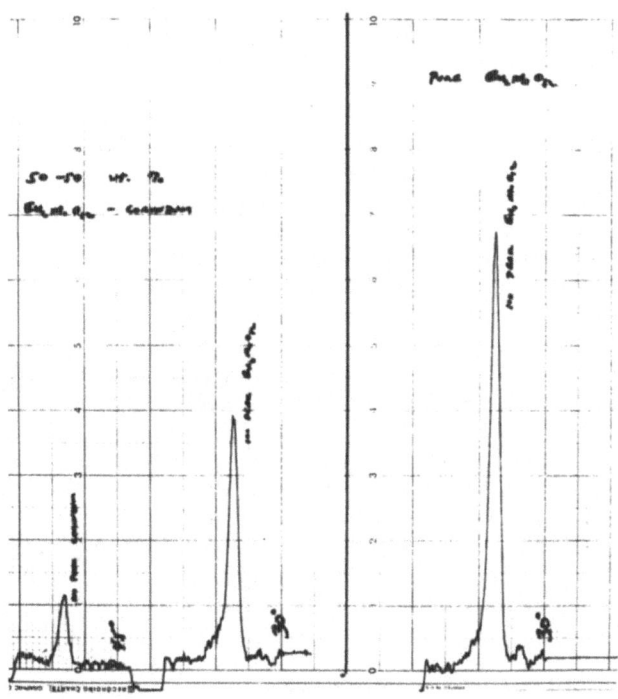

Fig. 5.   Measurement of $I/I_{corundum}$.   Left: 50-50 wt. %
Mixture.   Right: Pure Standard.

culated intensities of the components could be combined into "mixed" patterns, allowing for intensity changes due to
microabsorption, and determinative curves could be established as a function of quantity present. Also, calculated
patterns could be used strictly as a reference to check the
experimental patterns for preferred orientation, particle
statistics, extinction, etc. The experimental patterns
are then used to establish the determinative curves. Regardless how these calculated standards are used they should
yield increased confidence in the use of x-ray diffraction
techniques in quantitative phase analysis.

I/I_c PDF Card

3.0	22-1094	Dysprosium niobium oxide, $DnNbO_4$
4.7	22-1098	Europium molybdenum oxide, $Eu_2Mo_2O_7$
3.3	22-1097	Europium molybdenum oxide, $EuMoO_4$
1.6	22-1099	Europium niobium oxide, $EuNbO_4$
2.4	22-1095	Erbium niobium oxide, $ErNbO_4$
9.1	22-1101	Europium tungsten oxide, $EuWO_4$
1.5	22-1104	Gadolinium niobium oxide, $GdNbO_4$
3.4	22-1175	Neodymium niobium oxide, $NdNbO_4$
3.9	22- 885	Rhodium ammine chloride, $[Rh(NH_3)_5Cl]Cl_2$
4.0	22-1303	Samarium niobium oxide, $SmNbO_4$

Table II represents a selected portion of values of I/I_c as given in the Powder Diffraction File.

The computer programs needed for the calculation of powder patterns can be obtained by writing to D. K. Smith. These programs are written in FORTRAN IV for the IBM 360/50[+] and IBM 370/155[+]. The program also uses magnetic tape storage for the storage of data files (space group, atomic scattering factors, multiplicity factors, etc.).

ACKNOWLEDGEMENTS

This work, both the computer output and the experimental determination of I/I_c has been sponsored by the Joint Committee on Powder Diffraction Standards, a Pennsylvania nonprofit organization that collects and distributes standard x-ray powder diffraction patterns.

REFERENCES

Borg, I.Y. and D.K. Smith (1969) Calculated Powder Patterns. Part II. Six Potassium Feldspars and Barium Feldspar, Amer. Mineral. 54, 163–181.

Clark, C., D.K. Smith and G.G. Johnson, Jr. (1973) A FORTRAN IV Program for Calculating X-Ray Powder Patterns-Version 5, Bulletin of the Experiment Station, The Pennsylvania State University, University Park, Pa.

Gibbs, G.V. (1964) Private Communication.

Jahanbagloo, I.C. and T. Zoltai (1966) Calculated X-Ray Powder Patterns. University of Minnesota, Department of Geology and Geophysics, Minneapolis, Minnesota.

Jahanbagloo, I.C., V. Vand and G.G. Johnson, Jr. (1968) Calculated X-Ray Powder Patterns and Their Applications in Quantitative Interpretations. The Pennsylvania State University, Materials Research Laboratory, University Park, Pa., MRL Monograph No. 3.

Jietschko, W. and E. Parthe (1965) A FORTRAN IV Program for the Intensity Calculation of Powder Patterns. University of Pennsylvania, The School of Metallurgical Engineering, Philadelphia, Pa.

Joint Committee on Powder Diffraction Standards (1972) Inorganic Index to the Powder Diffraction File. p. 1421. Published by the Joint Committee on Powder Diffraction Standards, 1601 Park Lane, Swarthmore, Pa. 19081.

Published by the Joint Committee on Powder Diffraction Standards – a Pennsylvania nonprofit organization that collects and distributes standard x-ray powder diffraction patterns.

Klug, H.P. and L.E. Alexander (1954) X-Ray Diffraction Procedures, p. 410–439, John Wiley and Sons, New York.

Smith, D.K. (1963) A FORTRAN Program for Calculating X-Ray Powder Diffraction Patterns. Lawrence Radiation Laboratory, Livermore, California. UCRL-7196.

Yvon, K., W. Jeitschko and E. Parthe (1969) A FORTRAN IV
Program for the Intensity Calculation of Powder Patterns
(1969 Version). University of Pennsylvania, Laboratory for
Research on the Structure of Matter, Philadelphia, Pa.

_____ (1967) A Revised Program for Calculating
X-Ray Powder Diffraction Patterns. Lawrence Radiation
Laboratory, Livermore, California. UCRL-50264.

_____ (1968) Computer Simulation of X-Ray Diffractometer Traces. Norelco Rep. 15, 57-65.

SILICA ANALYSIS OF INDUSTRIAL HYGIENE SAMPLES BY X-RAY DIFFRACTION: INTERIM REPORT

D. T. Donovan, J. W. Knauber,
F. H. VonderHeiden
Bureau of Occupational Health, Pennsylvania
Department of Environmental Resources
Harrisburg, Pennsylvania

Following development of an x-ray diffraction packed powder internal standard procedure [1,2] for analyzing settled dust samples for free silica, a method was designed for respirable air-borne quartz. Silver membrane filters, because of their greater resistance to humidity, were being used for gravimetric determinations of inert or nuisance dusts. Further, it was determined that the x-ray diffraction pattern of the silver membrane was flat and even and that silver had no interfering peaks and should be adaptable to direct quartz analysis. Talvitie, Brewer, Knott and Oberg had shown the possibility of a more direct technic for the x-ray diffraction analysis of air-borne dusts. [3,4,5]

The packed powder procedure for settled dust by x-ray diffraction which is also used for coal dust samples entails combining approximately 0.5 g. dust with an internal standard, calcium fluoride, so that the ratio of the weight of sample to standard is constant (1:0.25). The mixture is mechanically ground for twenty minutes in tungsten carbide vials to ensure less than 325 mesh sizing. After moistening with 1-2 drops of collodion - iso amyl acetate (1:10), the mixture is packed immediately into phenolic linen sample holders in router milled cavities (300 mg. or 50 mg. capacity). The 300 mg. holder is 1" x 3/4" x 1 1/16"

deep. See Figure 1. A planar surface is attained by
smoothing off the surface of the samples with the edge of
a microscope slide.

The sample is automatically rotated from approxi-
ately $2\theta = 19°$ to $2\theta = 30°$ at constant angular speed, $0.4°/$
min., through the copper target x-ray beam (15 MA & 50
KV). The stripchart at the speed 0.2"/min. shows the
intensity of the rays diffracted by the sample as a function
of angular position, as shown in Figure 2. Quantitation is
accomplished by measuring the ratio of the intensity of the
quartz diffraction peak to that of the calcium fluoride peak
and comparing with a calibration curve -- Figure 3 prepared from standard mixtures of less than 325 mesh amorphous silica, calcium fluoride and varying amounts of
quartz. The areas of the quartz $(3.34 \mathring{A})$ and the calcium
fluoride$(1.93 \mathring{A})$ peaks are measured by drawing the best
average baseline and treating the peak area above the baseline as a triangle. Electronic scalers may also be used to
measure counts in peak areas. The peak areas are dependent upon the combination of goniometer and chart speeds
and these two variables must never be different from those
established with the standards.

Scanning from $19°$ to $30°$ (span differs for targets
other than copper) provides a range suitable for seeing the
major peaks of tridymite $(4.39 \mathring{A})$, cristobalite $(4.09 \mathring{A})$,
quartz $(3.34 \mathring{A})$, and calcium fluoride $(1.93 \mathring{A})$. Muscovite
and certain other feldspars, cassiterite and graphite also
have peaks at $3.34 \mathring{A}$.   When quartz is detected, the sample
should be scanned for secondary peaks of these interferences and if present the sample must be re-scanned in a
secondary peak area for quartz $(4.26 \mathring{A})$. In this case,
standardizations will have to be repeated for the secondary
quartz peak selected. The secondary peaks for muscovite,
cassiterite and graphite are $9.95 \mathring{A}$, $2.64 \mathring{A}$ and $1.68 \mathring{A}$
respectively. Quantitations of tridymite and cristobalite

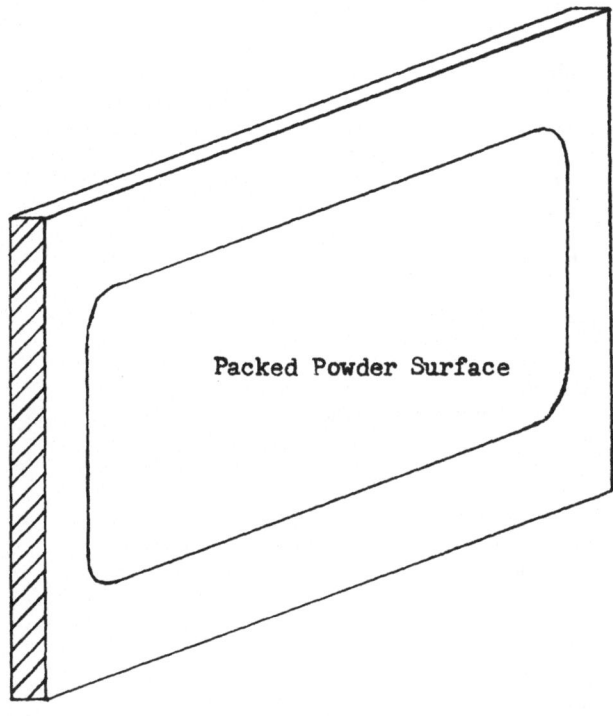

Figure 1 – Sample Holder for bulk and settled dust

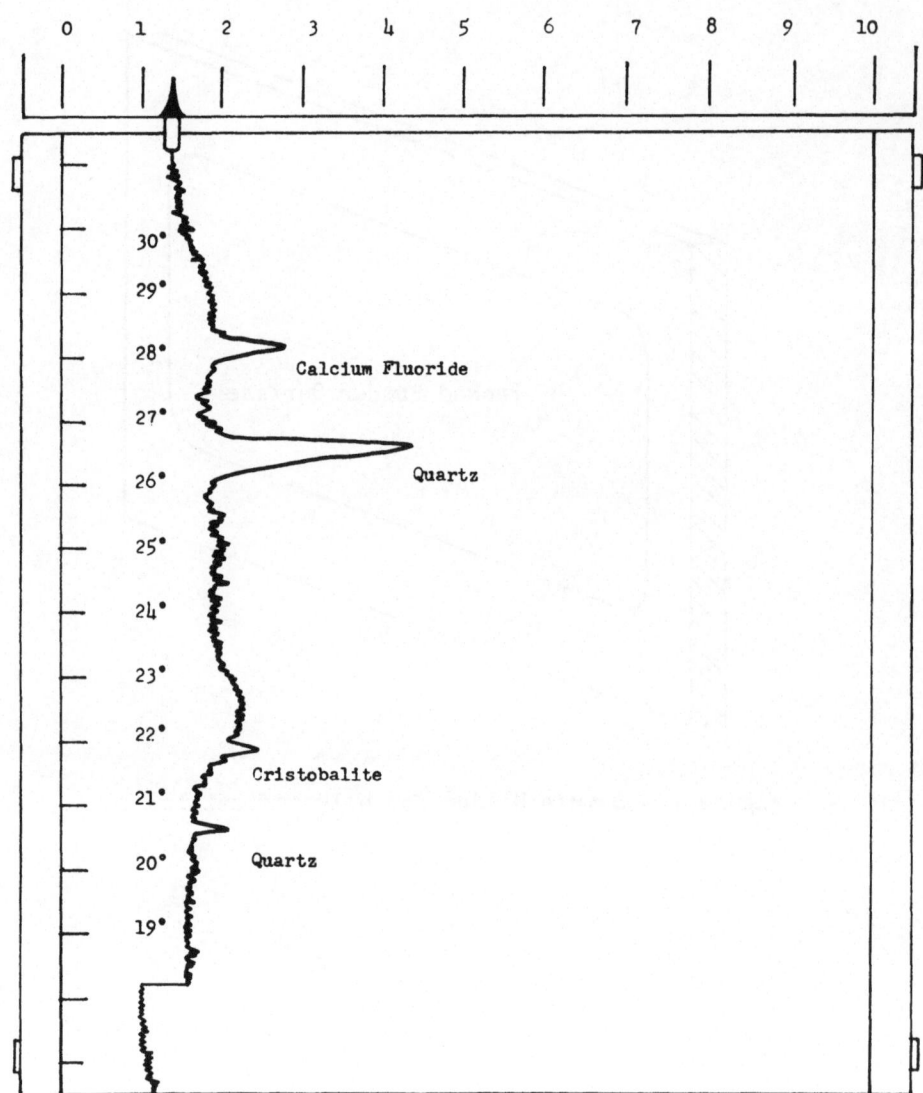

Figure 2.   Strip chart for bulk sample analysis

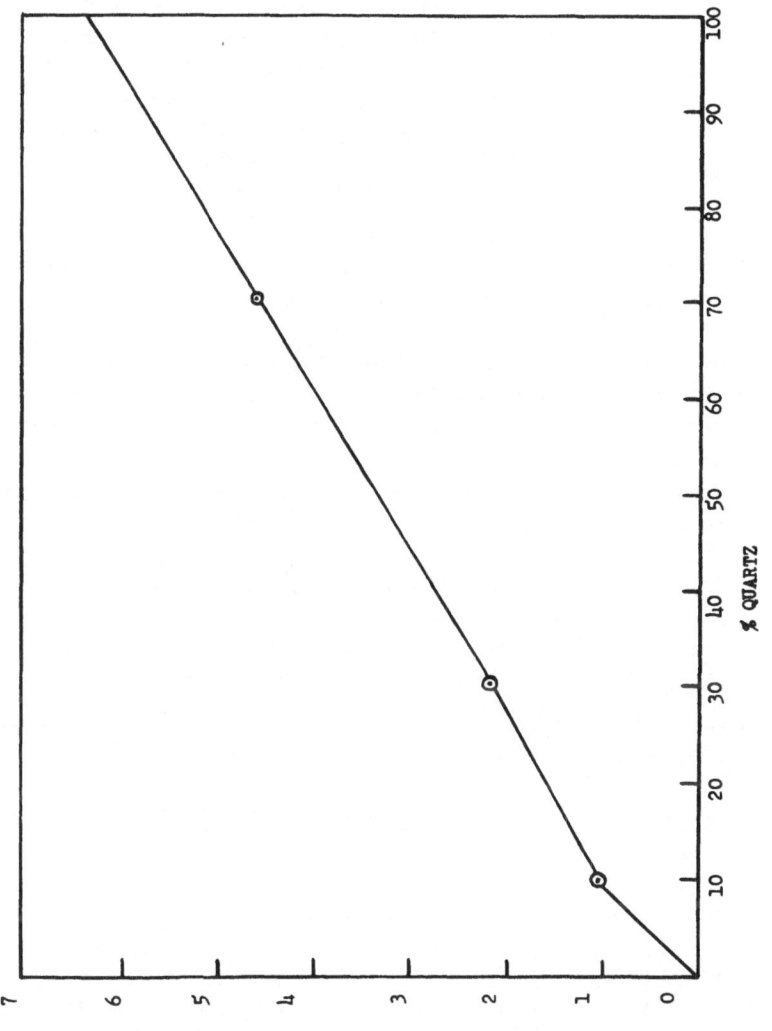

Fig. 3. Calibration curve for bulk samples.

are routinely made on all packed powder samples in our laboratory.

The practical limit of sensitivity for the internal standard technic is approximately 0.25 milligram or 0.5% of a 50-milligram sample. If the total weight of quartz is less than this, no quartz will be detected since the intensity of the peak will not be significantly above the level of the background.

Results using the x-ray diffraction procedure for settled silica samples compared favorably with the traditional wet chemical method by Talvitie[7] (Figure 4).

The transition from determining quartz in settled dust samples by x-ray diffraction with an internal standard to a silver membrane direct x-ray diffraction technic for air-borne quartz included using the silver peak (2.37 Å) as a measure of the extent of a sample's total x-ray absorption effect.[8,9]

In this direct approach, respirable air-borne dust samples are collected in a modified 25-mm Gelman in-line filter holder as designed by Knauber and Zullo on tared (always pre-weighed in laboratory) 25-mm silver filters (pore size 0.8 micron). The modification, shown in Figure 5, provides plenum chambers on both sides of the filter to get uniform air flow and even dust distribution over the deposit area of 3.8 square centimeters. The plenum chambers are fabricated and milled from nylon type rod. Without this modification of the in-line filter holder, the dust is concentrated in the center and at the edges of the filter.

Samples are collected with suction apparatus at a flow rate of 1.7 1/min. for approximately eight (8) hours. The in-line filter holder is placed in a vertical position with the inlet side down, as shown in Figure 6. A 10-mm nylon cyclone-type respirable mass separator also in a vertical position is connected to the in-line filter holder so that the air stream enters horizontally. The cyclone separator is a size selector which controls the size and

| SAMPLE NO. | MATERIAL | PER CENT FREE SILICA | | |
| | | Independent Laboratory | Pennsylvania | |
		Chemical	Chemical	X-Ray
1	Shale	56	61	57
2	Slate	32	36	25
3	Granite	29	38	34
4	Foundry dust	25	28	31
5	Trap rock	3	5	11
6	Clay	26	32	31
7	Magnetite ore	9	11	17

Fig. 4. Comparison of X-ray and chemical results for free silica.

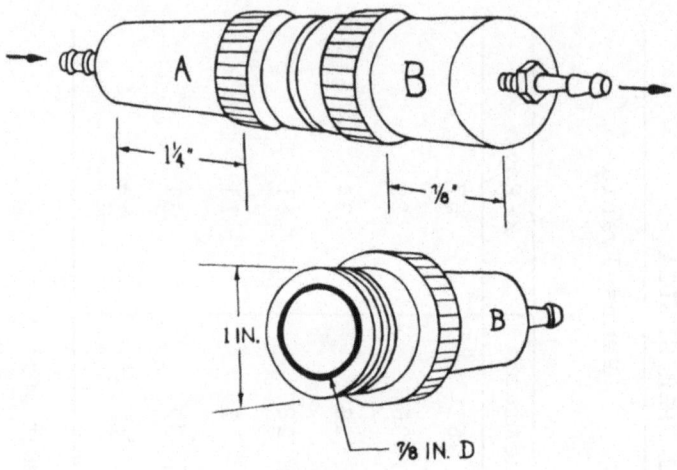

Fig. 5. Sampling Monitor.

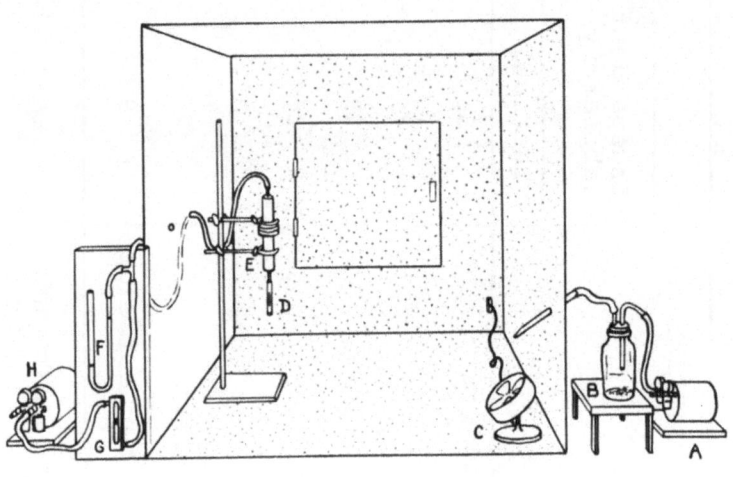

A - INLET PUMP   E - SAMPLING MONITOR
B - DUST RESERVOIR   F - MANIOMETER
C - FAN   G - ROTAMETER
D - CYCLONE   H - OUTLET PUMP

Fig. 6. Dust Chamber.

density of those particles which pass onto the filter.

Particles larger than an aerodynamic diameter of 10 microns drop into the lower section of the cyclone separator.

To prevent excessive handling, samples are usually submitted to the laboratory enclosed in the in-line filter holder. The pre-weighed silver membrane, following final weighing on a pedestal on the pan of a 5-place balance, [10] is placed in an aluminum jig, 1-1/2" x 1-1/8", shown in Figure 7, with a 3/4" diameter mask for holding the sample in the diffractometer and serving as a forward support. The mask provides a constant sample area and also prevents the unloaded peripheral portion of the membrane from diffracting the x-ray beam, so that the intensity of the major silver diffraction peak can be used as a measure of self absorption by the sample. The back piece has a 1" D plug which fits into the front piece and provides back support for the sample. The holder's width of 1-1/2" allows it to be aligned flush with the edges of the instrument clamp, so that the position of the sample in the x-ray beam is constant from sample to sample.

As in the procedure for settled dust samples, the air-borne samples are scanned from $2\theta = 19^\circ$ to $2\theta = 30^\circ$, and also from $2\theta = 36^\circ$ to $2\theta = 39^\circ$ to measure the area of the 2.37 $\overset{o}{A}$ silver peak, as shown in Figure 8. The area of the silver peak should be at least 90% of the area of the silver peak of a blank filter. If the area of the silver peak of the filter covered with sample is less than 90% of the area of the silver peak of a blank membrane, it must be assumed that the sample's true quartz content is greater than that determined from calibration. When the deposit is heavy enough to reduce the silver peak ratio below 90% (approximately 6 milligrams of dust) the quartz calibration is not linear due to excessive self absorption. There is some variation in the intensities of silver peaks from one batch of filters to another.

Calibration for this technic involved first grinding pure quartz to a size of less than 7 microns, dispersing

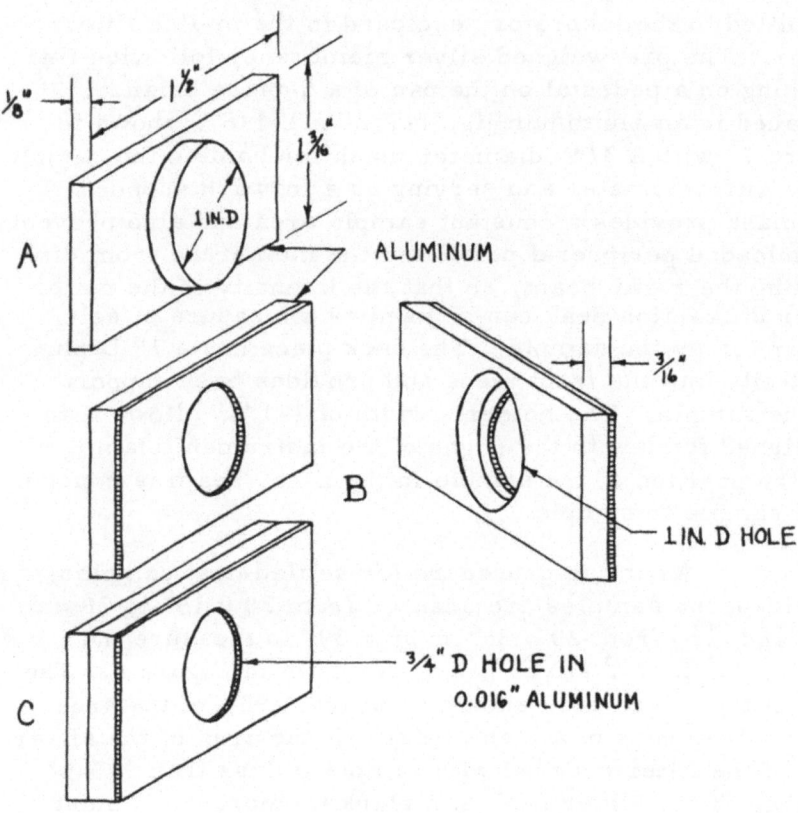

Fig. 7. Sample Holder.

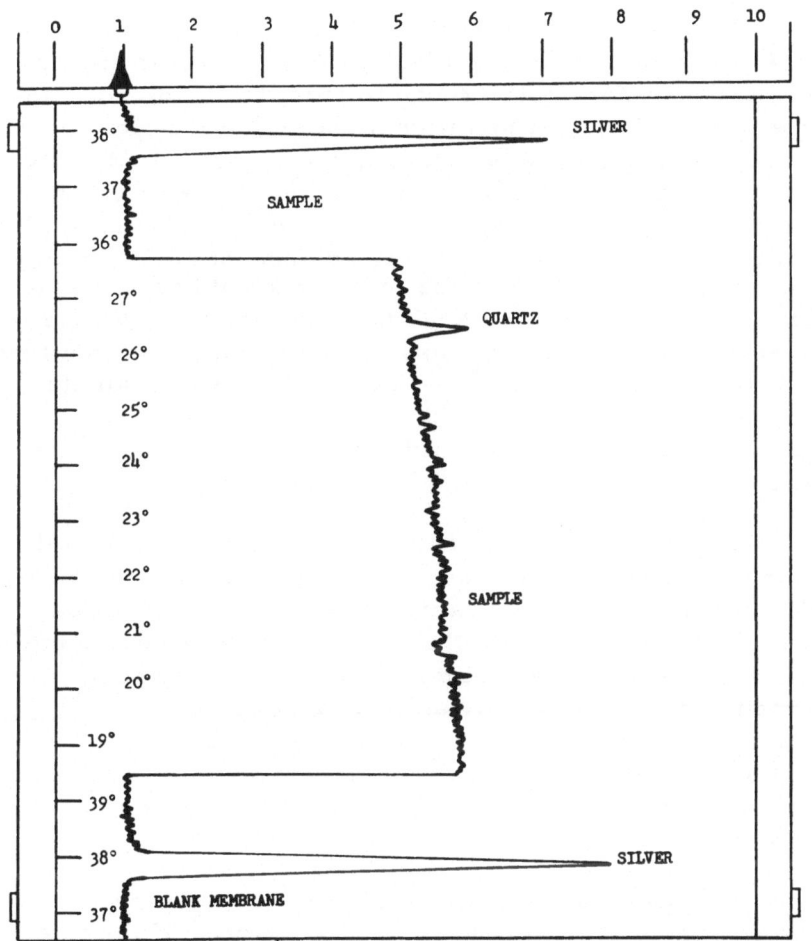

Fig. 8. Stripchart for airborne samples on silver membranes.

the material in a chamber, collecting it on silver mem - branes with cyclone separator and measuring the quartz peak intensities for amounts varying from 0.06 mg. to 3.45 mg. See Figure 9.

The next step involved preparing a mixture of reagent grade potassium chromate (density 2.73 g/cc and quartz (density 2.65 g/cc) both ground to less than 7 microns and dispersing and collecting the mixture in the identical manner. Potassium chromate was selected not only for its density but also because its linear x-ray absorption coefficient is considerably greater than that of the quartz (uK_2CrO_4=332; u quartz = 93). Any error due to an absorption difference between the standards and the mixture should be present. These samples were then analyzed first for their quartz content by the direct x-ray diffraction technic and then for CrO_4=colorimetrically, [11] and the total for each sample was compared with its weight gain as shown in Table I. The quartz calibration is valid with the limitations of the weight range (0.06 mg. - 0.18 mg.) studied.

Time does not permit the presentation of more self absorption data, but studies [9] comparing simultaneously collected light and heavy samples on silver membranes show as stated previously that a value of 90% would seem to be an acceptable safe minimum for the silver peak intensity ratio. No significant self absorption error should occur at this level. The light samples (1.5 - 2.5 mg.) were analyzed by the direct technic and the heavier samples were dusted from the silver membranes and analyzed by the packed powder technic.

The practical limit of sensitivity for the direct technic is approximately 0.06 mg. of quartz. If the total weight of quartz in the sample is less than this, no quartz will be detected since the intensity of the peak will not be sufficiently above the level of the background by the strip- chart method. This sensitivity would correspond to 0.075 mg. of quartz/m^3 in an 800 liter air sample of pure quartz.

Sample results are reported as milligrams of quartz per cubic meter of air.

TABLE I

QUARTZ - POTASSIUM CHROMATE SAMPLE RESULTS

Milligrams Sample	Milligrams SiO_2 + K_2CrO_4
0.48	0.53
0.50	0.51
0.51	0.50
0.82	0.88
1.05	1.14
2.49	2.48

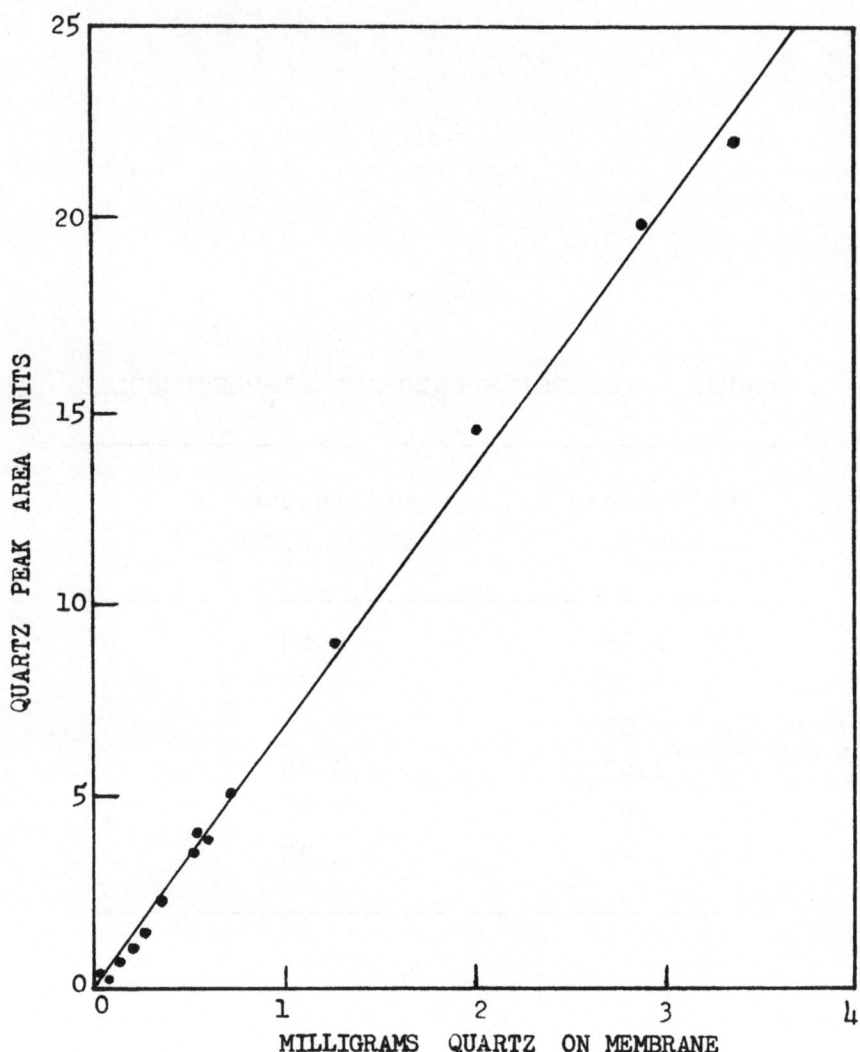

Fig. 9. Calibration for airborne samples on silver membranes.

A comparison of field samples (Table II) shows results of air-borne quartz samples and bulk settled dusts collected in the same areas. Although the data is not extensive the trend appears to be a lower percentage quartz for the settled dust samples.

Table III points up the effects of grinding and sieving on results from a number of samples. Each sample was analyzed first in the usual manner of the packed powder procedure. The remaining portion was screened to pass 325 mesh and divided in half. Each portion was then analyzed by the usual technic except one portion of the sieved material was not ground with the calcium fluoride internal standard -- merely mixed with it. In general, the quartz in the unscreened sample was higher than the average value of the screened portions. Grinding of one of the screened portions appears also to produce lower results.

In this interim report we have described our experiences with the use of x-ray diffraction for the analysis of quartz. It is readily discernible that further study of the technic is necessary. A rapid precise method for air-borne quartz without tedious sample preparation is vitally needed, considering the vast number of samples anticipated in the next few years to assure compliance with the Federal Occupational Safety and Health Act of 1970. We feel the direct x-ray diffraction technic is an answer.

Our next project will include gross weight or total particulate measurements coupled with free silica analysis of the bulk materials compared to respirable dust samples. The quartz determinations will be made by the packed powder internal standard x-ray diffraction technic and the spectrophotometric procedure for free silica by Talvitie[12] as well as the direct x-ray diffraction technic where applicable. Particle size differences will be studied in more detail.

TABLE II: QUARTZ DETERMINATION IN AIR-BORNE DUST (FIELD SAMPLES) BY X-RAY DIFFRACTION

INDUSTRY	LOCATION	% QUARTZ	
		AIR-BORNE	CORRESPONDING SETTLED DUST OR BULK SAMPLE
CEMENT	BREATHING ZONE - ROCK SILO	8	5
	B.Z. - PRIMARY CRUSHING	17	-
	B.Z. - PRIMARY CRUSHER OPERATOR	16	13
	B.Z. - CRANE OPERATOR	14	-
	B.Z. - DRILLER	36, 29	11, 13
	B.Z. - EUCLID OPERATOR	22	18
	B.Z. - EUCLID DRIVER	27	-
	DRILL OPERATOR	17	-
	B.Z. - PACKER	5*	LESS THAN 1
QUARRY	O.E.+ PRIMARY CRUSHER (STONE)	9*	-
	O.E. END LOADER (TRAPROCK)	42*	-
	O.E. DRILLER (TRAPROCK)	13	3
	O.E. DRILLER (TRAPROCK)	8	5
	O.E. EUCLID DRIVER (TRAPROCK)	22	5
BRICKS	MOLDING AREA	8	-
	BRICK DUMPING TABLE	3	-
STEEL	ROUGH CHIPPING OPERATOR	26	25**
METAL CASTING (BRASS)	O.E. - SHAKEOUT	42	42

* Possibly higher due to excessive sample self-absorption
** Bulk sample of fused sand
+ Operator's exposure

TABLE III: EFFECT OF SIEVING ON FREE SILICA CONTENT
          OF SETTLED DUST SAMPLES (PACKED POWDER
          INTERNAL STANDARD - X-RAY DIFFRACTION)
          IN PERCENT QUARTZ.

NOT SCREENED	SCREENED	
C	c_1	c_2
(WITH GRINDING)	(NO GRINDING)	(WITH GRINDING)
1	1	1
3	3	3
6	3	4
1	1	1
53	59	43
62	48	21
67	16	15
ND	1	ND
23	25	19
52	23	26
61	38	33
22	19	18
34	21	15
ND	ND	1
34	25	23

NOTE: The operating parameters for our x-ray
diffraction instrument are:

(a)   Goniometer speed -- 0.4°/min.
(b)   Collimating Slit -- 3° MR
(c)   Receiving Slit -- MR
(d)   Pulse height analyzer of base E-3V and a
window E=18V

The instrument is operated in compliance with the
safety precautions as prescribed in

Title 25 - Rules and Regulations
Part I - Pennsylvania Department of
Environmental Resources
Subpart D. - Environmental Health and Safety
Article V - Radiological Health
Chapter 235 - Other Radiation Sources

REFERENCES

1.   Klug, H.P. and Alexander, L.: X-Ray Diffraction
Procedures. p. 390. John Wiley and Sons, New York
N.Y., (1954).

2.   Cullity, B.D.: Elements of X-Ray Diffraction.
p. 269 and p. 378. Addison-Wesley, Reading,
Massachusetts (1959).

3.   Talvitie, N.A. and Brewer, L.W.: Separation and
Analysis of Dust in Lung Tissue. Amer. Ind. Hyg.
Assoc. J., 23:58 (1962).

4.   Crable, J.V. and Knott, M.J.: Quantitative X-Ray
Diffraction Analysis of Crocidolite and Amosite in Bulk
or Settled Dust Samples. Amer. Ind. Hyg. Assoc. J.,
27:449 (1966).

5.   Oberg M.: Evaluation of Quartz in Airborne Dust in
      the 0.5-2.0 Micron Size Range. Environ. Sci. Tech.,
      2:10 (1968).

6.   Perloff, A., Crystallography Section, National
      Bureau of Standards, U.S. Department of Commerce,
      Washington, D.C. -- personal communication.

7.   Talvitie, N.A.: Determination of Quartz in Presence
      of Silicates Using Phosphoric Acid. Anal. Chem.,
      23:623 (1951).

8.   Scoville, R.G.: Personal Air Sampling, Michigan's
      Occupational Health, 15:2 (1970).

9.   VonderHeiden, F.H. and Knauber, J.W.: A Technic
      for the Direct X-Ray Diffraction Analysis for Quartz
      in Airborne Dust Collected on Silver Membranes.
      Bureau of Occupational Health, Pennsylvania
      Department of Environmental Resources, Harrisburg,
      Pa., unpublished.

10.   Bureau of Occupational Health Committee on Analyt-
      ical Methods: Standard Procedure for Collection and
      Analysis of Inert or Nuisance Particulate. Bureau of
      Occupational Health, Pennsylvania Department of
      Environmental Resources, Harrisburg, Pa.,
      unpublished.

11.   Saltzman, B.E.: Microdetermination of Chromium
      with Diphenylcarbazide by Permanganate Oxidation.
      Anal. Chem. 24:1016 (1952).

12.   Talvitie, N.A.: Determination of Free Silica:
      Gravimetric and Spectrophotometric Procedure
      Applicable to Airborne and Settled Dust. Amer.
      Ind. Hyg. Assoc. J., 25:169 (1964).

APPLICATIONS OF X-RAY DIFFRACTION IN OCCUPATIONAL HEALTH STUDIES

Charles M. Nenadic and John V. Crable
National Institute for Occupational Safety
and Health, DHEW
1014 Broadway
Cincinnati, Ohio 45202

The Occupational Safety and Health Act of 1970[1] was promulgated for the express purpose of providing American workers with protection against personal injury and illness resulting from hazardous working conditions.

Section (2)(6)(3)[2] of the Act authorizes the Secretary of Labor to set mandatory occupational health standards. A list of some of the standards set under the authority issuing from Section (2)(6)(3) can be found in the Federal Register of August 13, 1971. This list contains standards for over 400 atmospheric pollutants. The responsibility for determining the criteria or background information for each of these standards rests in the National Institute for Occupational Safety and Health.[3] In order to develop the appropriate background information, it is also necessary to develop appropriate analytical procedures for each of the materials contained in the August 13, 1971 Federal Register list. Included in the list are a number of inorganic crystalline compounds for which the only practical method of qualitative and quantitative instrumental analysis is x-ray diffraction. Among these compounds are boron oxide, calcium oxide, cobalt fumes and dusts, copper fumes and dusts, iron oxide fume, magnesium oxide fume, vanadium oxide, zinc oxide, quartz, cristobalite, asbestos, and talc.

Many of these compounds are of special interest in occupational health because they have been implicated as

major factors in the etiology of such respiratory diseases
as silicosis, asbestosis, talcosis, and coal miner's
pneumoconiosis.

Silicosis is caused by the inhalation of one or more
of the polymorphic forms of silicon dioxide - quartz,
cristobalite, or tridymite. Talcosis is caused by the
inhalation of respirable particles of talc, a hydrated
magnesium silicate, found not only in the talc mining
industry but also in the cosmetic, paint, and rubber industries. Asbestosis is caused by the inhalation of one
of several fibrous mineral silicates, the more important
being chrysotile, amosite, crocidolite, and anthophyllite.
Coal miner's pneumoconiosis, as the name implies, is
caused by the inhalation of respirable coal dust.

In this country there are estimated to be 889,000
construction workers; 165,000 concrete, gypsum, and
plaster workers; 16,000 stone workers; 108,000 abrasive
and asbestos workers; 244,000 foundry workers; and
230,000 miners. Added together these groups constitute
a substantial number - over 1.5 million - of people exposed to the hazards of inhaling crystalline and other
pneumoconiosis-inducing particles.

Before a rational attempt at controlling the inci-
dence of these respiratory diseases can be made, the
causative agents involved must be identified and qualitative and quantitative analytical procedures developed
for them. The ability of x-ray diffraction to identify
and quantify inorganic crystalline compounds rather than
elements makes it the logical instrument to use in this
task. As with any other analytical technique, the full
potential of x-ray diffraction can only be achieved when
used in conjunction with other instrumental and wet chemical methods.

Sample Preparation

Occupational health and air pollution surveys generally require the collection and analysis of samples in
order to provide the data base from which decisions affecting control procedures or compliance with standards can
be formulated. Samples submitted for x-ray diffraction
analysis generally fall into one of five basic categories:
(1) a bulk sample of the parent material of some air-borne

dust, (2) membrane filter samples from personal air samplers,
(3) large membrane filter samples from hi-vol air samplers,
(4) samples of settled dust, and (5) biological samples.
The requirements of the analysis with respect to its accu-
acy, and the condition in which the sample is received, will
ultimately control the type and degree of treatment which
the sample receives. The simplest and most basic type of
preparation is the reduction of the sample to a powder.
The choice of a powder form is made to facilitate sample
manipulation, the addition of an internal standard, and
the selection of a representative portion of the sample for
analysis. The powder form generally will provide a greater
degree of random orientation and thus result in greater reproducibility. Many samples are initially collected in a
powder form - atmospheric dusts, for example - which may
eliminate the necessity for any sample preparation.

Methods for reducing samples to the required powder
form are numerous and include both automatic and manual
grinding equipment. Any method will suffice as long as the
grinding procedure does not unduly alter any of the material's intrinsic physical or chemical characteristics. The
authors have found a liquid nitrogen cooled grinder to be
extremely useful for grinding tough fibrous materials like
asbestos. The low temperature induced in the grinding
vials makes the material being ground more brittle and thus
more easily ground and also prevents any heat-induced
changes from taking place.

The purpose of the grinding operation is to produce a
powder sample whose particles reproducibly diffract x-rays
at maximum intensity levels.

The objectives of sample preparation for qualitative
analysis by x-ray diffraction are not as stringent as those
for quantitative analysis. Powder diffraction samples that
pass through a 200-mesh sieve may be perfectly satisfactory
for qualitative analysis purposes, but samples that are
prepared for quantitative analysis should be ground to a
mean particle size of approximately 5 microns if a high degree of accuracy and reproducibility is required.

Bragg[4] attained acceptable reproducibility of measurements made on a $Ca(OH)_2$ specimen when it had been ground to
a maximum particle size of 2.5 microns. Gordon and Harris[5]
found that the optimum diffracted intensity occurred when

the quartz particles had a mean size of 2 microns. They
claim that the error introduced by ignoring the effects
of particle size distribution when the actual distribution
is between 1 and 20 microns is on the order of 12%.
Williams[6] does not discuss the effects of particle size
but does state that the samples of quartz he used were
ground to less than 5 microns by ball milling.

Crable and Knott[7] used a mixer mill to dry grind
chrysotile asbestos samples to a maximum particle dimension of 3.5 microns. They found that chrysotile and crocidolite, minerals that are very susceptible to preferred
orientation, could be quantitatively analyzed if the
length of the fibers is reduced to 3.5 microns by grinding.[7,8]

Leroux et al.,[9] in discussing the analysis of industrial dust films deposited on membrane filters, used quartz
standards with a maximum particle size of 5 microns. Till
and Spears[10] used hand grinding in an agate mortar to reduce materials to an average particle size of 15 microns.
They were able to achieve satisfactory reproducibility with
that size particle. Leroux,[11] in developing a method for
depositing thin coatings on membrane filters, used only
particles less than 5 microns in diameter in order to obtain "consistent quantitative measurements."

Aoyagi,[12] quantitatively analyzing rock samples for
various minerals, ground samples to be analyzed for quartz
to less than 5 microns and samples to be analyzed for clay
minerals to less than 2 microns. Hayashi and Oinuma[13]
used hydraulic elutriation to obtain the 0.1- to 5-micron
fraction of standard samples of quartz, cristobalite, and
tridymite.

Quakernaat,[14] in developing a procedure to correct
for the effects of preferred orientation, used various clay
mineral standards that had been ground to a particle size
less than 2 microns.

Talvitie and Brewer[15] sieved and ground their silica
samples until they had achieved a particle size of less
than 5 microns.

Experimental work done by Klung and Alexander on the
reproducibility of diffraction measurements from quartz
samples composed of specific size ranges indicated that

reproducible x-ray measurements were obtained only when all particles were less than 5 microns in size.

If an accurate and precise quantitative analysis is desired, the grinding procedure should be strictly monitored and controlled. Matsumura and Hamada,[16] Brindley and Udagawa,[17] and Gordon and Harris[18] all experimentally determined the effects of grinding the specimen on the accuracy of the analysis. They found the intensity of x-rays diffracted from quartz to be affected both by the grinding time and by the final average particle size of the specimen. Grinding the specimen for too long a time produces an amorphous layer on t ie outside of the quartz particles, thereby significantly reducing the intensity and making standardized grinding times a must.

Foster and others[19] found not only that the grinding time was of importance in the accurate analysis of samples for cristobalite, but also that the order of the crystal lattice greatly affected the intensity of diffracted x-rays. Maximum accuracy is thus attained by using standards composed of the same material as that under analysis.

Once the dimensions of the sample particles have been reduced to the required size range, the next step is to expose them to the x-ray beam in a manner designed to provide the information required, whether that be a qualitative identification or a precise quantitative determination. Aside from particle size, the most important parameters in sample presentation are (1) the smoothness of the sample surface, and (2) the randomness of the particle orientation. Completely random orientation is the ideal situation but is rarely achieved. Some investigators have found it more informative to accentuate the preferred orientation if maximum random orientation cannot be obtained.[12]

One factor that is often overlooked in x-ray diffraction analysis work and yet one that can have a significant effect on the final result, especially where very small samples are concerned, is the sample holder. Among the more common holders used are glass microscope slides,[12,13] metallic bulk powder holders,[6,12] rotating sample holders,[20,21] pellets,[22] and membrane filters.[9,11,14,15,23-25]

The sample mount receiving the most current attention is the membrane filter. Membrane filters are manufactured out of a variety of materials most of which scatter substantial amounts of incident x-radiation. The scatter from most filter materials causes such a high background that filters made out of these materials are unsuitable as x-ray diffraction sample mounts.[15,25] The membrane material that has proven to be most satisfactory as a sample mount for x-ray diffraction analysis is silver.[9,24]

Silver membrane filters have not received general acceptance as air sample collectors for a number of reasons. First silver filters cost substantially more than the cellulose and PVC filters. Second they weigh much more than the comparable cellulose and PVC filters thereby reducing the sensitivity of any necessary gravimetric determinations. Third, silver filters do not hold an electrical charge.

Since an electrostatic charge is one of the mechanisms involved in holding a sample on the surface of most membrane filters, the silver filter's inability to hold a charge can result in loss of sample during shipment.

One way to avoid the problem of collecting samples on the silver membrane filters is to collect the samples on another type of filter and to transfer the sample to a silver filter. The transfer can be accomplished either by removing the sample ultrasonically from the collecting filter or by ashing the collecting filter. The sample is then dispersed in an aqueous solution and deposited on a silver filter.[15]

Qualitative Analysis

The identification of the crystalline components of a sample is necessary if the hazards of exposure to that material are to be properly evaluated. Historically, x-ray diffraction has been the only instrumental method available with the ability to rapidly identify and quantify crystalline, inorganic compounds and minerals. The development of newer instrumental procedures--infrared spectroscopy and differential thermal analysis--for the identification of minerals and inorganic compounds has by no means decreased the importance of x-ray diffraction in qualitative analysis, but, on the contrary, has enhanced it. Samples which formerly could not be identified with certainly by x-ray diffraction alone can now be definitely identified with the aid of information supplied by the other instrumental techniques.

The identification of inorganic compounds and minerals is accomplished in the majority of cases by the comparison of the powder diffraction pattern of an unknown with the diffraction patterns on file in the Joint Committee on Powder Diffraction Standards (JCPDS)[26] powder pattern indexing system.

A variety of indexing systems are marketed as JCPDS, including a computerized system supposed to be considerably more proficient at matching patterns than the standard human operator.

Chemically simple minerals and inorganic compounds (quartz, for example) generally are readily identifiable from their x-ray powder patterns. The more complex chain and sheet silicate minerals often generate diffraction patterns which tend to be more confusing than helpful. These minerals, of which talc and chrysotile are representative types, are often poorly crystallized, contaminated with unimportant, interfering minerals, and subject to varying degrees of isomorphous substitution. Poorly crystallized specimens exhibit diffraction patterns with broader, less intense peaks than well-crystallized specimens. The presence of interfering minerals results in the introduction of new peaks into the pattern and the increase in intensity of some of the old peaks as a result of the additive effect of more diffracted radiation. Isomorphous substitution, which consists of the replacement of some of the atoms of one of the components of the mineral with other similar but different-sized atoms, results in the alteration of some of the interplanar spacings, thereby shifting some peaks, causing others to disappear and new ones to appear.

The similarity of composition among many of the complex silicate minerals increases the probability of error when attempting to identify a diffraction pattern. For example, tremolite, actinolite, and ferroactinolite are minerals containing 0 to 20, 20 to 80, and 80 to 100 mole % $(OH)_2$ $Ca_2Fe_5Si_8O_{22}$, respectively. Figure 1 presents the diffraction patterns of a tremolite sample from the State of

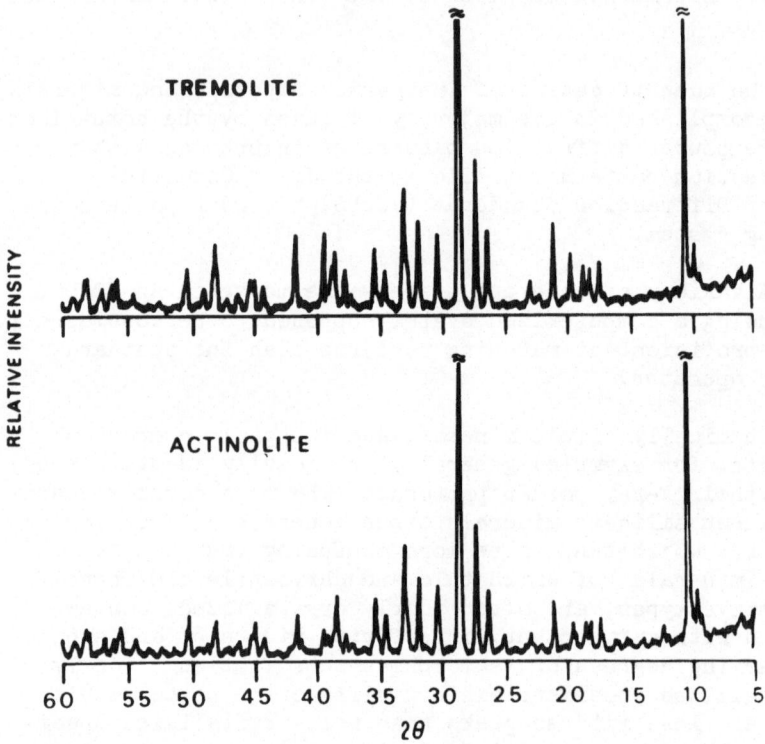

Figure 1. X-ray diffraction patterns of tremolite and actinolite.

New York and an actinolite sample from California. The similarity of the patterns is self-evident. Although it is usually possible to identify one of these minerals if it is the only silicate mineral present in the sample, two or more integrated in one sample can prevent the separation and identification of the components, especially if the sample cannot be altered or treated in any way.

Figure 2 presents the diffraction patterns of four talc samples: the talc profile found in the ASTM file: Montana talc; Texas talc; and New York talc. None of the samples are identical to the pattern constructed from the ASTM talc. Even though all the samples have been mineralogically identified as talc, the diffraction pat-

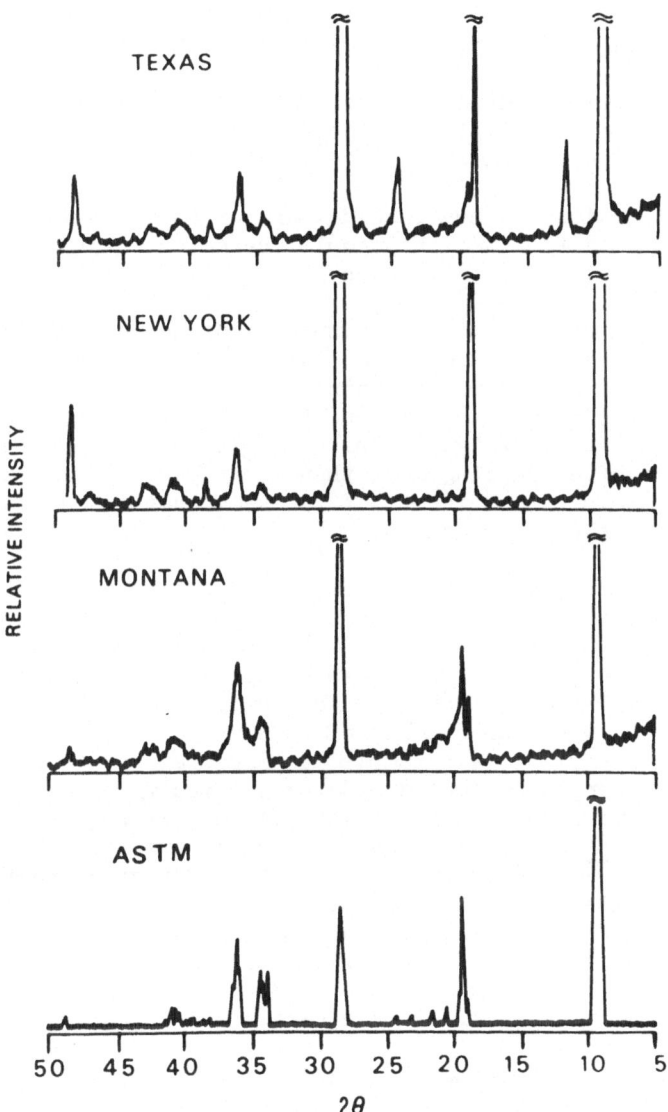

Figure 2. X-ray diffraction patterns of four talc samples originating from four different areas.

terns indicate that all are not pure talc, or, at most, only one is pure talc.

The pretreatment of silicate minerals in order to reduce them to readily identifiable forms has been described by several authors.[12,27,28,29]

Applications of heat, acids, and organic liquids have all been used to alter the sample characteristics and thus aid in the identification.

While pretreatment of a sample might be the preferred way to approach the problem of identification, it may not be possible because of the manner in which the sample was originally collected. Membrane filter samples often total less than 1 mg in weight. This, coupled with the common procedure of conducting more than one type of analysis on the same sample or aliquots of the same sample, may rule out any sample pretreatment.

The manner in which the sample was collected at the site can often control the number, type, and accuracy of any analyses to which it may later be subjected. The samples to be wary of are those collected in very small amounts by methods which tend to highly orient the sample particles. A common example of this type of sample is the membrane filter air sample mentioned above. The small amount of sample collected and the difficulty of effectively removing it from the filter often will require that the sample be analyzed as is. If the sample particles have been oriented to any degree, the diffraction pattern may prove misleading when compared to the patterns indexed in the JCPDS card file. The deviations in the diffraction pattern caused by preferred orientation can be serious enough to result in an erroneous identification, especially where many of these silicate minerals are concerned.

One way of checking the possibility of error is to prepare a number of standard samples in the same manner in which the sample was collected and then scan them and file the diffraction patterns. A file of diffraction patterns of highly oriented powder samples can be invaluable in the qualitative identification of a powder specimen.

The problem of preferred orientation is common because of the high degree of structural orientation in-

herent in the crystal building blocks of many minerals.
Many of the silicate minerals are platy or fibrous in
nature, meaning that the dimensions of the basic structural crystallite are much greater in some directions than
in the others. Consequently, when particles are formed
from a massive piece of the mineral, they assume the same
structural configuration as the basic unit cell has.

A word of caution is in order concerning the use of
the JCPDS card file. Minerals originating from different
deposits often exhibit significant variations in impurity
content and crystallinity. The point of origin does not
usually affect the composition of chemically simple minerals such as quartz as it does the complex silicate minerals. The chemically simple minerals are normally
obtainable in a highly pure state. On the other hand, the
complex silicate minerals are highly susceptible to variations in chemical composition and degree of crystalli-
zation. These facts must be considered when attempting to
make a correlation between a pattern obtained from an unknown specimen and the patterns indexed on the file cards.
Commonly there may be wide variations in relative intensities and some peaks missing and some new ones added.
There is no way to solve this problem other than to use
another confirmatory technique to provide more conclusive
evidence.

Quantitative Analysis

Direct Comparison

There are two basic quantitative x-ray diffraction
techniques: (1) the direct comparison method, in which
the intensity of one component of a sample is measured
and compared with measurements obtained from specimens composed only of the component material, and (2) the internal standard method, in which the ratio of the diffracted intensity of a specimen component to that of a
standard compound mixed with specimen is compared with
ratios of diffracted intensity measurements from
standards made up of the component and the standard compound. The labeling of all methods not included in the
internal standard category as direct comparison methods
is not entirely correct. To attempt to descriptively
identify each method would eventually prove to be more
confusing than to lump the many methods containing ele-

ments of the direct comparison method under the one
heading. The number of variations using these two basic
techniques is large. The direct comparison method in
general requires less sample preparation but may require
more data manipulation to achieve the accuracy desired.
Direct comparison measurements by their very nature require that measurements be made to determine the effect
of sample absorption of the incident x-radiation on the
accuracy of the analysis or that the sample be prepared
in a manner which reduces the effects of absorption or
the accuracy of the analysis.

Almost all direct comparison methods now in use
are derived from the basic expression for the intensity
of x-rays diffracted from one component of a multicomponent mixture developed by Klug and Alexander.[30] Mathematical refinements of the expression have resulted in
a considerable number of methods that are procedurally
different, but nevertheless closely related.

Leroux et al.[31] developed a direct absorption method
based on Klug and Alexander's work. Leroux manipulated
the equation until he was left with terms which would permit an experimental solution without the use of an internal standard. The equation required measurements of
both the intensity of x-rays diffracted by the sample and
the intensity of x-rays penetrating through the sample.
Using binary mixtures containing known amounts of quartz,
Leroux and the others were able to achieve an accuracy of
$\pm$ 5% in their analyses.

Williams,[6] starting from Klug and Alexander's basic
expression, derived an expression which required experimental measurement of the absorption effect on the sample.
Williams developed a sample holder which allowed him to
make the measurement quickly and with a minimum of sample
he was able to achieve an accuracy of $\pm$4% with his method.

Moore[32] proposed the theory that the effect of the
presence of any one mineralogical component on the
intensity of a diffraction maximum of any other given
component can be expressed by a single constant. Working
from Klug and Alexander's basic expression, Moore was
able to derive a mathematical expression employing these
"linear interaction coefficients." He discusses the
evaluation of these constants and their application to

the treatment of naturally occuring mineral systems.

Karlak and Burnett[33] developed a more general expression describing the effects the presence of one or more minerals has on a beam of x-rays diffracted from another mineral present in the same sample. The method is highly sophisticated and is more applicable to large numbers of routine samples than to varied numbers of totally different samples. They were able to demonstrate accuracies ranging from m $\pm$0.1 to $\pm$10% in their experimental work using samples composed of known amounts of inorganic compounds.

Standard Method No. 6,[34] developed and used by the Bureau of Mines, is a direct comparison method used to determine concentrations of quartz in coal dust. Cobalt (Co) radiation is used to minimize the effects of iron fluorescence. The absorption effects of the coal dust matrix were found to be relatively constant for all coal dusts. Muscovite interferences were eliminated by grinding the samples until the pattern of the soft muscovite was nearly extinguished. Calibration curves were prepared for 1-, 3-, and 5-mg samples containing 0 to 8% quartz.

The quantitative analytical techniques that are of particular interest in occupational health work are those which provide for the direct analysis of air samples collected on membrane filter samples. Membrane filters make excellent supports for small samples. The use of these filters as sample holders has reportedly resulted in detection limits approaching 2 micrograms for quartz[24] and 10 micrograms for chrysotile asbestos.[25]

Talvitie and Brewer[36] used a standard cellulose membrane filter as a sample mount for residues obtained from lung digestions. They developed standard curves by depositing known amounts of quartz and cristobalite on the filters and measuring the intensity of the diffracted x-radiation. They did not discuss the sensitivity of accuracy of their method.

Crable[23] used a direct comparison procedure for determining the amount of chrysolite, crocidolite, or amosite in dust samples deposited on membrane filters. He prepared standards by depositing known amounts of the asbestos

minerals on membrane filters and then comparing the integrated intensity measurements from the samples with those
of the standards. The absorption effect of the remainder
of the sample matrix is not taken into account. The accuracy of this method is highly dependent on both sample
preparation and the particular type of mineral under
study. Crable[23] found the accuracy of this method to vary
up to ±20%, being lowest at the lowest asbestos concentrations.

Bradley[37] investigated the membrane filter mounting
technique for the quantitative analysis of small samples
and found it to be satisfactory as long as certain weight
limits were not exceeded. Working with quartz, he found
that the x-ray response was nearly a linear function of
the density of deposition of the sample on the filter up
to a density of approximately 0.35 mg/cm^2 Bradley was able
to determine that the error caused by the variation in
absorbance of x-rays among different minerals does not
have a significant effect on thin-layered quartz samples.

Leroux and Powers[9] mathematically and experimentally
demonstrated that significant error resulting from sample
absorption could be avoided by the use of molybdenum (Mo)
radiation and thin samples supported by silver membrane
filters. For these conditions they found the intensity
diffracted from quartz in samples to vary linearly up to
a sample loading density of 200 µg/cm^2 on the filter.
Bergin[35] also developed a direct comparison method for
handling membrane filter samples. His method deals with
the absorption problem by measuring the decrease in intensity of x-rays transmitted through a thin filter specimen. By including these values in a formula derived for
the purpose, he was able to accurately calculate the concentration of a crystalline component of a sample without
adding an internal standard.

Crosby and Hamer[24] collected air samples on silver
membrane filters and then directly analyzed the samples
for quartz on the silver filters by x-ray diffraction.
They estimated their limit of detection to be 2 micrograms.

Richards[25] collected and analyzed samples for chrysotile asbestos directly on Solvinert filters.* He was able to attain a detection limit of 10 micrograms of chrysotile on the filter.

Internal Standard

The internal standard method of analysis has long been used to compensate for the difficulty of data manipulation in direct comparison measurements, instability in the electronic components of the x-ray instrument, and variations in the sample absorption. It is still the preferred method for most investigators. The theory behind quantitative analysis by the internal standard method is adequately explained by Klug and Alexander,[30] Cullity,[33] and Bragg.[4]

Perhaps the most critical aspect in the successful management of the internal standard procedure is the proper selection of the internal standard. Among the more important qualities sought in an internal standard are those listed below:

1. A strong diffraction peak close to a strong diffraction peak of the unknown.

2. A specific gravity similar to that of the sample matrix.

3. A particle size similar to that of the sample matrix.

4. Nonreactivity with the matrix.

5. Mass absorption coefficient similar to that of the matrix.

6. Hardness and brittleness similar to that of the matrix.

7. Availability in a pure form.

*Mention of commercial concerns or products does not constitute endorsement by the Public Health Service, DHEW.

8.   Particles that orient to the same degree as those of the matrix.

9.   Normal absence from the sample matrix.[10,15,34]

Talvitie and Brewer[15] used beryl as an internal standard for quartz in settled dust and bulk samples. They were able to achieve relatively good agreement between x-ray results for quartz and the results of a chemical determination for quartz on the same series of samples. Aoyagi[12] developed standard curves for quartz, cristobalite, calcite, and several of the clay materials using zinc oxide (ZnO) as the internal standard.

Till and Spears[10] determined quartz concentrations in sedimentary rocks using bochmite as an internal standard. They claim an analytical time of 7 minutes per sample using peak heights and a coefficient of variation of 1.9% for their method. Hayashi[20] used calcite as the internal standard and albite-oligoclase as the diluent for the quantitative determination of quartz in lung residue samples. Hayashi and Oinuma[13] constructed standard curves for quartz, cristobalite, and tridymite using two of the minerals as the internal standard and diluent, respectively, for the third.

Crable and Knott[7] used aquamarine as the internal standard in the quantitative determination of chrysotile. They were able to achieve accuracies approaching 100% when the chrysotile concentration was greater than 15%. Below that level the accuracy of the method decreased considerably.

The Bureau of Mines Standard Method No. 1 utilized calcium fluoride (CaF_2) or nickel oxide (NiO) as the internal standard for quartz.[39] This method requires greater than 20 mg of sample. The developers of the procedure were able to achieve an accuracy of $\pm$10% over a range of 0 to 100% quartz.

Crable and Knott[7] used quartz as the internal standard for the quantitative determination of crocidolite and amosite. Using sample mixtures containing known amounts of the two minerals, they were able to achieve an accuracy of $\pm$10% for all concentrations of crocidolite greater than 5% and for all concentrations of amosite greater than 10%.

Oberg[40] developed an experimental procedure for depositing an internal standard on a membrane filter air sample without unduly disturbing the sample. He used calcium fluoride (CaF_2) as the internal standard for quartz.

The quantitative analysis of quartz, cristobalite, and tridymite has been reported by numerous authors (see above), but other minerals, especially the fibrous and platy minerals, present a much more difficult problem to the analyst. Variations in composition, atomic substitution, and the orientation problems discussed with reference to qualitative identification are even greater problems in quantitative analysis.

Brindley and Kurtossy[20] investigated the problem of preferred orientation in the samples of kaolinite. They developed an orientation index which is a rough measurement of the extent to which the particles of a sample may be preferentially oriented.

Mossman and others[41] also considered the problem of handling fibrous and platy materials. Their work centered around an attempt to find an internal standard with the same or similar morphology as the material to be analyzed for. Their theory was that, no matter to what extent the sample was oriented, the internal standard would always be oriented to the same extent as the unknown, thereby maintaining the diffraction properties of each constant with respect to the other. They settled on zinc hydroxide, $Zn(OH)_2$, and pyrophyllite as internal standards for illite, kaolinite, and muscovite.

Quakernaat[14] investigated quantitative analysis of some clay minerals using an orienting internal standard as Mossman[34] did. Quakernaat selected molybdenite (MoS_2) as the internal standard for illite, pyrophyllite, and kaolinite. He was able to achieve an accuracy of $\pm 7\%$ with his technique.

1. Public Law 91-596, 84 Stat. 1590, U. S. Code
Title 29, Sec. 655

2. Occupational Safety and Health Act of 1970, Section 6(a), 84 Stat. 1593

3. Occupational Safety and Health Act of 1970, Section 22, 84 Stat. 1612

4. Bragg, R. H.: Quantitative Analysis by Powder Diffraction. In Handbook of X-Rays (E. F. Kaelble, ed.), McGraw-Hill Book Co., New York (1967)

5. Gordon, R. L., and G. W. Harris: Effect of Particle-Size on the Quantitative Determination of Quartz by X-Ray Diffraction. Nature 175:1135 (1955)

6. Williams, P. P.: Direct Quantitative Diffractometric Analysis. Anal. Chem. 31: 1842 (1959).

7. Crable, J. V., and M. J. Knott: Application of X-Ray Diffraction to the Determination of Chrysotile in Bulk or Settled Dust Samples. Amer. Ind. Hyg. Assoc. J. 27: 383 (1966)

8. Crable, J. V., and M. J. Knott: Quantitative X-Ray Diffraction Analysis of Crocidolite and Amosite in Bulk or Settled Dust Samples. Amer. Ind. Hyg. Assoc. J. 27: 449 (1966).

9. Leroux, J., and C. Powers: Direct X-Ray Diffraction Quantitative Analysis of Quartz in Industrial Dust Films Deposited on Silver Membrane Filters. Occup. Health Res. 21:26 (1970).

10. Till, R., and D. Spears: The Determination of Quartz in Sedimentary Rocks Using an X-Ray Diffraction Method. Clays Clay Minerals 17: 323 (1969).

11. Leroux, J.: Preparation of Thin Dust Coatings for Their Analysis by X-Ray Emission and Diffraction. Occup. Health Rev. 21:19 (1970).

12. Aoyagi, K.: Mineralogical Study of Sedimentary Rocks in the Oil Fields of Japan by the X-ray Diffraction Method, and Its Application to Petroleum Geology. Clay Sci. 3: 37 (1967).

13. Hayashi, H., and K. Oinuma: Rapid Method of Quantitative Mineralogical Analysis of the Silica Minerals from the Lungs of the Refractory Workers. Inc. Health 2: 172 (1964).

14. Quakernaat, J.: Direct Diffractometric Quantitative Analysis of Synthetic Clay Mineral Mixtures with Molybdenite as Orientation-Indicator. J. Sediment. Petrol. 40: 506 (1970).

15. Talvitie, N. A., and L. W. Brewer: X-Ray Diffraction Analysis of Industrial Dust. Amer. Ind. Hyg. Assoc. J. 23: 214 (1962).

16. Matsumura, Y., and A. Hamada: Change of Surface Properties of Quartz Particles by Grinding. Ind. Health 6: 220 (1968).

17. Brindley, G. W., and S. Udagawa: Sources of Error in the X-Ray Determination of Quartz, J. Amer. Ceram. Soc. 24: 643 (1959).

18. Gordon, R. L., and G. W. Harris: Effect of Particle Size on the Quantitative Determination of Quartz by X-Ray Diffraction. Nature 175: 1135 (1955).

19. Foster, P. K., I. R. Hughes, and K. McKenzie: X-Ray Diffraction and Thermal Expansion Properties of Cristobalite-Containing Ceramics. New Zealand J. Sci. 9: 249 (1966).

20. Brindley, G. W., and S. S. Kurtossy: Quantitative Determination of Kaolinite by X-Ray Diffraction. Am. Mineralogist 46: 1205 (1961).

21. DeWolff, P. M., J. M. Taylor, and W. Parrish: Experimental Study of Effect of Crystallite Size Statistics on X-Ray Diffraction Intensities. J. Appl. Phys. 30: 63 (1959).

22. Schliephake, R. W.: Procedure for the Routine X-Ray Determination of quartz in Mine Dusts from Bituminous Coal Mines. Gluckauf 99: 79 (1963)

23. Crable, J. V.: Quantitative Determination of Chrysotile, Amosite, and Crocidolite by X-Ray Diffraction. Amer. Ind. Hyg. Assoc. J. 27: 293 (1966)

24. Crouby, M. T. and P. S. Hamer: The Determination of Quartz on Personal Sampler Filters by X-Ray Diffraction. Ann. Occup. Hyg. 14: 65 (1971).

25. Richards, A. L.: Estimation of Trace Amounts of Chrysotile Asbestos by X-Ray Diffraction. Anal. Chem. 44: 1872 (1972).

26. Joint Committee on Powder Diffraction Standards: Powder Diffraction File. 1845 Walnut Street, Philadelphia, Pa. 19103

27. Hayashi, H.: Procedure of Mineral Analysis of Dusts in the Lung by X-Ray and Infrared Studies. Ind. Health 1: 37 (1964).

28. Brindley, G. W.: X-Ray Identification and Crystal Structures of Clay Minerals, The Mineralogical Society, London (1961).

29. Kaelble, E. F.: Qualitative Analysis by Powder Diffraction. In Handbook of X-Rays (E. F. Kaelble, ed.), McGraw-Hill Book Co., New York (1967).

30. Klug, H. P., and L. E. Alexander: X-Ray Diffraction Procedures, John Wiley and Sons, New York (1954).

31. Leroux, J., D. H. Lennox, and K. Kay: Direct Quantitative X-Ray Analysis. Anal. Chem. 25: 741 (1953).

32. Moore, C. A.: Quantitative Analysis of Naturally Occurring Multicomponent Mineral Systems by X-Ray Diffraction. Clays Clay Minerals 16: 325 (1968).

33. Karlak, R. F., and D. S. Burnett: Quantitative Phase Analysis by X-Ray Diffraction. Anal. Chem. 38: 1741 (1966)

34. Standard Method No. 6. U. S. Bureau of Mines, Pittsburgh, Pennsylvania.

35. Bergin, P.: Quantitative Diffraction Analysis of X-Ray Transparent Specimens. J. Sci. Instr. 41: 558 (1964)

36. Talvitie, N. A., and L. W. Brewer: Separation and Analysis of Dust in Lung Tissue. Amer. Ind. Hyg. Assoc. J. 23: 58 (1962).

37. Bradley, A. A.: The Determination of Quartz in Small Samples by an X-Ray Technique. J. Sci. Instr. 44: 287 (1967).

38. Cullity, B. D.: Elements of X-Ray Diffraction, Addison-Wesley Publishing Company, Reading, Massachusetts (1956).

39. Standard Method No. 1. U. S. Bureau of Mines, Pittsburgh, Pennsylvania.

40. Oberg, M.: Evaluation of Quartz in Airborne Dust in the 0.5- to 2-Micron Size Range. Environ. Sci. Technol. 2: 795 (1968).

41. Mossman, M. H., D. H. Freas, and S. W. Bailey: Orienting Internal Standard Method for Clay Mineral X-Ray Analysis. Clays Clay Minerals 27: 441 (1967).

DETERMINATION OF COMPOSITION OF SOLID SOLUTIONS USING X-RAY DIFFRACTION TECHNIQUES

Paul Cherin

Xerox Corporation
Research & Engineering Division
Xerox Square - W114, Rochester, N.Y. 14644

ABSTRACT

Although ordinarily a qualitative tool, x-ray diffraction techniques can be particularly powerful when dealing with continuous solid solutions. The very properties of the atoms that allow them to form these solutions often make it difficult to distinguish between them by chemical means. Even most physical analytical techniques have difficulty distinguishing among a physical mixture, a solid solution or a mixture of the two. This paper will be limited to substitutional solid solutions, where only two constituents are being altered. The principle behind the techniques to be discussed is based on the differences in effective atomic radii between the solvent atom and the substituting atom which results in an expansion or contraction of the lattice which is directly measurable. Several methods will be described which are particularly suited for investigating continuous solid solutions obeying Vegard's Law. When Vegard's Law is not obeyed, these methods can be suitably adjusted. Some of these methods can be extremely precise, permitting a determination of relative composition to 1 part in 10^9.

INTRODUCTION

Although X-ray diffraction techniques are ordinarily considered qualitative tools, there are situations where these techniques become particularly powerful quantitatively as well. This is especially true when dealing with continuous solid solutions, particularly substitutional solid solutions.

These solutions can be very difficult to study analytically. The very properties of the atoms, which are so similar so as to permit them to combine in this manner, make it difficult to distinguish among them by chemical means. I can relate several personal experiences where samples of CdSSe and CdZnS were sent to several reputable analytical services including our own in-house wet chemistry group. The results varied greatly for the same samples - by as much as 50%.

Physical analytical techniques are somewhat more useful. However, even physical methods have difficulty distinguishing among a physical mixture, a solid solution or a combination of the two. On the other hand, X-ray diffraction techniques can distinguish among these three cases and can yield a rapid determination of composition as well as a rather precise one.

DISCUSSION

The diffraction technique is based on differences in atomic radii, which cause observable changes in the lattice parameters. Figure 1 shows a two component system containing atoms of different radii. Note that the atoms are not located at their average lattice sites. If A were solvent atoms and B were substituting for A, the lattice would expand. On the other hand, if B were the solvent and A were substituting for B the lattice would contract. This change can be measured and related directly to changes in composition. It would appear that if the radii of A or B were very similar it would be difficult to measure a change in lattice size. However, this is not necessarily the case since atoms may differ in compressibility. Thus, the radii of A and B may be different in solution than when they are in the pure phase. And, in fact, studies by Warren, Averbach, and Roberts[1] have shown that the radii of atoms can, in

certain situations, change as a function of composition.
However, it is true that if substituting A or B causes no
discernible change in the lattice then the diffraction technique will not be fruitful.

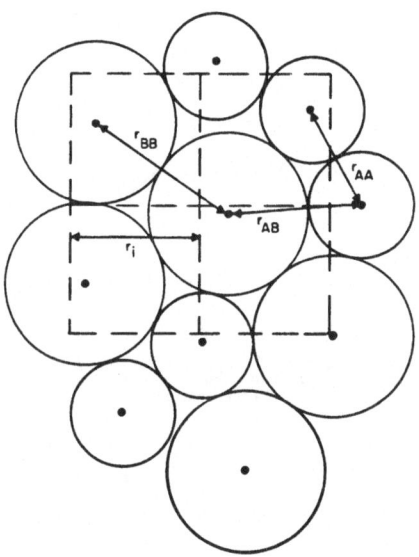

Figure 1. Two Component Solid Solution

Solid Solutions Obeying Vegard's Law

There are many, many techniques for measuring lattice
parameters. However, the simplest cases involve solid
solutions obeying Vegard's Law which states that there is
a linear relationship between the lattice parameters and
the composition as expressed in mole percent. This is an
empirical law and as such it is dependent on the precision
used to make the measurements.

Figure 2 is a schematic of three typical examples of
lattice parameter variation with composition. The central
line is an example of a solid solution obeying Vegard's
Law. The negative deviation from Vegard's Law usually
occurs when the electronegativity of the constituent atoms

A and B differ substantially. The positive deviation from
Vegard's Law occurs when the cohesive forces between the A
atoms and those forces between the B atoms are significantly greater than the interaction between the A and B atoms.

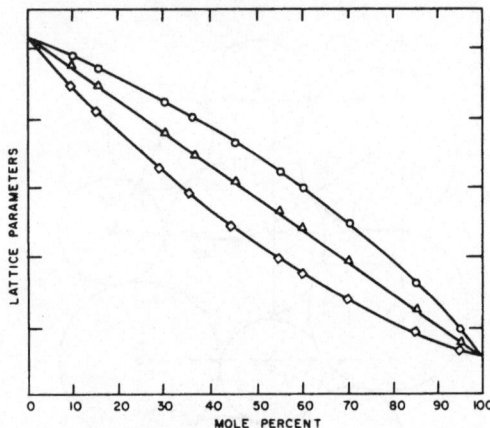

Figure 2.   Variation of Lattice Parameter with Composition:
Three Typical Examples.

There are several misconceptions associated with
Vegard's Law. It is usually thought that ideal solutions
and only ideal solutions, obey Vegard's Law. This is
neither a necessary nor sufficient condition for compliance
with the law. Zen[2] has shown this very nicely through
the use of equations (1) and (2).

$$a_s = a_B \left\{ 1 - \left[1 - \left(\frac{a_A}{a_B} \right)^3 \right] x_A \right\}^{1/3} \tag{1}$$

$$a_s = a_B \left\{ 1 - \left[1 - \left(\frac{a_A}{a_B} \right) \right] x_A \right\} \tag{2}$$

In the first equation, Zen has assumed ideality, that is, volumes are additives. Assuming a primative cubic unit cell, the volume has been converted into the cube of the cell edge, a^3. Equation (1) shows the unit cell edge of the solid solution, a_s, is related to the cube root of the mole fraction, i.e., X_A. In equation (2), he has simply assumed that the variation of the length of the unit cell edge is linear with the mole fraction, which is another way of stating Vegard's Law. It is clear that the two expressions are not equivalent. However, if a_A and a_B, which are the unit cell edges of the pure materials, are approximately equal, then both expressions become nearly identical.

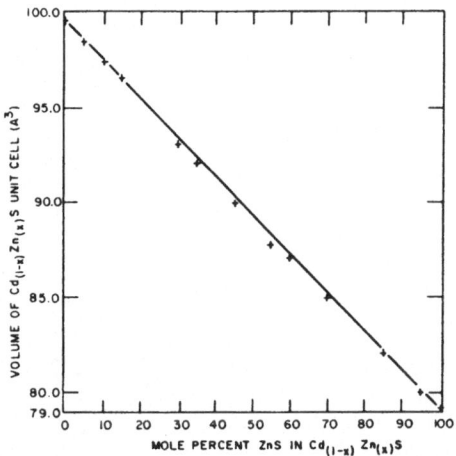

Figure 3. Unit Cell Volume Versus Composition

To demonstrate this in still another way, consider the case of $Cd_xZn_{(1-x)}S$ system, which is hexagonal and has a Wurtzite structure. Figure 3 shows a plot of the unit cell volume as a function of composition. Because the variation of volume shows a slightly negative deviation from linearity, this is considered a nonideal system. However, Figures 4 and 5 show that the unit cell edges c and a vary linearly with composition. Thus, both unit cell edges for this system obey Vegard's Law even though the system itself is not ideal.

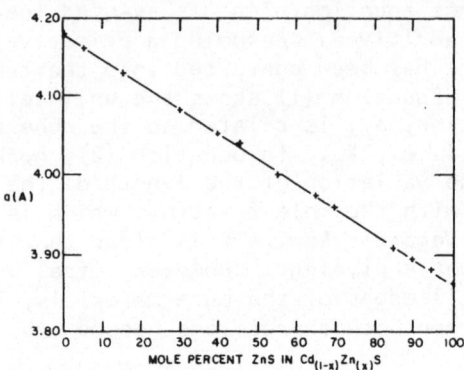

Figure 4.   Variation of the Length of the a-Axis With
Composition.

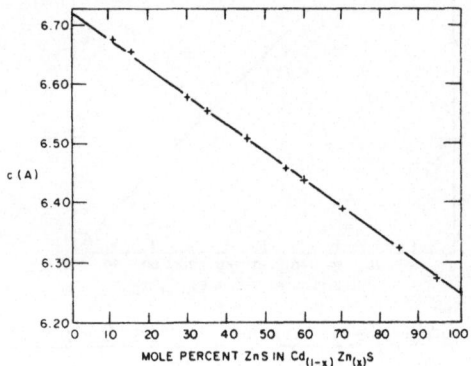

Figure 5.   Variation of the Length of the c-Axis with
Composition.

General Case

When a system is known to obey Vegard's Law and the
lattice parameters of the pure materials are known, the
composition of a solid solution may be easily determined.
A straight line can be generated analytically or graphical-
ly from the known parameters. By measuring the parameters

of the unknown solid solution the composition can be determined using the analytical or graphical information. One additional step is recommended, however. It is wise to redetermine the lattice parameters of the pure materials even though they are readily available in the literature, because whenever the lattice parameters are determined invariably systematic errors are incurred. Since these errors will be constant for the unknown solution as well as the pure phase materials, the composition determination will be unaffected.

Frequently information concerning compliance with Vegard's Law is not readily available in literature. In these cases one can investigate radii, compressibility and electronegativity data. If these are favorable, one can assume compliance as a first approximation and determine the composition as described above.

In many cases, however, deviations from Vegard's Law are sufficient to demand a more extensive study. For these general cases, a series of solid solutions of known composition must be prepared. This must be done using a complimentary technique as, for example, fluorescence. If this technique is used, it is necessary to prepare a series of physical mixtures containing known concentrations of the constituents. The samples must be mixed very thoroughly and the particle size kept uniform. It is often useful to dissolve the mixture in a solvent, assuming no volatile constituents are present and sufficient quantities dissolve; it is necessary to obtain sufficient fluorescence intensity for decent statistical precision.

After obtaining the fluorescent data, a plot is drawn of the intensity ratio of the varying constituents versus the mole ratio. Now a series of solid solutions is prepared for fluorescence in exactly the same manner. It is necessary to check these solid solutions by X-ray diffraction techniques to ascertain that these are single phases. The intensity ratios are determined for these solid solutions exactly as for the physical mixtures and compared with the plots previously prepared for the known mixtures. In this way, the composition of the solid solutions can be determined. The next step requires the determination of the lattice parameters of these known solutions as a function of the known composition.

It may appear that the determination of lattice parameters is a seemingly unnecessary step.   It may seem sufficient to determine the intensity ratios of the constituents of the unknown solutions with fluorescence and determine the composition in that way.   Diffraction techniques
are recommended because with fluorescence we cannot distinguish between a solid solution and a physical mixture.
The diffraction technique is inherently more precise than
the fluorescence technique, especially when one of the constituents is in low concentration or the amount of sample
present is small.   It is much easier to prepare a sample
for diffraction than it is for fluorescence techniques.

Lattice Parameter Determination

Although there are literally hundreds of techniques
used to determine lattice parameters, some of which yield
very high precision (several authors[3,4] have claimed a
precision of one part in 10^9), all are based on the principle
that measurements made at high angles are inherently more
accurate.   This can be seen with the aid of Figure 6.   A
constant error $\Delta\theta$ causes a significantly larger error in
$\sin\theta$ and consequently in the lattice spacing, d, at low
diffraction angles than it does at high angles.   Thus, all
techniques have been devised to take advantage of this
situation.

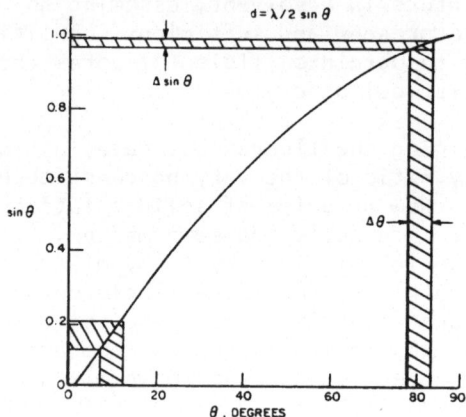

Figure 6.   Effect of an Error in θ on $\sin\theta$ as a Function of
θ.

Figure 7 illustrates the typical Debye-Scherrer method, which allows diffraction maxima to be recorded at fairly high diffraction angles. However, diffracted radiation at very high angles can be physically blocked by the collimator. This technique has other problems which include absorption, beam divergence, spectral dispersion, etc.

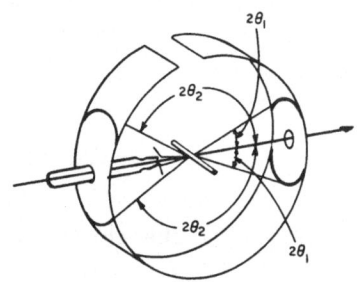

Figure 7. Debye-Scherrer Method.

The symmetrical focusing technique is illustrated in Figure 8. The x-ray beam comes into the camera through a slit, which provides very little interference for diffracted radiation. The maximum 2θ angle that can be recorded exceeds 176°. The sample is placed along the circumference of the camera as is the film. The sample focuses the diffracted x-ray beam to a line which enhances the intensity and sharpness of the diffracted lines permitting more precise measurements. Absorption errors are eliminated. It is necessary to choose the proper x-ray wavelength in order to maximize the number of reflections at high angles.

When diffractometry techniques are used, film is replaced by an electronic recorder such as a scintillation counter permitting a more precise measurement of the diffraction angle. However, there are limitations on the maximum angle of diffracted radiation that may be recorded. For example, a Norelco diffractometer has a maximum 2θ angle of approximately 150°. This is significantly less than is observable by the symmetrical focusing technique.

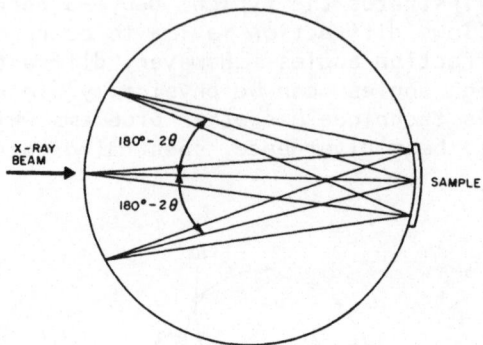

Figure 8.   Symmetrical Focusing Technique.

The above are probably the three most widely used
methods for measuring diffraction angles, enabling a determination of the lattice parameters. It is possible to
increase ones accuracy, when using these basic methods,
by emphasizing high angle reflections which are inherently
more accurate. Figure 9 shows a plot of lattice parameter
versus $\cos^2\theta$. Since the value of the lattice parameters is
more accurate at $\theta = 90°$, extrapolation is made to that
point. The slope of the line is generally negative, because the major systematic errors involved, such as absorption, tend to reduce the observed value of a. The slope of
the line is a measure of the severity of the errors and of
the precision. The more nearly horizontal the slope the
greater the precision.

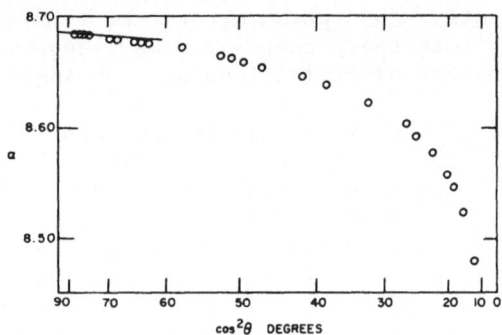

Figure 9.   Lattice Parameter Versus $\cos^2\theta$.

In most cases, the points do not fall as neatly on the line as this textbook case and since this curve is not linear for the Debye-Scherrer technique it can be difficult to use. Therefore, many people use the technique shown in Figure 10. Taylor and Sinclair[5] and independently Nelson and Riley[6] have shown that

$$1/2 \left(\frac{\cos^2\theta}{\sin\theta} + \frac{\cos^2\theta}{\theta} \right)$$

represents a mean between two different functions which are proportional to the error caused by absorption and by beam divergence. These are the major sources of error when using the Debye-Scherrer technique. Thus, a versus this function is usually a straight line and therefore results in a more accurate extrapolation.

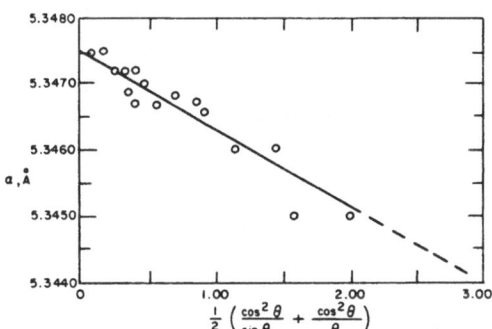

Figure 10.   Lattice Parameter Versus $1/2\left(\frac{\cos^2\theta}{\sin\theta} + \frac{\cos^2\theta}{\theta}\right)$.

Whenever straight lines are present it is possible to use least squares methods. The classical approach shown below is called Cohen's Method.

$$\frac{\lambda^2}{4a_o^2} (h^2 + k^2 + \ell^2) + D \sin^2 2\theta_i = \sin^2\theta_i \qquad (3)$$

$$A \alpha_i + D \delta_i = \sin^2\theta_i \qquad (4)$$

where $A = \dfrac{\lambda^2}{4a_o}$; $\alpha_i = (h^2 + k^2 + \ell^2)$; $\delta_i = \sin^2 2\theta_i)$

Min R = $(A\alpha_i + D\delta_i - \sin^2\theta_i)^2$ (5)

   Equation (3) is essentially the Bragg equation for cubic crystals where $1/d^2$ is replaced by its relation to the lattice parameter, $1/a^2$ and the Miller indices h, k, and ℓ. $D\sin^2 2\theta$ is the accumulation of systematic errors which causes a deviation from Bragg's Law. Equation (5) is min- imized by differentiating with respect to the coefficients A and D setting the resulting expressions equal to zero and then solving for the coefficients, which give the best values for the lattice parameter.

   For the noncubic case the expression is a bit more complex(7)

$$(1/d)^2 = S_{11}h^2 + S_{22}k^2 + S_{33}\ell^2 + S_{12}hk + S_{23}k\ell + S_{31}h\ell \qquad (6)$$

$$\sum_i^N w_i \left\{ (1/d_i)^2 - (S_{11}h_i^2 + S_{22}k_i^2 + \ldots\ldots\ldots\ldots + S_{31}h_i\ell_i) \right\}^2 \qquad (7)$$

where

$$S_{11} = b^2c^2\ (\sin^2\alpha)/v^2$$

$$S_{12} = 2abc^2(\cos\alpha\cos\beta - \cos\gamma)/v^2$$

$$S_{22} = c^2a^2\ (\sin^2\beta)/v^2$$

$$S_{23} = 2a^2bc\ (\cos\beta\cos\gamma - \cos\alpha)/v^2$$

$$S_{33} = a^2b^2\ (\sin^2\gamma)/v^2$$

$$S_{13} = 2ab^2c(\cos\alpha\cos\gamma - \cos\beta)/v^2$$

Equation continued...

$$W_i = 1 / \left\{ \left[\frac{d(1/d^2)}{d(2\theta)} \right] \triangle(2\theta_i) \right\}^2 = \left\{ \frac{\lambda}{8 \sin 2\theta_i \, \triangle (2\theta_i)} \right\}^2 \qquad (8)$$

abc, α, β, γ, are the lattice parameters.

The minimization is essentially the same. However, because there are now 6 parameters to determine, more data are needed and it is necessary to use relatively low angle reflections. However, they are properly weighted as shown in Equation (8).

Ultra-Precise Methods

Aside from the three basic methods for determining lattice parameters there are also specialized methods, some of which can yield ultra-precise results. Consider the Bond[8] method (a schematic of the experimental technique is shown in Figure 11), which is useful for single crystals only. The crystal obeys the Bragg conditions in two positions separated by π-2θ degrees. One can either measure the angle between the two collector positions or can measure the angle through which the crystal has been rotated (R_2-R_1). This technique essentially eliminates eccentricity, absorption and zero-position errors. Other errors due to crystal tilt, axial divergence, Lorentz-polarization and spectral dispersion still exist. However, with proper corrections precision good to 1 part in 10^6 may be achieved.

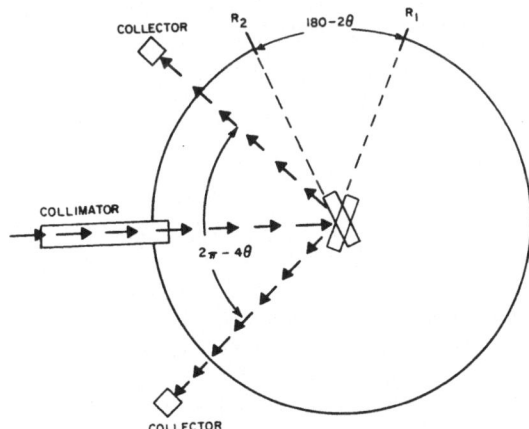

Figure 11. The Bond Method.

For still greater precision one can use a double crystal arrangement and rocking curve as developed by Hart[9].
This method requires a single crystal sample and a reference
crystal. The diffracting planes of these crystals must be
parallel to the morphological crystal planes. A schematic
drawing is shown in Fig. 12. Radiation from two sources is
passed through a narrow slit and onto a thin reference
crystal. The beams are positioned so that the Bragg conditions are obeyed for the reference crystal and the x-rays
are diffracted onto the sample crystal. The sample crystal
is then rocked through an angle β until the Bragg conditions
are obeyed. This will occur in two positions and the x-rays
are diffracted into the two detectors. A typical rocking
curve for an almost perfect crystal of Si is shown in
Figure 13.

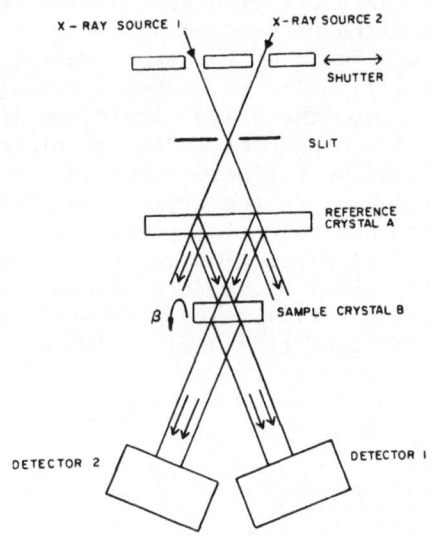

Figure 12. **The Hart Method**

The separation between the peaks $\Delta\beta$ can be measured
and can be related to the Bragg angle, θ_B, of the sample
crystal, since the Bragg angle of the reference crystal is
known. The angular separation $\Delta\beta$ can be determined very
precisely. This method eliminates the error due to spectral
dispersion, which was one of the major errors of the Bond

method. The unit cell dimension can be determined to 1 part in 10^9. The absolute value of lattice parameters can not be determined that accurately, because the wavelength is known only to 1 part in 10^5. However, this is a systematic error which is constant for all cases and will not affect the composition determination, which theoretically may still be determined to 1 part in 10^9.

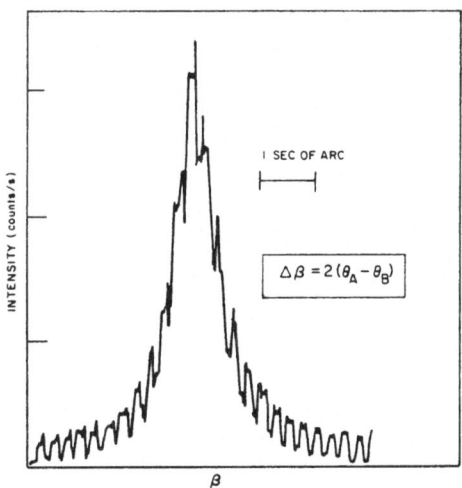

Figure 13.   A Typical Rocking Curve for a Nearly Perfect Crystal of Si.

It should be emphasized that the Bond and Hart methods are specialized techniques suitable only for very good single crystals of a solid solution. Special equipment is required and great care is necessary in collecting experimental data. For example, very good temperature control is necessary. Most situations do not require a composition determination to 1 part in 10^9; often 1 part in 10^4 is sufficient. There are several rapid methods that are particularly suitable for determining compositions of solid solutions and will yield this precision.

Rapid Methods for Solid Solutions

Consider a solid solution of known crystal structure. It is possible to approximate the composition without having to set up a series of standards simply by measuring the lattice parameters of the solid solution. This can be done with the aid of the following expressions [10].

where
$$S_X - S_o \approx x \, (r_X - r_B) \tag{9}$$

$$\bar{S}_o = (U_B - U_A)\bar{a}_o + (V_B - V_A)\bar{b}_o + (W_B - W_A)\bar{c}_o$$

$$\bar{S}_X = (U_X - U_A')\bar{a}_X + (V_X - V_{A'})\bar{b}_X + (W_X - W_{A'})\bar{c}_X$$

$$(U_X - U_{A'}) \approx (U_B - U_A), \quad (V_X - V_{A'}) \approx (V_B - V_A), \quad \text{etc.}$$

$\bar{S}_o$ is the translational vector between atoms A and B in pure phase AB. $\bar{S}_X$ is the translational vector between atoms X and A in the solid solution where X is being substituted for B. U, V, and W are the positional coordinates of the atoms in the unit cell. r_X and r_B are the radii of atoms X and B respectively and x is the mole fraction of constituent X. When the solid solution is dilute and the ionic radii, the electronegativity and compressibility of atoms B and X are not very different than one can assume $U_B - U_A$ is approximately equal to $U_X - U_A'$, etc. Since the radii, structure and parameters of the pure phases are known, the mole fraction x can be determined from equation (9).

In effect, we have assumed Vegard's Law holds. This is a reasonable assumption under the limitations previously prescribed. Earlier we described even simpler methods for dealing with systems that obeyed Vegard's Law. Therefore, this method is useful, whenever the second phase AX is not stable or crystallizes in a phase which is different than AB. In that case the method described earlier could not be used.

Harrison and Curtis[11] utilize a very rapid method for hexagonal and tetragonal solid solutions obeying Vegard's Law, in which it is necessary to measure only one diffraction line utilizing equations (10-12).

$$\frac{c_X}{a_X} = \frac{xc_A/a_A + (1-x)c_B/a_A}{x + (1-x)a_B/a_A} \tag{10}$$

$$\frac{\sin^2\theta}{\lambda^2} = \frac{1}{c_X^2}\left\{\frac{c_X^2}{a_X^2}\left(\frac{h^2+hk+k^2}{3}\right) + \frac{\ell^2}{4}\right\} \tag{11}$$

$$x = \frac{c_X - c_B}{c_A - c_B} \tag{12}$$

Assume a value for x; the c_X/a_X ratio may be calculated with the aid of equation (10), since the lattice parameters of the pure phases are known. Using equation (11) the value of c_X can be calculated from the c_X/a_X ratio and the measured value of θ. The value of x can now be recalculated using equation (12); this can be substituted back into equation (10), and the process can be repeated iteratively until the final value of x is determined.

This method relies on the dependence of the c_X/a_X ratio on the composition. However, generally this ratio is somewhat independent of the composition. Figure 14 presents a plot of the c_X/a_X ratio as a function of composition for the wurtzite system $Cd_xZn_{(1-x)}S$, which was given as an example by Harrison and Curtis (11). The variation with composition is not great. Compare this variation with that shown in Figure 5. The slope of the line in Figure 5 is 40 times that which is shown in Figure 14. Thus, this technique is rapid, but will not yield high precision, because small errors in the value of c_X/a_X will yield relatively larger errors in the composition.

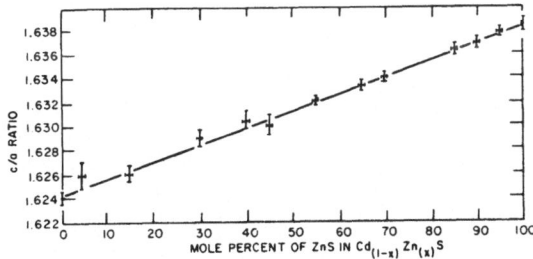

Figure 14.   c/a as a Function of Composition for the $Cd_xZn_{(1-x)}S$ System.

The "inner ring" method developed by Cherin, Davis, and Bielan[12] is useful. Figure 15 illustrates the meaning of the term "inner ring". The diffracted lines at the highest 2θ angle of the Debye-Scherrer photograph are called the inner ring. There are other rings at high angles which may be utilized. In fact, the radiation should be chosen to maximize the number of inner rings. A set of solid solution standards of known compositions as previously described, is necessary for this method. One measures the diameter of the inner ring, S, of the standard solid solutions and relates this to the known composition as is shown in Figure 16, which shows the variation of the diameter as a function of the composition for the $Cd_x Zn_{(1-x)} S$ system. The nearly linear character of the lines is no accident. For a continuous solid solution crystallizing in the cubic phase and obeying Vegard's Law a plot of $cos^{-1} S/4$ versus the mole fraction is linear. The same is true of hexagonal or tetragonal systems, when the c/a ratio is independent of the composition. These are usually reasonable assumptions so that this plot is generally a useful one. After the standard curves have been drawn it is necessary to measure only the ring diameter of the unknown solid solution.

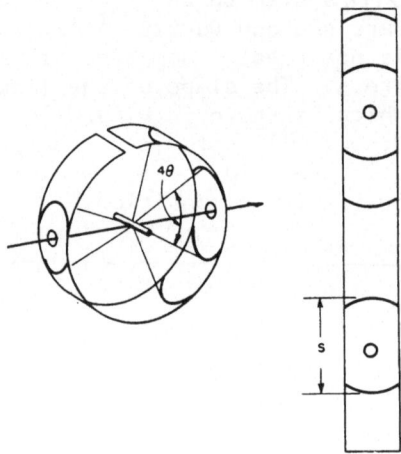

Figure 15.   Inner Ring Method.

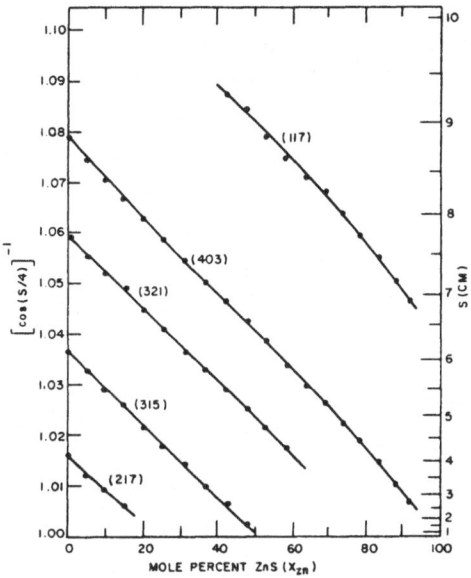

Figure 16. <u>Diameter of Inner Ring as a Function of Composition for the Cd$_x$Zn$_{(1-x)}$S System.</u>

As you can see in the Cd$_x$Zn$_{(1-x)}$S system the variation is nearly linear; there are several high angle reflections with which to work. The sensitivity of this method can be appreciated by noting that the diameter of the (403) ring, for example, varies by about .25mm for every change of 1 mole percent near the insensitive region, where the concentration of ZnS is very low. A change of less than 0.1mm can be routinely measured. Where the concentration of ZnS is very high the variation in diameter is almost 1mm per change of 1 mole percent in concentration. By judiciously choosing the right inner rings of the unknown solid solution one can easily determine small changes in composition. The use of more than one ring is recommended to minimize the effect of random and systematic errors. Compositions can usually be determined to 1 part in 10^4.

CONCLUSION

In conclusion, determining compositions is basically a question of determining lattice parameters, although it is possible to circumvent this relatively lengthy procedure. It has been shown that precision can be quite high with the usual trade off of more effort yielding greater precision.

REFERENCES

1. B. Warren, B. Averbach and B. Roberts, J. Appl. Phys. 22, 1493 (1951).
2. E. Zen, American Mineralogist 41, 523-4 (1956).
3. H. H. Wills, Proc. Royal Soc. A 309 281-96 (1969).
4. S. Propoviz, J. Appl. Cryst. 4, 240 (1971).
5. A. Taylor and H. Sinclair, Proc. Phys. Soc. A 57, 126 (1945).
6. J. B. Nelson and D. P. Riley, Proc. Phys. Soc. A 57, 160 (1945).
7. P. Cherin, E. Lind and E. A. Davis, J. Electrochem. Soc. 117, 233 (1970).
8. W. L. Bond, Acta. Cryst. 13, 814 (1960).
9. M. Hart, Proc. Roy. Soc. A 309, 281 (1969).
10. L. K. Frevel, Analytical Chem. 42, 1583 (1970).
11. F. W. Harrison and B. J. Curtis, Brit. J. Appl. Phys. 13, 247 (1962).
12. P. Cherin, E. A. Davis and C. Bielan, Analytical Chem. 40, 611 (1968).

NMR OF LESS-SENSITIVE SPIN - 1/2 NUCLEI

Robert R. Sharp

University of Michigan

Ann Arbor, Michigan 48104

INTRODUCTION

Nearly every stable element has one or more magnetic isotopes which can, at least in principle, be observed by nuclear magnetic resonance. Only argon, among the naturally abundant representative and outer transition elements, possesses no isotope with a nuclear magnetic moment. Magnetic nuclei have properties that fall conveniently into two classes according to the magnitude of nuclear spin. Those with spins greater than one-half (shown hatched in figure 1) have nonspherical nuclear charge distributions which give rise to electric quadrupole moments. Resonances of these nuclei are generally broad, often beyond the practical limits of detection, because of the highly efficient electric quadrupole contribution to the relaxation rates. Chemically interesting information can in favorable cases be obtained from these resonances however (1-6) although the quadrupolar broadening tends to obscure chemical shift differences and spin-spin multiplet structure.

Spin-1/2 isotopes in mobile diamagnetic liquids typically show narrow resonances characteristic of high resolution n. m. r. The Periodic Table in figure 1 contains these isotopes in unhatched squares along with a number which represents the relative n. m. r. sensitivity at constant field for the most abundant isotope (based on

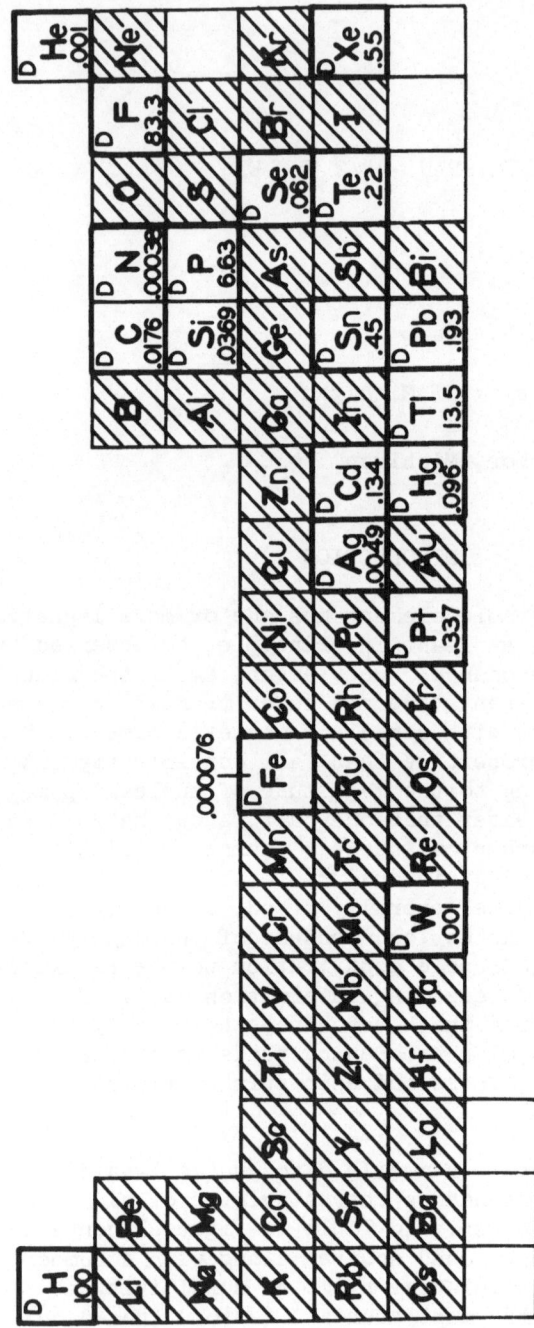

Figure 1. Periodic table of the elements according to nuclear spins of naturally abundant isotopes. Blank squares are positions of elements with no naturally abundant magnetic isotopes. Elements with at least one spin-1/2 isotope lie in unhatched squares. Elements for which the only naturally abundant magnetic isotopes have spin greater than one-half are shown in hatched squares.

$^1H = 100$).[*]   A glance at the sensitivities shows why the
great bulk of n.m.r. research prior to 1970 has involved
^1H, ^{19}F, and ^{31}P. The recent introduction of Fourier
transform techniques in n.m.r. has permitted a quantum jump
of more than an order of magnitude in instrumental sensitivity, so that ^{13}C in natural abundance is now routinely
observable using commercial spectrometers (see the following paper). Fourier transform n.m.r. is also being
used to observe resonances of some heavier nuclei and gives
promise of increasing substantially the chemical utility
of these resonances in the future. Nevertheless, an F.T.
capability is not essential for observing less-sensitive
nuclei, and certain less sophisticated alternative tech-
iques are discussed below.

Although the heavier spin-1/2 nuclei are undeniably
demanding from a sensitivity standpoint, their spectra
offer certain advantages relative to ^1H spectra which
should in large part compensate for the sensitivity difficulties. Chemical shift ranges (in p.p.m.) generally
increase with atomic number and span ranges the order of
7000-17000 p.p.m. for ^{195}Pt, ^{205}Tl and ^{207}Pb (8-16). For
heavier isotopes very subtle changes in the chemical surroundings give rise to easily identifiable chemical shifts,
so that spectra obtained under relatively low resolution
($\Delta H/H_o \approx 10^{-5}$) are sensitive probes of the chemical environment. Early measurements by Figgis (14) on the ^{205}Tl
resonance in aqueous solutions of Tl (III) salts illustrate the detailed information obtainable from metal resonances. Complex chemical shift variations exceeding
1000 p.p.m. in absolute magnitude are observed as excess
Br^- and Cl^- are added to solutions containing Tl (III).
The thallium isotopes resonate as singlets in aqueous solutions due to rapid exchange of the metal between several
species of complex halide ion. The resonance position is
a weighted sum of shifts in the various complex ions. From
an analysis based on solution composition at minima and
inflection points of the chemical shift data, Figgis infers
that the predominant species are $Tl_2X_9^{-3}$ at intermediate
halide concentration and TlX_6^{-3} at higher halide concentration. Chemical shifts of other metal resonances have

---

[*] Sensitivities in Table I are the product of the natural
abundance in percent and the intrinsic sensitivity at constant field (as given in reference 7).

been used in a similar manner to study rapid chemical exchange equilibria in solutions containing ^{119}Sn (17,18),
^{199}Hg (19), ^{111}Cd (20), and ^{95}Pt (8). Such large chemical
shifts are especially valuable for distinguishing geo-
metrical and diastereoisomers. cis and trans isomers of
Pt (II) complexes show ^{195}Pt shifts in the range 20-500
p.p.m. (8,9), while ^{31}P shifts (in the ligands) are somewhat smaller, 3-5 p.p.m. (21,22). Most resonances of
heavier isotopes are not understood well enough at the
present time to provide an easily applied method of
structure analysis, however. Only for the phosphorus
resonance (among elements outside the first row of the
Periodic Table) is sufficient information available to
identify functional groups by simple empirical correlations
with tabulated shifts in model compounds (23,24). Nevertheless, chemical shift data for many nuclei form a useful
complement to other conventional means (polarography,
absorption and Raman spectroscopy, etc.) for monitoring
solution species.

Further information is obtainable when the resonant
spin-1/2 isotope is scalar (or spin-spin) coupled to a
second magnetic nucleus; when both nuclei relax slowly[1]
multiple structure is observed. This structure differs
in two important respects from that normally seen in proton spectra. In the first place coupling constants, like
chemical shifts, are often much larger for heavy than for
light nuclei and frequently exceed one kilohertz in magnitude for directly bonded elements. Secondly, several
heavy elements consist of two or more naturally abundant
spin-1/2 isotopes (e.g., ^{115}Sn, ^{117}Sn and ^{119}Sn, ^{203}Tl and
^{205}Tl, ^{199}Hg and ^{201}Hg, ^{111}Cd and ^{113}Cd). The spectrum of
any nucleus bonded to one of these elements normally con-
sists of two or three superposed multiplets arising from
coupling to the various spin-1/2 isotopes, as well as a
simpler parent spectrum arising from molecules that contain only nonmagnetic metal nuclei. Fortunately the coupling constants to pairs of isotopes of a given element fall
in the same ratio as the resonance frequencies of the pair.
Therefore the n.m.r. spectra of nuclei bonded to any heavy
element can be described in terms of a single coupling con-

---

[1]More precisely, the transverse relaxation time T_2 of both
nuclei must be much longer than $(2\pi J)^{-1}$, where J is the
spin-spin coupling constant (25).

stant per pair of bonded atoms. When compounds contain
two or more identical heavy elements bonded to each other
the metal resonances display complex multiplet structure
which is particularly useful for molecular structure determination. Perhaps the most elegant example of this
situation is found in the work of Schneider and Buckingham
(15), who have used the large (2.56 kHz) ^{203}Tl-^{205}Tl coupling in TlOEt to confirm polymeric structure in this liquid.
The thallium isotopes ^{203}Tl and ^{205}Tl resonate as a septet
and quintet respectively, whereas a single pair of non-
identical coupled nuclei would exhibit only a pair of
doublets with identical splitting. The observed multiplets
are consistent however with a tetramer in which each Tl
nucleus is spin-coupled to three nearest neighbors. Random
isotopic distribution of Tl isotopes in the tetramer produces the observed spectra composed of overlapping multiplets. The existence of several magnetic isotopes present
in natural abundance is the key to a molecular structure
determination of the polymer. If only a single Tl isotope
were naturally abundant, the predicted resonance for the
tetramer would be a singlet.

N.m.r. linewidths have not been used extensively for
interpreting spectra of spin-1/2 resonances, although for
the heavier isotopes they are potentially a valuable source
of chemical information. In many systems the natural linewidth is less than the magnetic field inhomogeneity and is
measureable only by means of pulsed methods. Substantial
line broadening is frequently observed however for one of
the following causes. Rapid exchange of the resonant nucleus between two or more chemically shifted environments produces broadening when the rate of exchange (in sec^{-1}) is
comparable to the chemical shift difference between the
two sites. Alternatively broadening may be due to a partial
collapse of spin-spin multiplet structure when spin-spin
coupling is interrupted either by chemical exchange reactions
or by rapid relaxation of one or the nuclei[1]. Whatever its
origin, broadening is usually more pronounced for heavy than
for light nuclei because the former are usually associated
with much larger chemical shifts and coupling constants
(see below). In the work of Figgis (14) cited above, the
thallium resonance in aqueous halide solution was broadened,
probably due to chemical exchange of the metal between several chemically shifted sites. The thallium linewidth shows
a strong and complex dependence on added halide concentration

and was used, along with chemical shifts, to monitor variations in chemical equilibria. In favorable cases (20),
linewidth data can be analyzed quantitatively to obtain
rate constants for chemical exchange.

1. CHEMICAL SHIFTS

Experimental and theoretical n.m.r. investigations
have very largely involved proton spectra and have provided
a tested, reasonably quantitative basis for interpreting
proton chemical shifts. It is indeed unfortunate that
these studies are almost wholly inapplicable to the interpretation of chemical shifts in heavier atoms. The reason
for this qualitative difference in chemical shift origins
for protons and heavier nuclei is readily apparent from
Ramsey's (26) perturbation formulation of the theory of
magnetic shielding. According to Ramsey's theory the
shielding constant σ is a sum of two terms of opposite sign:

$$\sigma = \sigma_d + \sigma_d. \tag{1}$$

Chemical shifts represent differences in the shielding constant in different molecular environments. The first term
σ_d in equation (1) is diamagnetic and describes an increase
in magnetic shielding due to free precession of electrons
close to the resonant nucleus. The second term σ_p results
from a mixing of excited electronic states with orbital
angular momentum into the ground state wave function.
Chemical shifts of protons and heavy nuclei differ in that
the former originate primarily from changes in σ_d while the
latter are determined by changes in σ_p. Proton shielding
provides a measure of electron density in the hydrogen 1s
orbital and is also sensitive to local magnetic fields
caused by precession of electrons in nearly molecular orbitals. Chemical shifts of heavier atoms are determined
principally by factors that influence σ_p, namely, by deviation from spherical symmetry in the electronic wave
function and by the energy difference of ground and excited
states. The diamagnetic term is quite large for heavy
atoms (27,28) but is not strongly dependent on the valence
electrons (e.g., σ_d changes relatively little upon ionization (27)), while a variety of calculations (29-33) indicate that changes in σ_p are comparable to the observed
shifts.

Unfortunately the calculation of chemical shifts in heavy atoms is a very difficult matter since σ_p is derived using second order perturbation theory and contains a sum over excited states for which accurate energies are almost never known. Thus the interpretation of ^{19}F, ^{13}C and ^{31}P chemical shifts is essentially correlative in nature. These resonances have been investigated extensively, and the requisite data bases are available for practical applications of measured chemical shifts to problems of molecular structure. Such large data bases are not presently available for other spin-1/2 nuclei. Periodic or other semi-theoretical relations between nuclear shielding scales of different elements may eventually provide the most satisfactory basis for interpreting chemical shifts of less common resonances. Useful relations of this type are not widely accepted by n.m.r. spectroscopists at present, although Jameson and Gutowsky (34) have pointed out a few very general trends which appear to be firmly supported by the existing data. These workers draw their conclusions from a simplified M.O. expression for σ_p, equation (A-1), which is given in the appendix. According to this expression, chemical shifts depend on (1) the average energy, ΔE, of excited states (2) $<r^{-3}>$ for p and d valence orbitals and (3) the unbalance, P_u and D_u, of electron population in different p and d orbitals (this term measures asymmetry in the electronic surroundings). The second parameter, $<r^{-3}>_{p,d}$, is strongly periodic and generally determines the range of chemical shifts encountered for various nuclei. Chemical shifts, like $<r^{-3}>_{p,d}$, tend to become larger across a given row or down a given column of the Periodic Table (34,35). For example, the known range of chemical shifts for silicon is about 160 p.p.m. (36,37), for tin is 1860 p.p.m. (17,37), and for lead is 15000 p.p.m. (16). But beyond this simple generalization very few correlations have been found which are applicable to chemical shifts of a wide variety of heavy elements.

Chemical shift scales for several spin-1/2 nuclei are shown in figures 2-7. For four of these elements, F, P, Pb, and Sn, sufficient information is available to construct absolute chemical shift scales[1], which shown experimental

---

[1]These scales are based on the definitions of σ_p' and σ_d' proposed by Deverell (38) and Flygare (39), which differ slightly from σ_p and σ_d defined by Ramsey.

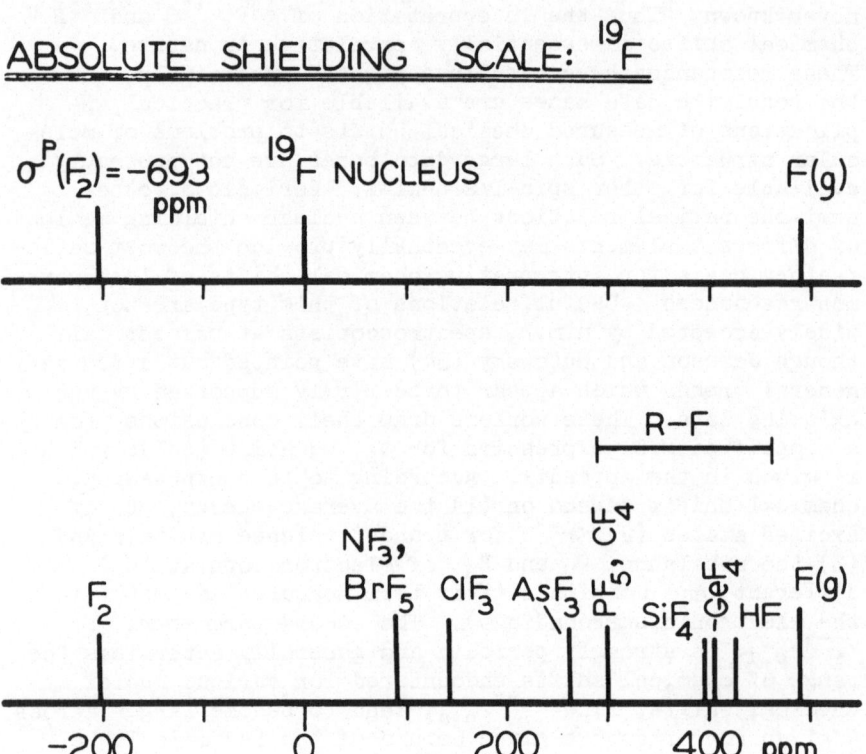

Figure 2.   Nuclear shielding of ^{19}F. Values for the absolute
shielding constants are taken from reference 38. Chemical
shifts of the binary fluorides are given in reference 40.
Chemical shifts of alkyl fluorides (R-F) are compiled in
reference 7.

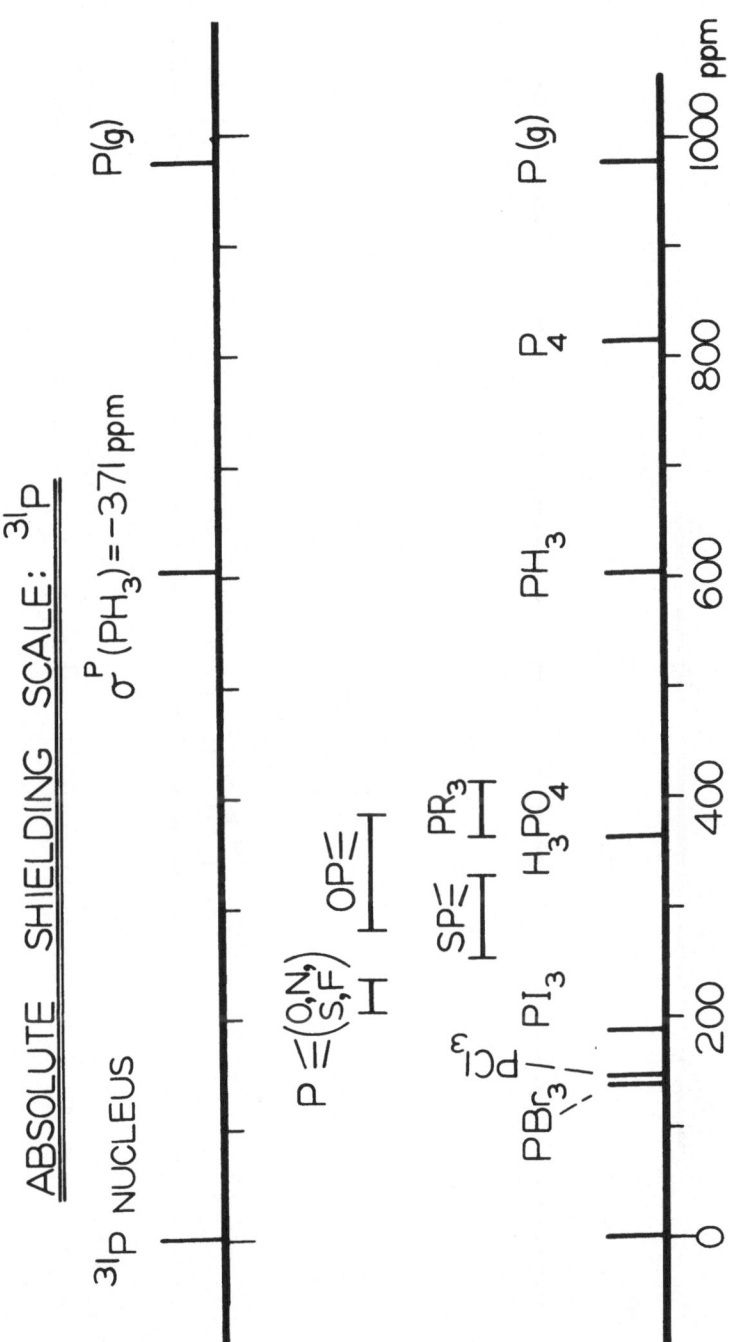

Figure 3. Nuclear shielding of ^{31}P. Values for the absolute shielding constants are taken from reference 88. Compilations of ^{31}P shifts are taken from references 7 and 42.

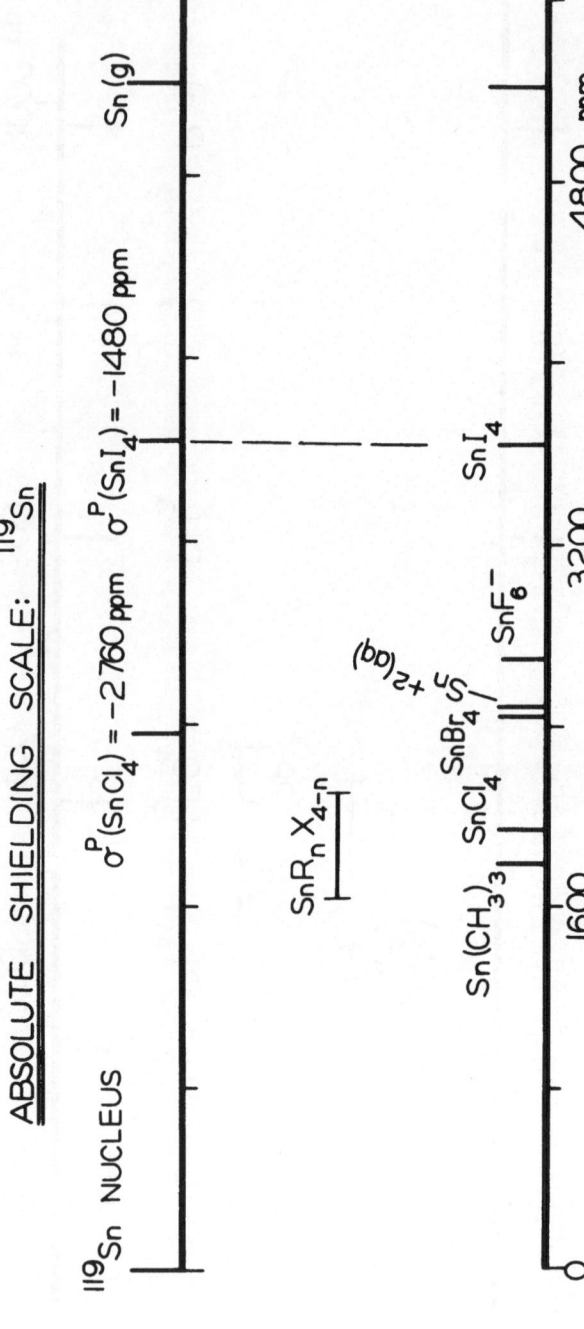

Figure 4. Nuclear shielding of ^{119}Sn. Values for the absolute shielding constants are taken from reference 74. Chemical shift ranges for ^{119}Sn are taken from references 17 and 37.

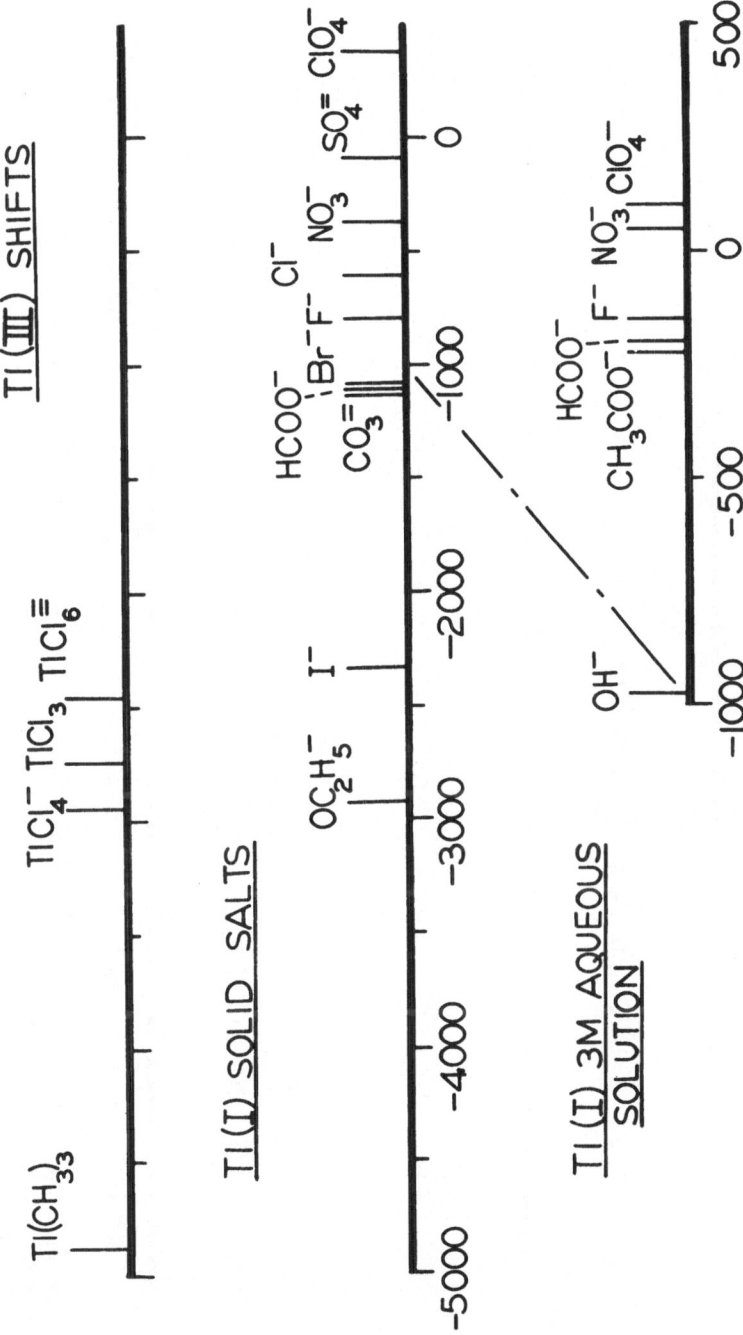

Figure 5. Nuclear shielding of ^{205}Tl. Chemical shifts are taken from references 11-15.

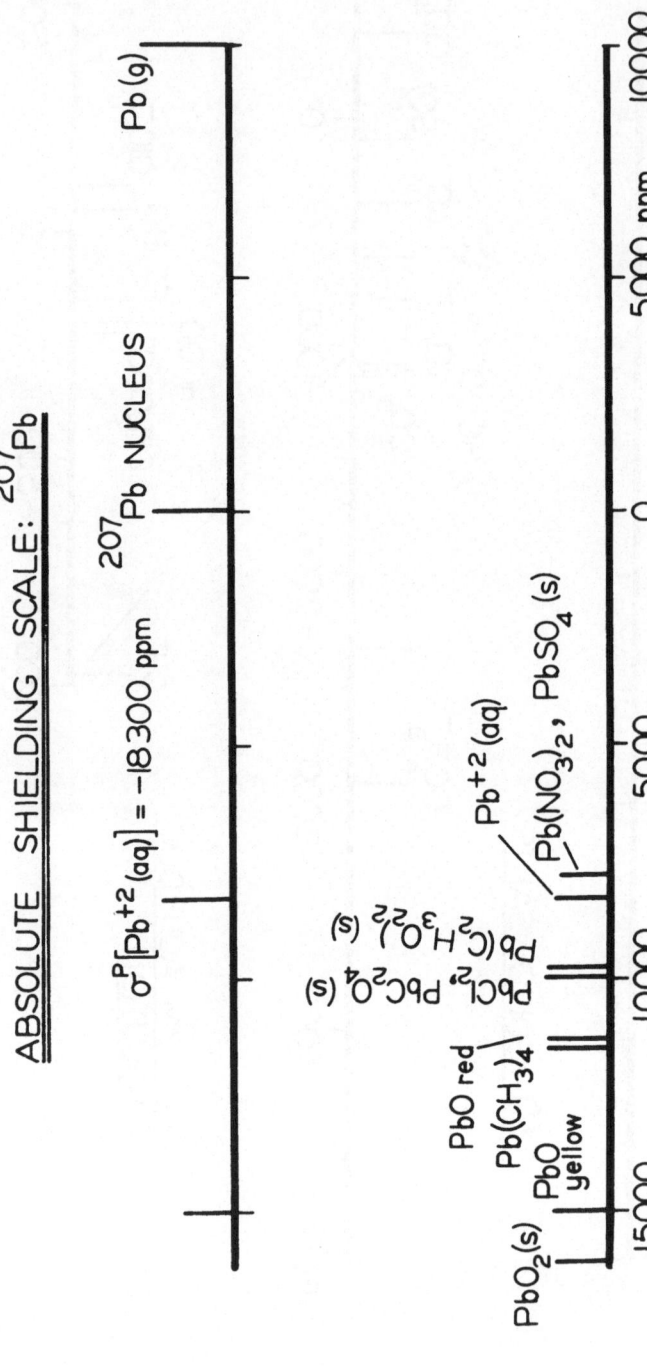

Figure 6. Nuclear shielding of ^{207}Pb. Absolute shielding constant is taken from reference 87. Chemical shifts are taken from reference 16.

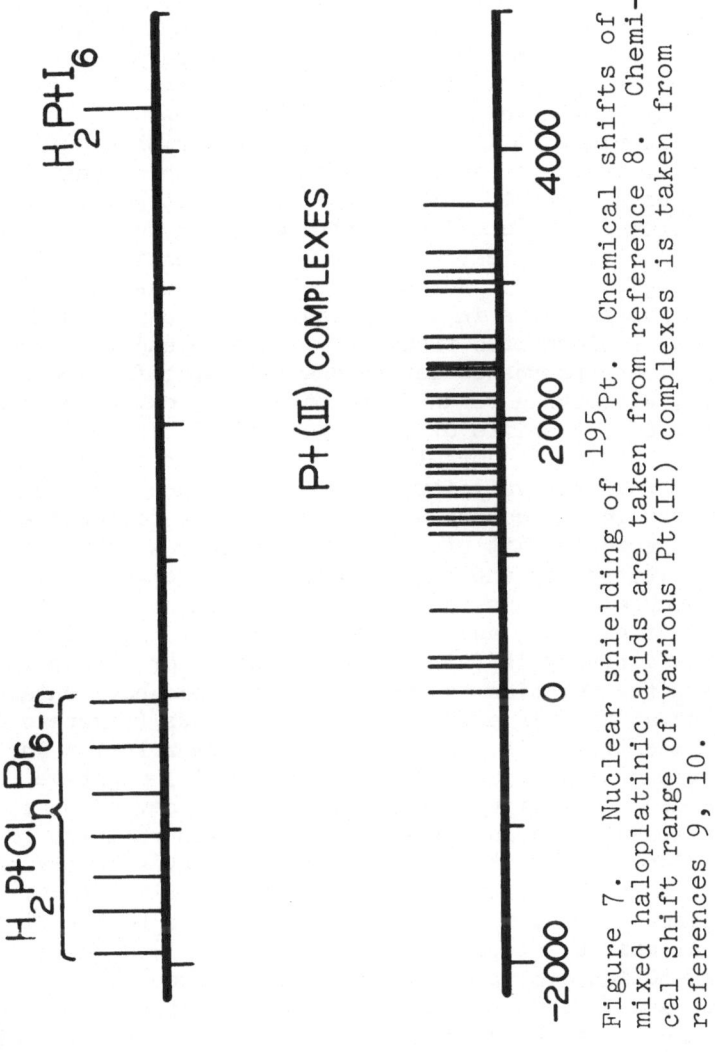

Figure 7. Nuclear shielding of ^{195}Pt. Chemical shifts of mixed haloplatinic acids are taken from reference 8. Chemical shift range of various Pt(II) complexes is taken from references 9, 10.

shift ranges in relation to the absolute values of σ_d and
σ_p. Compounds of certain elements, fluorine in particular,
scan a range of chemical shifts which represents a substantial fraction of the total σ_p. For these resonances
it seems reasonable to expect more direct and accurate correlations between measured shifts and the parameters in
equation (A-1) than is the case when chemical shifts reflect only subtle variations in the total σ_p. Shifts of
binary fluorides do indeed correlate well with electronegativity of the attached element (40). Unfortunately
phosphorus appears to fall in the latter category, and the
correlation of its chemical shifts with the usual chemical
parameters (e.g., electronegativity or oxidation state) is
not straightforward. For example the additive shift of
various substituents in triply connected phosphorus shows
no easily rationalized trend (41). The +3 and +5 oxidation
states of phosphorus do not resonate in specific spectral
regions, although a rough correlation with oxidation state
is observed for shifts of Tl(I) and Tl(III) compounds (figure 5). One simple correlation that appears to have at
least approximate validity for many resonances is that of
additivity in the contributions of directly bonded elements
to the chemical shift of the central atom. When one halide
is substituted for another the additivity relation is often
quite accurate. Thus in the mixed tin and carbon tetrahalides (figure 8), and in mixed chloroplatinic acids (figure 7) additivity is fairly precise while for the series
PCl_xI_{3-x} the relation is approximately valid but contains
some curvature. In many series of compounds (figure 8)
simple additivity is replaced by a concave relation in
which the mixed compounds show a substantial negative (i.e.,
deshielded) deviation from linearity. The deviation appears always to be in a low-field direction relative to the
symmetrically substituted compounds. Ramsay's theory provides a straightforward interpretation of this effect. The
paramagnetic term, σ_p, is negative and leads to low-field
chemical shifts. According to equation (A-1) σ_p depends
on the unbalance of p-orbital population about the resonant
nucleus and is thus a measure to the deviation from spherical symmetry of the electron distribution. Mixed ligands
increase the electronic asymmetry and thus lead to larger
values of σ_p. Van Wazer and Letcher (42) have analysed
data for ^{31}P and have concluded that deviations from additivity are largest (at least for phosphorus) when one
substitutent is a π-bonding ligand and the other is not.

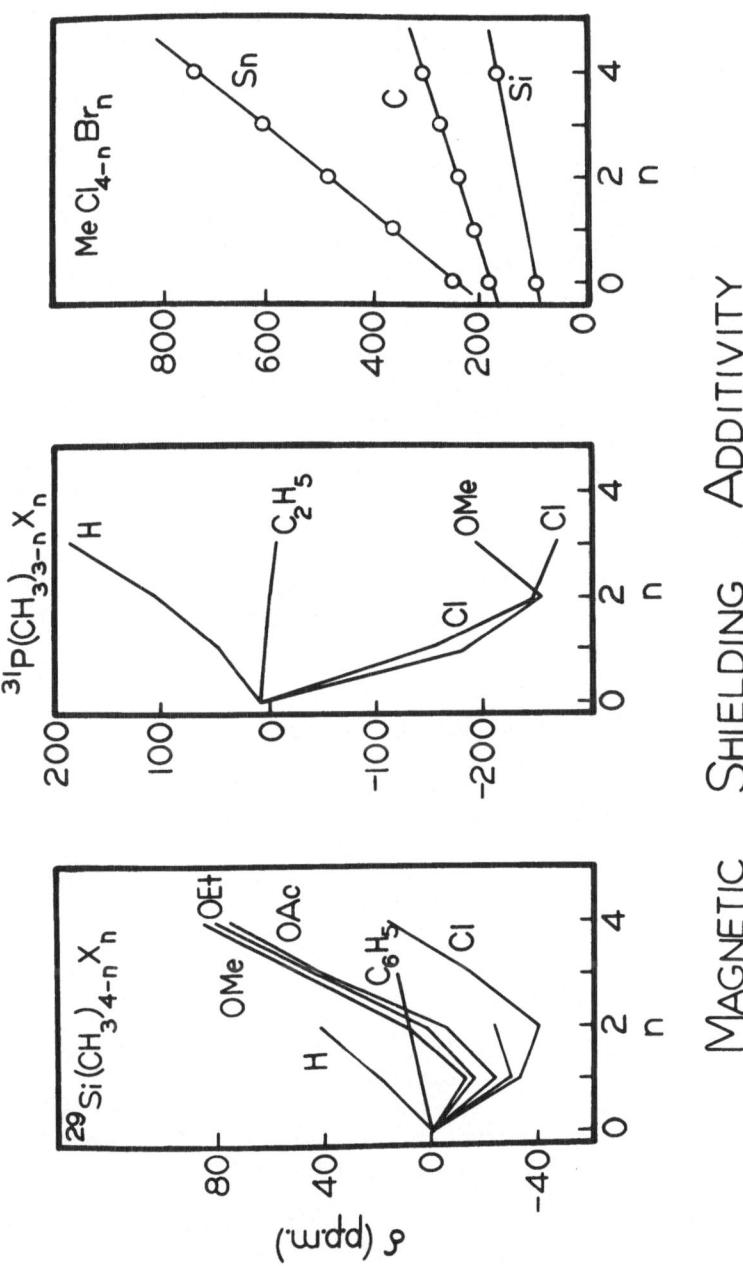

MAGNETIC SHIELDING ADDITIVITY

Figure 8. Effect of progressive ligand substitution on the chemical shift of the central atom.

In other respects, chemical shifts of elements that are closely related in the Periodic Table are often qualitatively dissimilar even when analogous series of compounds are compared. An interesting example of such dissimilarity has been discussed by Lauterbur (36,43) for ^{13}C and ^{29}Si chemical shifts in compounds of the series $(CH_3)_xM(OCH_3)_{4-x}$. The effect of increasing alkoxy-substitution on chemical shifts is essentially opposite in sign for the two nuclei. Additivity is still observed for each nucleus however, with relatively minor concave deviations from linearity. One can rationalize this behavior in terms of d-orbital participation in the bonding of silicon, but in fact there is no hard evidence that other factors in equation (A-1) do not dominate this effect. Indeed, one might expect the ^{29}Si shifts to be larger in absolute magnitude than the ^{13}C shifts if σ_p for ^{29}Si contained both d and p orbital contributions, whereas the ^{13}C shifts are found experimentally to span a slightly larger range of values. A similar inversion in the predicted magnitudes of chemical shifts for ^{13}C, ^{29}Si and ^{119}Sn is observed in the mixed tetrahalides (figure 8) of these elements. Although the shifts vary in a linear manner for each of the resonances, the absolute range of ^{29}Si shifts is less than that of ^{13}C or ^{119}Sn. The cause of these and other anomalies is not well understood at present.

2. SPIN-SPIN COUPLING

For a variety of reasons, nuclear spin-spin coupling is even more difficult to understand from a theoretical standpoint than are chemical shifts. Spin-spin coupling depends on the electronic configuration about (at least) two nuclei and often involves the inner electronic shells of heavy atoms. Nevertheless, the basic chemical parameters that influence coupling constants appear to have been discussed systematically by several authors (7,44-46).

Transmission of spin information between neighboring nuclei results from spin polarization of electrons in the bonding orbitals. That is, the bonding electrons of given spin, say $|\alpha>_e$, are energetically stablized in the presence of an opposite nuclear spin $|\beta>_n$ and destabilized near the parallel nuclear spin state, $|\alpha>_n$. Electrons in a bonding orbital therefore unpair to a very slight extent due to interactions with nuclear magnetic moments and cause split-

tings of the nuclear energy levels (figure 9). The mechanism
of coupling between a nucleus and its electronic environment determines the magnitude and sign of the coupling
constant, as well as the nature of chemical influences on
spin-spin coupling.

Most authors seem to agree that spin coupling to heavy
nuclei is usually dominated by the Fermi contact term (44,45,
47), which is proportional to unpaired spin density at the
nucleus. Since all orbitals except the s have nodes at the
nucleus, spin-spin coupling seems to reflect primarily the
degree of s electron participation in chemical bonding.
This interpretation has been reinforced by numerous measurements of coupling constants to heavy nuclei such as [199]Hg
(15,48-50), [205]Tl (51,52), [195]Pt (53,54), [119]Sn (55-57),
[103]Rh (58), and [183]W (59) as well as to the lighter nuclei
[29]Si (60-61) and [12]C (62). For example a surprisingly
accurate correlation exists between $J(^{13}C-^{1}H)$ and the
fractional s character of sp^1, sp^2, and sp^3 hybridized
carbon for directly bonded nuclei (62). Two other interactions (spin-spin dipolar coupling and spin-orbital coupling) may also be appreciable under appropriate circumstances. Grim and coworkers (63-65) have suggested that
d and p orbitals may transmit spin information in π-bonded
complexes of platinum and tungsten. On the other hand, the
data that support this interpretation may simple reflect
enhanced spin coupling through σ orbitals due to increased
bond strength of π-bonded ligands.

It should be emphasized that spin-spin coupling is
interpreted in a more straightforward manner when nuclei
are directly bonded than when coupling is transmitted
through several bonds. For this reason the discussion below
is limited to directly bonded nuclei. When comparing coupling constants of different nuclei, it is also convenient
to define a reduced coupling constant, $K_{ij} = 2\pi J_{ij}/h\gamma_i\gamma_j$,
where J_{ij} is in Hz and γ_i, γ_j are gyromagnetic ratios of
the coupled nuclei. K_{ij} is independent of nuclear properties and depends only on the electronic portion of the
coupling constant.

Jameson and Gutowsky (45) have proposed a simple
classification based on electron configuration and bond
character that explains many qualitative features of spin-
spin coupling of heavy nuclei. Their system recognizes
three classes involving different degrees of s-electron
participation in bonding orbitals. They predict that the

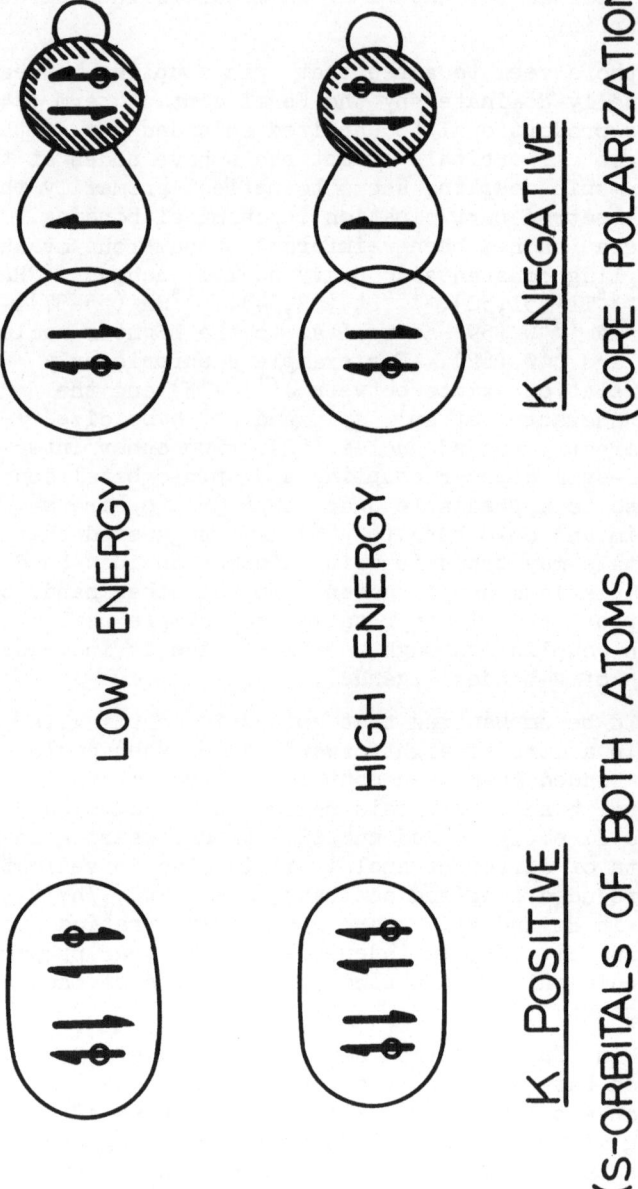

Figure 9. Pictorial representation of the electron-coupled nuclear spin-spin interaction.

contribution to K_{ij} will be large and positive for elements
with an open s configuration in the ground state of the
free atom. This class includes alkali metals and hydrogen.
A smaller, but still positive, contribution to K_{ij} is predicted when s-orbitals are filled in the ground state atom
but are utilized in the formation of bonding orbitals (e.g.,
sp^3 hybridized carbon). When the bonding orbitals involve
only p-electrons, as is approximately true in the negative
oxidation states of group VI A and VII A elements, and
possibly of triply connected elements in group V as well,
spin coupling is transmitted by polarization of the core
electrons. In this case the contribution to K_{ij} is relatively small and negative. Of course the observed coupling constant is determined by the contribution of both
atoms forming the bond. If either contribution is negative,
then K_{ij} is negative, while the magnitude of K_{ij} is given
by the product of the two contributions. A further effect
superimposed on that of bond type is the Periodic trend of
the Fermi contact term: electron density at the nucleus of
a valence s orbital increases across a row or down a family
of the Periodic Table.

The periodic nature of K_{ij} implied by this classification is borne out by a wealth of data (45). A few examples will suffice to illustrate the predictive value and
the limitations of these rules.

1. We first consider coupling in the $^{29}Si-^{1}H-^{19}F$
    triad. Hydrogen, having a (ls) atomic ground
    state, should give a large positive contribution.
    Silicon, when sp^3 hybridized, will give a mod-
    erate positive contribution. Fluorine, having
    no readily accessible open s states, gives a
    small negative contribution. Taking the products
    of these contributions and making allowance for
    the difference in atomic number one would predict:

    $K(^{29}Si-^{1}H)$ = fairly large, positive
    $K(^{29}Si-^{19}F)$ = moderate, postive
    $K(^{1}H-^{19}F)$ = moderate, negative

    Comparison of these predictions with measured
    values is quite satisfactory and is shown in
    Table I (values of K are given in units of
    $10^{20}cm^{-3}$).

electron configuration	Ground state open s configuration	accessible bonding orbital with open s contribution	Bonding orbitals with little or no s character (core polarization)
examples	H, alkali metals	sp^1, sp^2, sp^3 carbon d^2sp^3 sulfur	$X(-I)$ (X = halogen)
contribution to K_{ij}	positive large	positive moderate	negative small

Z – dependence: increase with Z across a row of the Periodic Table
                 increase with Z down a given family of the Periodic Table

Examples	Predicted	Observed, K_{ij}
K(Si–H) I x II	large, positive	+161 (66)
K(H–F) I x III	moderate, negative	(−) 54 (68)
K(Si–F) II x III	moderate, negative	−122 (69)

Table I. Gutowsky's and Jameson's (45) classification of coupling constants
          between directly bonded elements.

2. s electron density at the nucleus increases down a family of the Periodic Table. An increase in coupling constant is accordingly observed for analogous compounds containing elements in the carbon family when hybridization remains constant (see Table II).

3. Because of the inaccessibility of open s configurations in the halides one would expect $K(Sn-X)$ to be smaller in absolute value than $K(Sn-Sn)$ and probably comparable to $K(Sn-H)$. Experimental sign information unfortunately is lacking for tin-halide coupling constant, but agrees with prediction for $K(Sn-H)$ and $K(Sn-Sn)$:

$$K(Sn-H) = +431 \quad (55)$$
$$K(Sn-F) = 370 \quad (73)$$
$$K(Sn-Cl) = 1072 \ (74)$$
$$K(Sn-Sn) = +2668 \ (57) \quad K(Sn-I) = 1048 \ (74)$$

The observed magnitudes are also in general agreement with prediction, although the expected increase from $K(Sn-Cl)$ to $K(Sn-I)$ does not occur.

4. Identical directly bonded nuclei in symmetrical compounds have identical contributions to K_{ij}, which should therefore always be positive. This is often not the case for directly bonded phosphorous nuclei (75):

$$K(P-P) = -55 \qquad \text{in } H_2P-PH_2$$
$$= -115 \qquad \text{in } F_2P-PF_2$$

These coupling constants are relatively small however, and may be dominated by an interaction other than Fermi Contact term.

Specific correlations between coupling constants and parameters such as Taft σ^* constants (76-8) and electronegativity (79) have also been discussed in the literature. These parameters are usually intended to reflect s electron participation in bonding orbitals.

$|\psi_s(0)|^2$ central atom	compound	$K(X-H)$ (10^{20} cm^{-3})
1.38 a.u.[a]	$^{13}C\,^1H_4$	+41.4[c]
2.06[b]	$^{29}Si\,^1H_4$	+85.0[d]
4.92[b]	$^{73}Ge\,^1H_4$	233[e]
8.16[b]	$^{119}Sn\,^1H_4$	+431[f]
13.8[b]	$^{207}Pb\,^1H(CH_3)_3$	947[f]

(a) Reference 46.   (b) Reference 45.   (c) Reference 71.
(d) Reference 60.   (e) Reference 72.   (f) Reference 55.

Table II. Z-Dependence of the Spin-Spin Coupling Constant K(X-H)

3. EXPERIMENTAL ASPECTS

The only spin-1/2 isotopes that have sensitivities[2] within two orders of magnitude of that for protons are ^{19}F, ^{31}P, and [205,203]Tl. The remaining isotopes are normally observed using one of several techniques for signal enhancement. The most generally applicable of these is Fourier transformation, which gives signal enhancements of order $(\Delta/\delta)^{1/2}$, where Δ is the range of chemical shifts to be observed and δ the desired resolution. Fourier transform n.m.r. spectrometers have been commercially available since about 1970, but are usually purchased as a specialized unit for observing ^{13}C in natural abundance. The major modifications required to observe other nuclei with a ^{13}C spectrometer is a probe (or coil insert) tuned to the appropriate resonance frequency and an r.f. oscillator and preamplifier, all of which are normally supplied by the manufacturer. Fourier transform n.m.r. and the use of high field magnets have been described in detail elsewhere (80-1). Commercial Fourier transfrom spectrometers are quite costly however, and less expensive methods have been devised by which satisfactory high resolution spectra can be obtained in chemically interesting systems. Straightforward signal averaging without Fourier transformation by means of multiple scan data accumulation can be accomplished with commercial units priced from about $4000 (Northern Scientific SAC, Model 560) to about $7500 (Fabri-Tek, Model 1020). These hard-wired, on-line computers give a signal-to-noise enhancements equal to the square root of the number of scans.

In the special case that the less sensitive nucleus is spin-coupled to a nucleus of higher sensitivity (usually ^{1}H or ^{19}F) the elegant INDOR technique provides substantial signal enhancement (82-84). The experimental procedure is outlined in figure 10. The ^{1}H spectrum of $(CH_3)_4Pb$ contains of an intense singlet arising from molecules which contain nonmagnetic lead isotopes, [204]Pb, [206]Pb and [208]Pb. The remaining molecules contain [207]Pb (spin 1/2, 21.1% abundant), and the protons resonate as a doublet with a splitting of 61.4 Hz. In the INDOR experiment, one of the proton satellites is observed while a second variable r.f. frequency is swept across the [207]Pb spectrum. "Tickling" the [207]Pb transitions by ν_2 leads to a partial collapse of the proton

---

[2]See footnote 1.

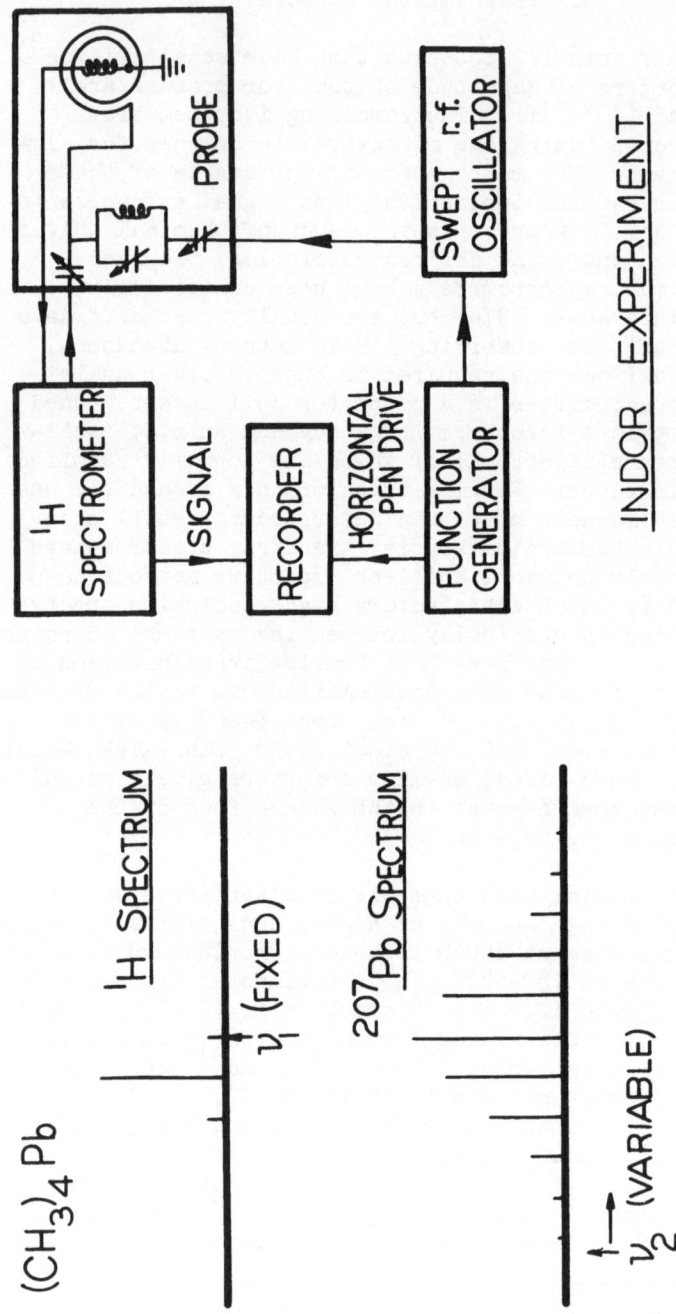

Figure 10. Block diagram of an INDOR spectrometer and a description of the experimental procedure.

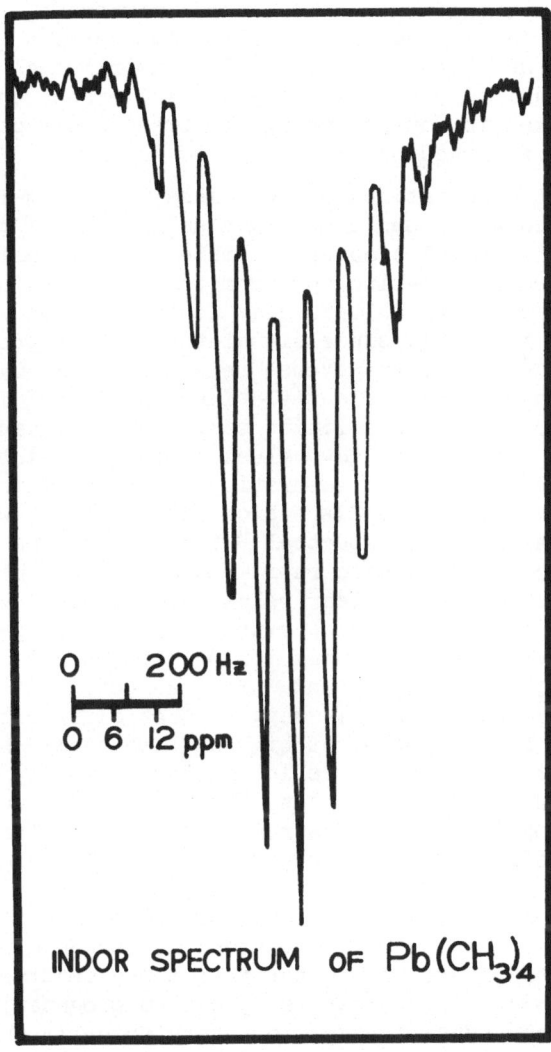

Figure 11.   INDOR spectrum of ^{207}Pb in Pb(CH$_3$)$_4$ (reference 82).

doublet and a decrease in signal intensity at ν_1. As ν_2
is swept, the inverted ^{207}Pb multiplet is recorded at ν_1
(figure 2). This method leads to substantial improvements
in spectral quality when the natural abundance of an isotope
is not too low to measure satellite intensities quantitatively in the parent spectrum. The experimental modifications required are relatively simple, except that care
must be taken in double tuning the transmitter coil to accept two r.f. frequencies.

Direct observations of less sensitive spin-1/2 nuclei
have occasionally been undertaken using special probes with
standard high resolution spectrometers and by using variable frequency wide-line spectrometers. The main limitation on sensitivity in this case is the longitudinal nuclear
relaxation time, T_1, since signal-to-noise is proportional
to $(T_2/T_1)^{1/2}$. The transverse relaxation time T_2 is usually determined by magnetic field inhomogeneity. T_1 is often
much longer than T_2 and limits sensitivity by causing the
onset of saturation at low values of the irradiating H_1
field. The problems of saturation may be minimized by observing spectra in the dispersion mode (36,85), which has
more intensity in the spectral "wings" than does the absorption mode. Adiabatic rapid passage is useful for very
long values of T_1. Nuclear relaxation of heavy spin-1/2
nuclei has not been studied extensively in the past and there
is little experimental basis for predicting T_1's of isotopes
like ^{199}Hg, ^{195}Pt, ^{119}Sn and ^{77}Se. Recent data on ^{31}P (86),
^{119}Sn (33), and ^{207}Pb (87) suggest that different relaxation
mechanisms are important at high and low atomic numbers.
The nuclear magnetic dipole interaction normally dominates
T_1 for lighter nuclei such as ^{1}H, ^{13}C and ^{19}F, but this
contribution depends very strongly on internuclear separation
and falls off rapidly for heavy nuclei with large convalent
radii. Fortunately other relaxation mechanisms (spin-
rotation, chemical shift anisotropy and scalar coupling) become more efficient for large atoms and produce relaxation
times comparable to or shorter than those observed for light
isotopes (33). These results appear to prognosticate relatively good sensitivities for heavy spin-1/2 nuclei even
when nuclear dipole relaxation is very inefficient.

ACKNOWLEDGEMENT

Acknowledgement is made to the National Science Foundation and to the Donors of the Petroleum Research Fund, ad-

ministered by the American Chemical Society, for support in the preparation of this manuscript.

REFERENCES

1. L. Petrakis and F. E. Dickson, Appl. Spect. Rev. 4, 1 (1970).
2. C. Deverell, Prog. in N.M.R. Spectroscopy (Pergamon, New York, 1969), Vol. 4, p. 235.
3. R. A. Kovar and G. L. Morgan, J.A.C.S. 92, 5067 (1970).
4. G. R. Eaton and W. N. Lipscomb, N.M.R. Studies of Boron Hydrides and Related Compounds (W. A. Benjamin, New York, 1969).
5. J. D. Memory, Anal. Chem. Nitrogen and its Compounds (Wiley-Interscience, New York, 1970), Pt. I, p. 29.
6. A. A. Van Geet, Paper Presented at the Eastern Analytical Symposium, Atlantic City, New Jersey, Nov. 2, 1972.
7. J. W. Emsley, J. Feeney and L. H. Sutcliffe, High Resolution Nuclear Magnetic Resonance Spectroscopy (Pergamon, New York, 1966).
8. A. Von Zelewsky, Helv. Chem. Acta 51, 803 (1968).
9. A. Pidcock, R. E. Richards, L. M. Venanzi, J.C.S. (8), 1970 (1968).
10. W. McFarlane, Chem. Comm., 393 (1968).
11. R. Freeman, R. P. H. Gasser, R. E. Richards, and D. H. Wheeler, Mol. Phys. 2, 75 (1959).
12. R. Freeman, R. P. H. Gasser, and R. E. Richards, Mol. Phys. 2, 301 (1959).
13. R. P. H. Gasser and R. E. Richards, Mol. Phys. 2, 357 (1959).
14. B. N. Figgis, Trans. Far. Soc. 55, 1075 (1959).
15. W. G. Schneider and A. D. Buckingham, Disc. Far. Soc 34, 147 (1962).
16. L. H. Piette and H. E. Weaver, J. Chem. Phys. 28, 735 (1958).
17. J. J. Burke and P. C. Lauterbur, J.A.C.S. 83, 326 (1961).
18. A. Fratiello, Shirley Peak, R. E. Schuster, Don D. Davis, J. Phys. Chem. 74, 3730 (1970).
19. P. D. Godfrey, M. L. Heffernan, and D. F. Kerr, Aust. J. Chem. 17, 701 (1964).
20. R. J. Kostehnik, Ph.D. Thesis, 1971, University Microfilms, Ann Arbor, Mich.
21. F. H. Allen and S. N. Sze, J.C.S. A (12), 2054 (1971).

22. B. T. Heaton and A. Pidcock, J. Organomett. Chem. 14, 235 (1968).
23. Reference 7, Chapter 12.
24. H. Finegold, Annals N.Y. Acad. Sci. 70, 875 (1958).
25. A. A. Abragam, The Principles of Nuclear Magnetism, (Oxford, New York, 1961), Chapter 8.
26. N. F. Ramsey, Phys, Rev. 78, 699 (1950); 86, 243 (1952).
27. K. M. S. Saxsena and P. T. Narasimhan, Int. J. Quant. Chem. Symp. 1, 731 (1967).
28. R. A. Bonham and T. G. Strand, J. Chem. Phys. 40, 3447 (1964).
29. A. Saika and C. P. Slichter, J. Chem. Phys. 22, 26 (1954).
30. M. Karplus and T. P. Das, J. Chem. Phys. 34, 1683 (1962).
31. J. A. Pople, Mol. Phys. 7, 301 (1964).
32. J. D. Halliday, R. E. Richards, and R. R. Sharp, Proc. Roy. Soc. A. 313, 45 (1969).
33. M. Karplus and J. A. Pople, J. Chem. Phys. 38, 2803 (1963).
34. C. J. Jameson and H. S. Gutowsky, J. Chem. Phys. 40, 1714 (1964).
35. C. Deverell and R. E. Richards, Mol. Phys. 16, 421 (1969).
36. G. R. Holtzman, P. C. Lauterbur, J. H. Anderson and W. Koth, J. Chem. Phys. 25, 172, (1956).
37. B. K. Hunter and L. W. Reeves, Can. J. Chem. 46, 1399 (1968).
38. C. Deverell, Mol. Phys. 18, 319 (1970).
39. W. H. Flygare, J. Chem. Phys. 41, 793 (1964).
40. H. S. Gutowsky and C. J. Hoffman, J. Chem. Phys. 19, 1259 (1951).
41. J. R. Van Wazer, C. F. Callis, J. N. Shoolery and R. C. Jones, J.A.C.S. 78, 5715 (1956).
42. J. R. Van Wazer and J. H. Letcher, Top. in Phos. Chem. 5, 169 (1967).
43. P. C. Lauterbur, Determination of Organic Structures by Physical Methods (Academic, New York, 1962), Vol. 2, Chapter 7.
44. J. A. Pople and D. P. Santry, Mol. Phys. 8, 1 (1964).
45. C. J. Jameson and H. S. Gutowsky, J. Chem. Phys. 51, 2790 (1969).
46. J. N. Murrell, Prog. in N.M.R. Spectroscopy (Pergamon, New York, 1970), Vol. 6, p. 1.

47. D. W. Davies, The Theory of the Electric and Magnetic Properties of Molecules, (Wiley, New York, 1967).
48. W. McFarlane, J.C.S., A, 2280 (1968).
49. K. A. Mc Lauchlan, D. H. Whiffen and L. W. Reeves, Mol. Phys. 10, 131 (1966).
50. R. L. Keiter and S. O. Grim, Chem. Comm. (9), 521 (1968).
51. J. P. Maher and D. F. Evans, J.C.S., 637 (1965).
52. J. P. Maher, M. Evans and M. Harrison, J.C.S. Dalton Trans. (2), 188 (1972).
53. A. Pidcock, R. E. Richards and L. M. Venanzi, J.C.S., A, 1707 (1966).
54. F. H. Allen and A. Pidcock, J.C.S., A, 2700 (1968).
55. N. Flitcroft and H. D. Kaesz, J.A.C.S. 85, 1377 (1963).
56. H. C. Clark, J. T. Kwon, L. W. Reeves and E. J. Wells, Can. J. Chem. 42, 941 (1964).
57. W. McFarlane, J.C.S. A (7), 1630 (1968).
58. T. H. Brown and P. J. Green, J.A.C.S. 92, 2359 (1970).
59. R. L. Keiter and J. G. Verkade, Inorg. Chem. 8, 2115 (1969).
60. E. A. V. Ebsworth and J. J. Turner, J.C.P. 36, 2628 (1962).
61. E. A. V. Ebsworth and S. G. Frankiss, Trans. Far. Soc. 59, 1518 (1963).
62. J. N. Shoolery, J. Chem. Phys. 31, 1427 (1959).
63. S. O. Grim, D. A. Wheatland and W. McFarlane, J.A.C.S. 89, 5573 (1967).
64. S. O. Grim and D. A. Wheatland, Inorg. Chem. 8, 1716 (1969).
65. S. O. Grim, R. L. Keiter and W. McFarlane, Inorg. Chem. 6, 1133 (1967).
66. W. McFarlane, J.C.S. (8), 1275 (1967).
67. Von H. Elser and H. Dreeskamp, Ber. Buns. Gesell. Phys. Chem 73, 619 (1969).
68. I. Solomon and N. Bloembergen, J. Chem. Phys. 25, 261 (1956).
69. S. S. Danyluk, J.A.C.S. 86, 4504 (1964).
70. Rolf B. Johannesen, J. Chem. Phys. 47, 3088 (1967).
71. N. Muller and D. E. Pritchard, J. Chem. Phys. 31, 768 (1959).
72. H. Dreeskamp, Z. Naturforsch. 19A, 139 (1964).
73. P. A. W. Dean and D. F. Evans, J.C.S. A, 1154 (1968).
74. R. R. Sharp, J. Chem. Phys. 57, 5321 (1972).
75. E. G. Finer and R. K. Harris, Prog. in N.M.R. Spectroscopy (Pergamon, New York, 1971), Vol 6, p. 74.

76. K. Kawakami, T. Saito and R. Okawara, J. Organomet. Chem. <u>8</u>, 377 (1967).

77. J. Duffermont and J. C. Maire, J. Organomet. Chem. <u>7</u>, 415 (1967).

78. M. Gielen, M. R. Barthels, M. DeClercq and J. Nasielski, Bull. Soc. Chim. Belg. <u>80</u>, 189 (1971).

79. R. Ditchfield, M. A. Jensen and J. N. Murrell, J.C.S. A, 1674 (1967).

80. T. C. Farrar, Paper presented at the Eastern Analytical Symposium, Atlantic City, New Jersey, Nov. 2, 1972.

81. E. D. Becker and T. C. Farrar, Pulse and Fourier Transform N.M.R. (Academic, New York, 1971).

82. P. J. Banney, D. C. Mc William and Peter R. Wells, J. Magn. Res. <u>2</u>, 235 (1970).

83. E. B. Baker and L. W. Burd, Rev. Sci. Inst. <u>34</u>, 238 (1963).

84. E. B. Baker, L. W. Burd and G. N. Root, Rev. Sci. Inst. <u>34</u>, 243 (1963).

85. Reference 25, Chapter 3.

86. A. D. Jordan, R. G. Cavell and R. B. Jordan, J. Chem. Phys. <u>56</u>, 483 (1972).

87. R. R. Sharp, J. Mag. Res., in the press.

88. K. T. Gilley, J. Chem. Phys. <u>56</u>, 1573 (1972).

APPENDIX

Jameson and Gutowsky's expression for σ_p is

$$\sigma_p = -\left(\frac{2e^2h^2}{3m^2c^2}\right)\left(\frac{1}{\Delta E}\right)\left(\frac{1}{\langle r^3\rangle_p}\ P_u\ +\ \frac{1}{\langle r^3\rangle_d}\ D_u\right) \quad (A-1)$$

where $P_u = p_{xx} + p_{zz} - 1/2\ (p_{xx}p_{zz} + p_{yy}p_{xx} + p_{yy}p_{zz})$

$$+ 1/2\ (p_{xy}p_{yx} + p_{xz}p_{zx} + p_{zy}p_{yz})$$

$$\Psi_{M.O.} = a_x p_x + a_y p_y + a_z p_z + a_z 2d_z 2 + \ldots$$

$$+ \text{ s-orbitals} + \text{ orbitals on other atoms}$$

$$p_{\mu\nu} = \sum_s n_s\ (a_\mu a_\nu^*)$$

ΔE in equation (A-1) is the "average energy denominator" often used in the closure approximation of perturbation theory. Physical constants in the first set of parentheses have their usual meaning, r is the electron radius for a p or d orbital, Ψ is an L.C.A.O. molecular orbital and $p_{\mu\nu}$ defines an orbital population.

NMR OF SOLUTIONS OF ALKALI METAL IONS AND OTHER NUCLEI WITH SPIN GREATER THAN 1/2

Anthony L. Van Geet

Department of Chemistry
State University of New York
College at Oswego, Oswego, New York   13126

To exhibit nuclear magnetic resonance, an element needs to have at least one isotope with non-zero spin. Out of 81 stable elements, only argon does not have a suitable isotope. If one eliminates resonances over 100 times weaker than hydrogen, 32 elements remain. Of these, 24 have a spin greater than 1/2, and their spectra exhibit quadrupole broadening in varying degrees. Fortunately, the nuclei with broad lines also exhibit very wide chemical shifts. Shifts as large as 1% are observed for cobalt. Good resonances are found especially in the following rows of the periodic system: 1, 3A, 3B, 5A, 5B, 7.

In solids, the magnetic nuclei are held in place, and they interact, leading to very broad lines, which may be hard to detect. The chemical shift of these lines is often much smaller than the line width, and is then difficult to determine accurately. We will restrict ourselves to solutions.

In solution, the molecules tumble, and the magnetic interactions between different molecules are averaged out. As a result, very sharp resonances are obtained in favorable circumstances.

In ionic solution, the magnetic resonance of the ions provides information about the immediate environment. For example, the magnetic resonance of the sodium ion provides a unique probe for the immediate environment of the ion.

Solvation, complexation and ion pairing may be studied in
this way.

Heteronuclear spin-spin splitting is sometimes observed. For example, the ^{11}B spectrum of the BH_4^- ion in
0.1 M NaOH is a quintet, while the ^1H spectrum is a quartet,
corresponding to the magnetic quantum numbers 3/2, 1/2, -1/2,
-3/2 of the ^{11}B nucleus. The peaks of the ^1H spectrum are of
equal height, unlike the quintet of the ^{11}B spectrum. The
peak separation, J_{BH} = 81 Hz is the same in both spectra.

For many of the elements, especially the metals, the
spectrum usually consists of a single, symmetric peak,
because the molecule or ion contains only one resonating
nucleus.

A resonance peak is characterized by its chemical shift
and its width. Both provide useful information, but the
chemical shift is more easily interpreted.

Quadrupole Broadening

The orientation of nuclei with spin 1/2 cannot be influenced by electric fields, since these nuclei have a
spherical charge distribution, and possess no electric dipole
or quadrupole moment. Reorientation can only occur by interaction with fluctuating magnetic fields. Quite often these
magnetic interactions are weak, and reorientations are in-
frequent. Thus, the relaxation time T for reorientation is
long, and the line width $\nu_{1/2} = 1/(\pi T)$ is narrow for spin 1/2
nuclei.

Nuclei with spin greater than 1/2 have a non-spherical
charge distribution.

This charge distribution behaves as a combination of a
point charge, +, and a quadrupole, $-+\!+-$. The quadrupole is
represented by a second-rank tensor, eQ. It can interact
with an electric field gradient ∇E = eq, which is also a
second rank tensor. The energy of interaction is e^2qQ.

Molecular motions cause a random fluctuation of the
field gradient. The probability to find the field gradient
unchanged during time t can be represented by $\exp\{-t/\tau\}$,
where τ is the correlation time of the molecular motion

responsible for relaxation. The dominating molecular motion is usually reorientation of the ion, molecule or complex to which the nucleus belongs. It leads to reorientation of the nucleus with a relaxation time of $1/T = K(e^2qQ/\hbar)^2 \tau_c$, where K is a constant depending only on the spin quantum number I of the nucleus. For example[1], K = 1/10 for I = 3/2. These electric interactions are more intense than the magnetic interactions for a spin 1/2 nucleus, resulting in a larger line width. The line width is determined by the value of the quadrupole coupling constant e^2qQ/h. Broad lines will occur when the field gradient eq is larger, as will be the case in an asymmetric environment. In this way, the line width provides information about the symmetry of the environment.

Chemical Shift

The chemical shift of proton spectra is determined primarily by the screening effect of the surrounding electrons. This screening effect is also the cause of diamagnetism, and produces an upfield shift. This effect is not of much direct importance for most other nuclei.

The degree of symmetry of the surrounding electrons has an indirect effect on the screening, however. If the symmetry is spherical, the electron cloud will rotate freely, producing maximum screening. If the cloud has axial symmetry, as in the case of a p_z or σ_z orbital, it will rotate freely about its axis of symmetry (z-axis), but rotation about the x or y axis is inhibited, and the screening is reduced. This reduction is called the paramagnetic term and leads to a downfield shift. For example, the shift of F_2 occurs 550 ppm downfield from the F^- ion in water[2]. Electrons in a p or d orbital have angular momentum. The magnetic field caused by the angular momentum contributes to the paramagnetic term.

For the alkali metal and halide ions, the observed shifts are caused by the electronic repulsion which occurs between the ion and its nearest neighbors. For example, an H_2O molecule is attracted to a Na^+ molecule by the electrostatic ion-dipole interaction $-ep/r^2$. Some overlap between the H_2O and Na^+ orbitals occurs, leading to an equal repulsive force. This overlap means that the H_2O is donating electrons to the Na^+ and vice versa. The Na^+ can

accommodate this donated electron only in a higher orbital.
The partial filling of the 3p orbital causes a downfield
shift. Good electron donors such as pyridine and OH⁻ cause
large downfield shifts.

The range of chemical shift values depends on the size
of the ion. The large, "soft" cesium ion has a range of
almost 500 ppm, while for lithium it is only 5 ppm. However,
the lithium resonance is very sharp.

ALKALI METALS

The alkali metals have good resonances, but potassium
is weak. Lithium and cesium give sharp lines. Sodium is
perhaps the most interesting nucleus because of its widespread occurrence and biological importance. In dilute
aqueous solution, it has a natural line width of 6 Hz.

Only lithium forms covalent compounds. Brown[3-5] has
studied alkyllithium compounds by ^{6}Li and ^{7}Li NMR. These
compounds form tetramers in solution, $(LiR)_4$, in which the
lithium atoms are arranged in a tetrahedron. The ^{7}Li
resonance of a mixed methyllithium-ethyllithium solution in
ether consists of a single, sharp line. Chemical exchange
occurs between the alkyl groups. At low temperature, the
line broadens and can be partially resolved into 4 peaks,
corresponding to various methyl ethyl lithium compounds, such
as $MeEt_3Li_4$.

The other alkali metals form ionic solutions only. Ion
pairing with the anion may occur. The ions may be in close
contact, or separated by the solvent. The interaction energy
for ion pair formation is $-e^2/r$, while it is $-ep/r^2$ for
solvation. In this, p is the dipole moment of the solvent,
and r is the separation of the centers. For lithium, r is
small and solvation dominates in aqueous solution. For
cesium, r is large, and contact ion pairing is preferred
over solvation. The equilibrium constant for the anion-
solvent competition

$$M^+(H_2O) + Z^- = M^+Z^- + H_2O \qquad (1)$$

is larger than one for cesium, and less than one for lithium.
Sodium occupies an intermediate position, with K = 1.

These conclusions follow from the chemical shifts of

the metal ion[6-8]. A downfield shift occurs when a neighboring molecule or ion dents the metal ion. Orbital overlap occurs in the dent, causing repulsion. Thus, the shift provides information about the nearest neighbors. Only a single peak occurs because the neighbors are exchanged rapidly.

Sodium

Sodium NMR provides information about the structure of ionic solutions. It also provides a unique tool for the study of sodium transport in biological systems.

In aqueous solution, the sodium ion has little preference regarding its neighbors, and the equilibrium constant for the anion-solvent competition (Eq. 1) is one. The ions ClO_4^-, Cl^-, Br^-, I^-, OH^- are equally well accepted, and the neighbors are statistically distributed. The sodium ions are surrounded by Z^- and H_2O molecules in proportion to the mole fraction of each:

$$X = NaZ/(NaZ + H_2O) = Z^-/(Z^- + H_2O) \qquad (2)$$

As a result, the chemical shift varies linearly with the mole fraction for all concentrations up to saturation[6], and can be represented by

$$\delta = X\delta_z \qquad (3)$$

It is likely[6,9] that the constant δ_z corresponds to the shift of the crystal NaZ.

The chemical shift depends on the ion Z^-, and is increasingly negative in the order:

$$ClO_4^-, \quad NO_3^-, \quad THF, \quad H_2O, \quad Cl^-, \quad Br^-, \quad I^-, \quad OH^- \qquad (4)$$

The ions ClO_4^- and NO_3^- are poor electron donors, while OH^- is a good donor[10]. Water and tetrahydrofuran (THF) occupy intermediate positions.

Usually the aqueous sodium ion is not very selective regarding its neighbors. However, when these neighbors are other solvent molecules such as THF or ethylene diamine, it has definite preferences. For THF, the competitive equilibrium constant[11] of Eq. 1 is K = 1/10, while for ethylene diamine K is slightly greater than one.

Van Geet[11] used the preference for H_2O over THF to determine the hydration number of the sodium ion. A solution of sodium tetraphenylborate, $NaBPh_4$ in THF was titrated with water. The chemical shift showed an indistinct break corresponding to a hydration number 3-4. The $NaBPh_4$ appears to form solvent separated ion pairs[13] in THF, and possibly also in water[6], since the shift is independent of $NaBPh_4$ concentration in these solvents within experimental error.

Sodium Complexes. Sodium forms weak complexes with some anions[14]. More interesting are complexes with neutral molecules such as the crown complexes[15] and enzymes. These complexes are useful for liquid membrane electrodes. The crown complexes are large cyclic polyethers. Dibenzo-18-crown-6 (DBC) has 18 atoms in its ring, 6 of which are oxygen. The central hole accommodates a Na^+ or K^+ ion, which forms an electrostatic bond with the ether oxygens. The hole is not large enough to accommodate Cs^+, but two crown molecules can form a sandwich with the Cs^+ in between. These complexes form stable, crystalline salts such as Na-crown-CNS. The complexation constant[16] for the formation of this complex is $K_s = 500 \ M^{-1}$.

Addition of crown compound to a solution of $NaBPh_4$ in THF results in drastic quadrupole broadening of the ^{23}Na resonance[17]. The ring of the sodium complex is an imperfect plane of symmetry, while the uncomplexed sodium is in an environment approaching spherical symmetry.

The chemical shift of the complex is rather surprising[18]. Usually a strong interaction leads to a downfield shift because of the electron overlap which occurs in the dent made in the sodium ion. However, the sodium ion fits almost perfectly in the ring, and the ion is not dented. As a result, the complex has a large upfield shift[18]. This shift shows that the interaction is entirely electrostatic.

In biological membranes, sodium is transported by complex formation with enzymes such as monactin[18], which has a complexation constant $K_s = 10^3 \ M^{-1}$. Valinomycin forms a much weaker sodium complex with $K_s = 13 \ M^{-1}$. Unlike monactin, valinomycin has a downfield shift. Apparently the sodium does not fit very well and is dented. The resulting repulsion leads to the lower complexation constant.

The ^{23}Na line width of these complexes is much narrower than for the crown complexes. In the enzymes, the ^{23}Na is surrounded by the oxygen ligands in imperfect spherical symmetry, while in the crown complexes the oxygens are essentially planer.

Quadrupole Effects. The effect of the electric quadrupole on the line width has been discussed. The sodium ion has 4 spin states. In a magnetic field, the energy separation between these states is equal, and a single resonance is observed in solution.

In the ordered environment of a lyotropic nematic phase this is no longer so[19], and splitting occurs[20]. The electric fields at the nucleus are no longer random, and the quadrupole has preferred orientations. The resulting nuclear-quadrupole splitting leads to a triplet, since there are 4 spin states. The splitting is quite large, over 10000 Hz.

OTHER NUCLEI

Halides

The magnetic resonance of the halides is less attractive than the alkali metals, except for fluorine, which has a spin of 1/2. Chlorine is rather weak, while bromine and iodine have large quadrupole moments, resulting in broad lines. In aqueous solution at infinite dilution, the line widths are 10, 400 and 1060 Hz for ^{35}Cl, ^{79}Br and ^{127}I respectively[21-23].

In covalent compounds, the width is much larger. In CCl_4 it is[22] about 10 kHz. Covalent bromine and iodine are broad beyond detection. Fortunately, covalent ^{35}Cl also exhibits very large chemical shifts[22].

In aqueous solution, the chemical shift of the halide ions[24] is similar to the alkali metal ions. The soft cesium ion causes a strong downfield (upfrequency) shift[24] of the ^{35}Cl ion. The effect of lithium is unusual[24].

Chloride complexes have also been studied by ^{35}Cl NMR, and resonances[25] have been obtained for the paramagnetic complexes with Co(II).

^{35}Cl resonance has also been used as a probe to study

proteins[26]. Mercury ions will bind to reactive or exposed
SH-groups. The Hg tag then provides a coordination site
for chloride, and the ^{35}Cl resonance is broadened. Rapid
exchange occurs between the free and bound Cl-ions, and the
line width is the mole average of the free and bound resonances.

Cobalt

Cobalt(III) complexes have been studied by ^{59}Co NMR.
The resonance is strong and also broad because of the large
quadrupole moment. Paramagnetic cobalt (II) cannot be
directly detected. The paramagnetic electrons broaden the
resonance beyond detection.

The chemical shifts are very large[27], and are sometimes measured in percent. The shifts correlate very well
with the optical absorption spectra, but provide somewhat
more resolution. This provides the possibility to distinguish
similar compounds such as structural isomers which cannot
be distinguished optically[28,29].

REFERENCES

(1) A. Abragam, "Principles of Nuclear Magnetism", London,
     Oxford University Press, 1961, p. 314, Eq. 137.

(2) A. Saika and C. P. Slichter, J. Chem. Phys., 22, 26
     (1954).

(3) L. M. Seitz and T. L. Brown, J. Am. Chem. Soc., 88,
     2174 (1966).

(4) T. L. Brown, Acc. Chem. Res., 1, 23 (1968).

(5) T. L. Brown, Pure Appl. Chem., 23, 447 (1970).

(6) G. J. Templeman and A. L. Van Geet, J. Am. Chem. Soc.,
     94, 5578 (1972).

(7) C. Deverell and R. E. Richards, Mol. Phys., 10, 551
     (1966).

(8) J. D. Halliday, R. E. Richards, and R. R. Sharp, Proc.
     Roy. Soc. Lond. A., 45, 313 (1969).

(9) J. Burgess and M. C. R. Symons, Quart. Rev., Chem. Soc.,
     22, 276 (1968).

(10) R. H. Erlich and A. I. Popov, J. Am. Chem. Soc., 93, 5620 (1971).

(11) A. L. Van Geet, J. Am. Chem. Soc., 94, 5583 (1972).

(12) E. G. Bloor and R. G. Kidd, Can. J. of Chem., 46, 3425 (1968).

(13) G. W. Canters, J. Am. Chem. Soc., 94, 5230 (1972).

(14) G. A. Rechnitz and S. B. Zamochnick, J. Am. Chem. Soc., 86, 2954 (1964).

(15) C. J. Pederson, J. Am. Chem. Soc., 92, 386 (1970).

(16) E. Shchori, J. Jagur-Grodzinski, Z. Luz, and M. Shporer, J. Am. Chem. Soc., 93, 7133 (1971).

(17) A. M. Grotens, J. Smid and E. deBoer, Chem. Comm., (1971), 759.

(18) D. H. Haynes, B. C. Pressman, A. Kowalsky, Biochemistry, 10, 852 (1971).

(19) D. M. Chen and L. W. Reeves, J. Am. Chem. Soc., 94, 4384 (1972).

(20) H. G. Hecht, "Magnetic Resonance Spectroscopy", John Wiley, N. Y. 1967, p. 92.

(21) C. Deverell, D. J. Frost and R. E. Richards, Molec. Phys., 9, 565 (1965).

(22) C. Hall, R. E. Richards, G. N. Schulz and R. R. Sharp, Molec. Phys., 16, 529 (1969).

(23) C. Hall, Quart. Rev., Chem. Soc., 25, 87 (1971).

(24) C. Deverell and R. E. Richards, Molec. Phys., 16, 421 (1969).

(25) A. H. Zeltmann, N. A. Matwiyoff, and L. O. Morgan, J. Phys. Chem., 73, 2689 (1969).

(26) T. R. Stengle and J. D. Baldeschwieler, J. Am. Chem. Soc., 89, 3045 (1967).

(27) J. W. Emsley, J. Feeney and L. H. Sutcliffe, "High Resolution Nuclear Magnetic Resonance Spectroscopy", Vol. II, Pergamon Press, 1966, p. 1081.

(28) S. Fujiwara, F. Yajima and A. Yamasaki, J. Magn. Resonance, 1, 203 (1969).

(29) F. Yajima, A. Yamasaki, S. Fujiwara, Inorg. Chem., 10, 2350 (1971).

LIQUID CHROMATOGRAPHY OF METAL-ORGANIC COMPLEXES

Hans Veening

Department of Chemistry
Bucknell University
Lewisburg, PA   17837

INTRODUCTION

Many types of metal-organic compounds have been shown
to be sufficiently stable and volatile to permit their
separation and quantitative determination by gas chromatography (GC) (1-6). For example, metals can be converted
into neutral, volatile β-diketonates and analyzed by GC;
fluorinated ligands are usually employed for chelation (1-3).
Likewise, it has been shown that volatile arene metal carbonyls can be satisfactorily separated and determined by
GC (4,5). In other cases, pyrolytic GC permits determination
of certain complexes when elution without decomposition is
not possible (6).

There are of course many metal complexes which cannot
be separated by GC because they do not survive the high
column temperatures. It was this fact which prompted us to
consider the technique of high speed liquid chromatography
(LC) for the elution of these compounds. One of the significant advantages of LC over GC is that LC columns are
operated at or near room temperature, thus permitting the
elution of thermally unstable substances. The interchange
of solvents can also provide special selectivity effects in
LC, since the relative retention of the two solutes is
strongly influenced by the nature of the eluent used. This
report represents a review of our work which has dealt with
the liquid chromatographic separation of two different

classes of metal-organic compounds: non-volatile metal-β-
diketonates, and non-volatile arene metal carbonyls.

APPARATUS AND INSTRUMENTAL CONDITIONS

The liquid chromatograph used for this work was as-
sembled from commercial and homemade parts. It consisted
essentially of a reservoir for storing the moving phase; a
high pressure pump supplied with a suitable damping device
to eliminate undesirable pump pulsations; a precolumn to
assure attainment of equilibrium between the moving and
stationary liquids; a sample injection device; a thermostatted separation column; a UV detector; and a recorder.
A Zeiss PMQII or Beckman DB spectrophotometer was used as
the UV detector. For the Zeiss PMQII, 7.5 µl flow cells
with a path length of 10 mm were used. This detector has
been previously evaluated (7).

A variety of columns and packings were used for this
work (8-11). All columns were constructed from borosilicate
glass (2.3, 2.7, or 3.5 mm i.d.) and were thermostatted by
means of a water jacket. Column lengths varied from 10 to
100 cm depending on the separation.

For the separation of metal-β-diketonates, a ternary,
two-phase liquid-liquid system was used (10); it was prepared from distilled water, absolute ethanol and 2,2,4-tri-
methylpentane (TMP). The two liquid phases were prepared
separately according to their equilibrium composition. The
solid support consisted of diatomaceous earth of average
particle size 5 to 10 and 10 to 20 µm. The stationary phase
was coated on the support by pumping it through the column
until the air was displaced. Eluent was then pumped through,
displacing the excess stationary liquid.

For the separation of arene metal carbonyls, chemically
bonded as well as uncoated supports were used. The following
packings were used for various portions of the work:
(a) Carbowax-400 on 36 to 75 µ Porasil-C; (b) Carbowax-400
on 36-75 µ Corasil I; (c) uncoated Corasil I (Waters Assoc.,
Inc.). The eluent was TMP, and was pumped at a flow rate of
1.85 ml/min.

METAL COMPLEXES

The following acetylacetonates (AA) were obtained commercially: $Be(AA)_2$, $Al(AA)_3$, $Cr(AA)_3$, $Fe(AA)_3$, $Co(AA)_3$, $Ni(AA)_2$, and $Cu(AA)_2$. $Ru(AA)_3$ was synthesized according to standard procedures (1). (2,3-Dimethylnaphthalene)tricarbonylchromium (DMNTC) was prepared and characterized as reported previously (12). (2,5-Dimethoxytriptycene)tricarbonylchromium (DTTC) was synthesized and purified as described (11).

LC SEPARATION OF METAL ACETYLACETONATES

Most transition metal ions complex easily with acetylacetone (HAA) to form stable, neutral, bis- and tris- chelates. The structures for the ligand and the resulting bis- and tris- metal complexes are shown below. These chelates

Acetylacetone: $CH_3\text{-}\overset{O}{\overset{\|}{C}}\text{-}CH_2\text{-}\overset{O}{\overset{\|}{C}}\text{-}CH_3$

HAA

$M(AA)_2$ $M(AA)_3$

are generally soluble in polar organic solvents and display limited solubilities in water. Their aromatic-type structures cause them to be strong ultraviolet absorbers.

It was found that separations of these compounds could best be achieved by means of a ternary, two-phase, liquid-liquid system (10). In such a system, the polarity difference of the phases can be easily controlled by changing the quantitative composition of the phases. This ternary system has been thoroughly studied and evaluated by Huber (13). Water and TMP were chosen as the immiscible pair. Metal acetylacetonates are not very soluble in either solvent and a one-sided distribution is thus avoided. Ethanol was the

third component; it mixes easily with both water and TMP
and serves to establish the polarity difference between the
phases. This polarity difference decreases as the ethanol
fraction in the two phases is increased. A number of equilibrium compositions of the liquid-liquid system were investigated. The one which was found to be the most satisfactory
contained 34.3% water, 64.1% ethanol, and 1.6% TMP in the
more polar phase; and 0.7% water, 2.2% ethanol and 97.7% TMP
in the less polar phase*. The more polar phase was used as
the stationary liquid in the column.

Initially it was found that several of the metal acetylacetonates produced asymmetric peaks. This effect is shown
in Figure 1 for Ni(AA)$_2$. The peak is highly distorted and
indicates a nonlinear isotherm. Figure 2 shows the anomalous
results obtained for Al(AA)$_3$. Curve 1 represents the peak
produced when a freshly prepared solution of Al(AA)$_3$ in the
stationary liquid was injected into the column. Curves 2
and 3 represent the elution profiles for one-week and two-
week old solutions of Al(AA)$_3$ respectively. It was specu-
lated that the aluminum chelate underwent hydrolysis in the
stationary phase on standing, producing what were apparently
three different species. It is possible that mixed hydroxo-
acetylacetonatoaluminum(III) species form on standing, and
that this these species separate on the column. Thus, it is
likely that dissociation and hydrolysis of the metal complexes
in the stationary phase were responsible for the observed
results.

The suppression of these side reactions was accomplished
by considering the equilibria involved in the classical liquid-liquid distribution of metal complexes (10), and by
adding a small trace of the ligand (HAA) to the moving phase.
The amount of ligand added was such that at equilibrium there
was 0.8% HAA in the moving phase and 0.2% in the stationary
phase. The mass action effect of the excess ligand suppressed
the side reactions and resulted in symmetrical elution peaks.

A number of metal acetylacetonate separations were carried out using the liquid-liquid system described, plus a
trace of ligand. Figure 3 shows a four component separation
of Be(AA)$_2$, Fe(AA)$_3$, Cr(AA)$_3$, and Co(AA)$_3$. It can be seen
that the Fe(AA)$_3$ peak is symmetrical when the mixture is eluted in the presence of a trace of ligand. It was observed in

---
* percentages are by weight.

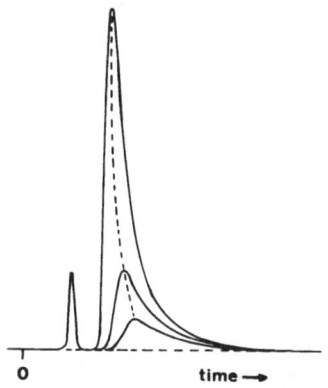

Figure 1.   Asymmetric Elution Peaks for Ni(AA)$_2$.
Reprinted from (10) with permission of the
American Chemical Society.

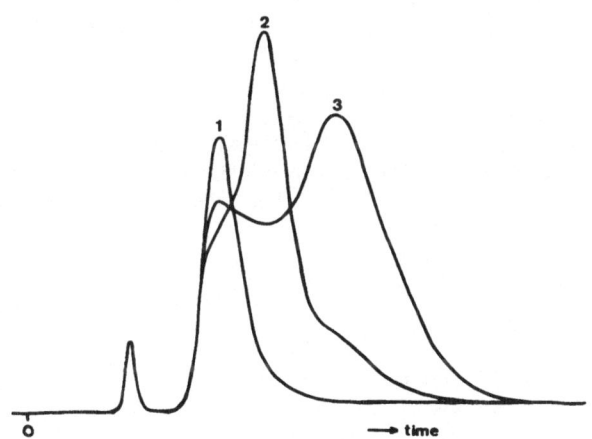

Figure 2.   Successive Elution Profiles for Al(AA)$_3$.
Curve 1:   Fresh solution of Al(AA)$_3$.
Curve 2:   One-week old solution of Al(AA)$_3$.
Curve 3:   Two-week old solution of Al(AA)$_3$.
Reprinted from (10) with permission of the
American Chemical Society.

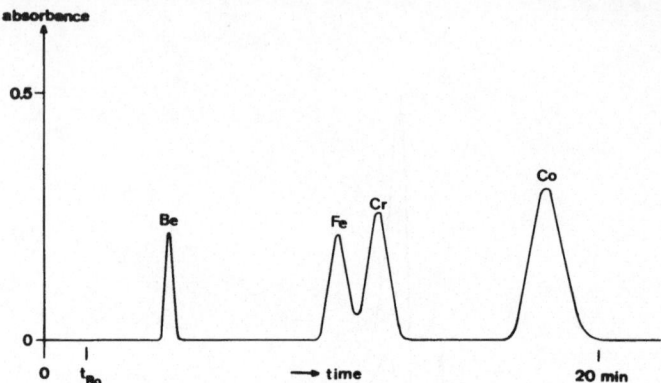

Figure 3.   Separation of Be(AA)$_2$, Fe(AA)$_3$, Cr(AA)$_3$ and
            Co(AA)$_3$ (trace of ligand in the moving phase)
            Column: 500 x 2.7 mm; particle size 10-20 μm;
            Fluid velocity: 2.6 mm sec^{-1}
            Reprinted from (10) with permission of the
            American Chemical Society.

earlier experiments, however, that the Fe(AA)$_3$ peak was distorted and asymmetrical if eluted without the excess trace
of ligand.   Figure 4 shows a rapid four-component separation of Be(AA)$_2$, Cu(AA)$_2$, Ru(AA)$_3$, and Co(AA)$_3$ accomplished
in less than 6 minutes.   Figure 5 shows a six component separation of Be(AA)$_2$, Cu(AA)$_2$, Al(AA)$_3$, Cr(AA)$_3$, Ru(AA)$_3$ and
Co(AA)$_3$ accomplished in 25 minutes.   The partition coefficients for a number of metal-acetylacetonates in the liquid-
liquid system described have also been determined by static
methods (10).

     It can be seen from these preliminary studies that high
resolution liquid chromatography can be a potentially powerful technique for the separation and determination of metal-
β-diketonates, as well as for equilibrium studies involving
the dissociation and stabilities of these compounds in various solvents.

                LC SEPARATION OF ARENE METAL CARBONYLS

     It is appropriate to review briefly the general struc-
ture of arene metal carbonyls, in particular the structure

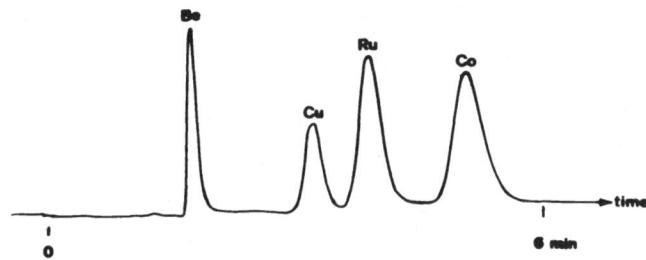

Figure 4.   Rapid Separation of Be(AA)$_2$, Cu(AA)$_2$, Ru(AA)$_3$,
            and Co(AA)$_3$.
            Column: 500 x 2.7 mm; particle size: 5-10 μm;
            Fluid velocity: 3.0 mm sec^{-1}.
            Reprinted from (10) with permission of the
            American Chemical Society.

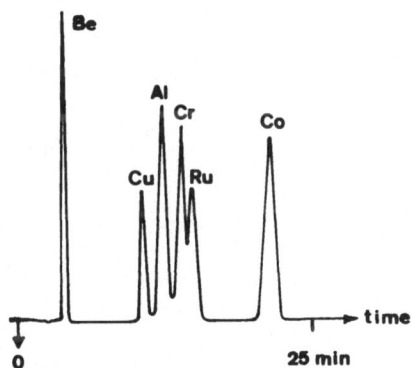

Figure 5.   Separation of six Metal Acetylacetonates.
            Column: 500 x 2.7 mm; particle size: 5-10 μm;
            Fluid velocity: 1.8 mm sec^{-1}.
            Reprinted from (10) with permission of the
            American Chemical Society.

of arene tricarbonylchromium complexes. A structural model
of the parent compound, benzene tricarbonylchromium (BTC) is
shown below. This compound, first synthesized by Fischer

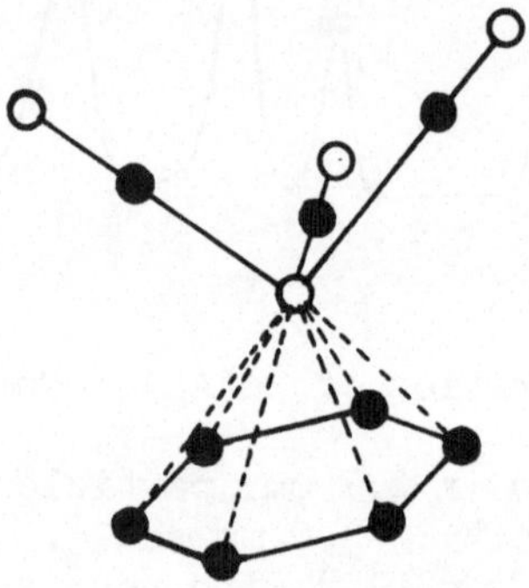

BTC

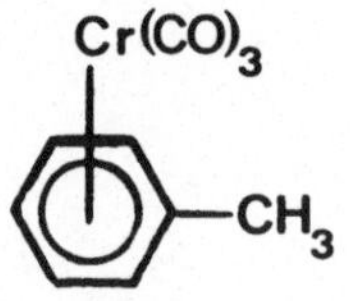

TTC

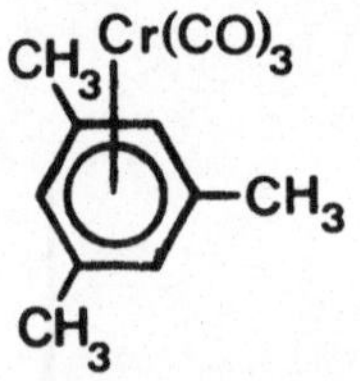

MTC

and Öfele in 1957 (14), is a complex in which a chromium(0)
atom is π-bonded to a benzene molecule (below the chromium)
and to three carbonyl ligands (above the chromium). Many
derivatives of this compound are possible, i.e., the structures for toluene- (TTC) and mesitylene- (MTC) tricarbonylchromium are also shown as examples. Naphthalene, anthracene
and other multiring tricarbonylchromium derivatives have also
been synthesized. In general these compounds are stable under
atmospheric conditions, and they are soluble in most organic
solvents. The single ring tricarbonylchromium complexes are
usually thermally stable and can be sublimed at reduced pressure, whereas the multiring compounds decompose when heated
to higher temperatures.

In 1969, we reported the first GC separation of several simple arene tricarbonylchromium complexes (4). It was shown that these complexes survived high temperature packed GC columns without decomposition (5). These studies were continued in later investigations and quantitative methods were developed (6). While GC is an effective and useful procedure for the analysis of mixtures of arene metal carbonyls, it is limited only to those complexes which are volatile and thermally stable. The potential value of LC for the separation and quantitative determination of the less stable complexes thus became apparent, and it was shown that it was possible to elute arene metal carbonyls from LC columns (8).

Tricarbonylchromium complexes of substituted naphthalenes were first reported in 1967 (12). It was found that naphthalenes containing two methyl substituents on only one ring react with hexacarbonylchromium to give a mixture of two unique isomers, in which the tricarbonylchromium group is bonded respectively to two different aromatic rings (I and II).

(I)   **(II)**

DMNTC

It was found that these isomers could be separated fairly easily by LC (9). The chromatogram obtained at 26.5°C is shown in Figure 6 and indicates a satisfactory separation of the isomers obtained on a 1000 x 3.5 mm column containing chemically bonded Carbowax-400 on Porasil-C using UV detection at 350 nm. The moving phase was TMP, using a flow rate of 1.85 ml/min. The two isomers were found to have capacity ratios of 3.9 and 4.9 respectively. The areas under the curves indicate that the mixture was 73% isomer I, in good agreement with the value obtained from NMR data (12). By lowering the temperature to 7.5°, complete resolution was achieved; however, the analysis time was significantly longer (Fig. 6). It was also found that repeated recrystallization of a synthetic mixture of the isomers from benzene/hexane solvent resulted in the enrichment of isomer II in solution. By this preparative method it was possible to isolate small

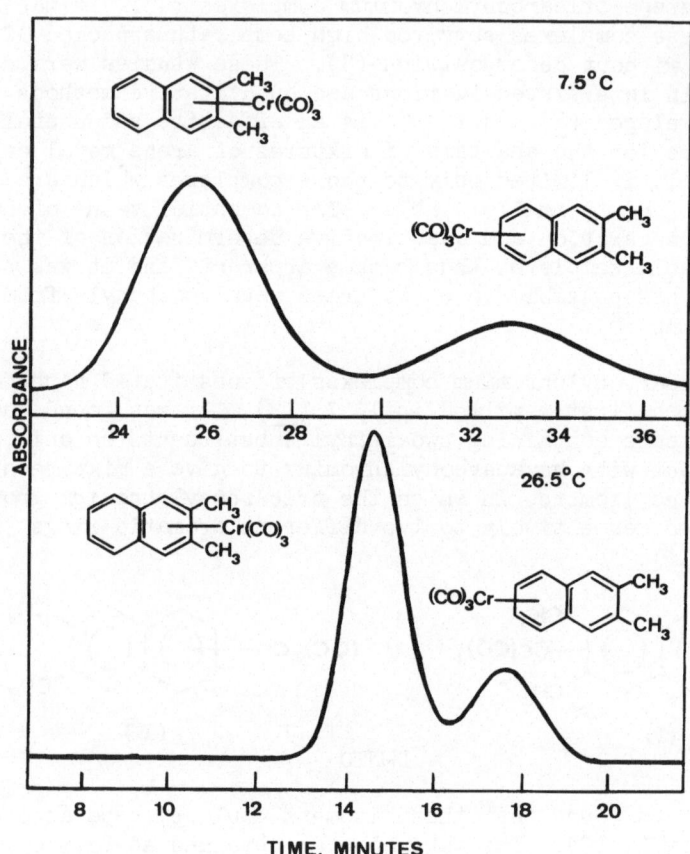

Figure 6.   Separation of DMNTC Isomers at 7.5 and 26.5°C.
            Column: 1000 x 3.5 mm; packing: Carbowax-400
            on Porasil-C; flow: 1.85 ml min^{-1} TMP; wave-
            length 350 nm.
            Reprinted from (9) with permission of the
            Journal of Organometallic Chemistry.

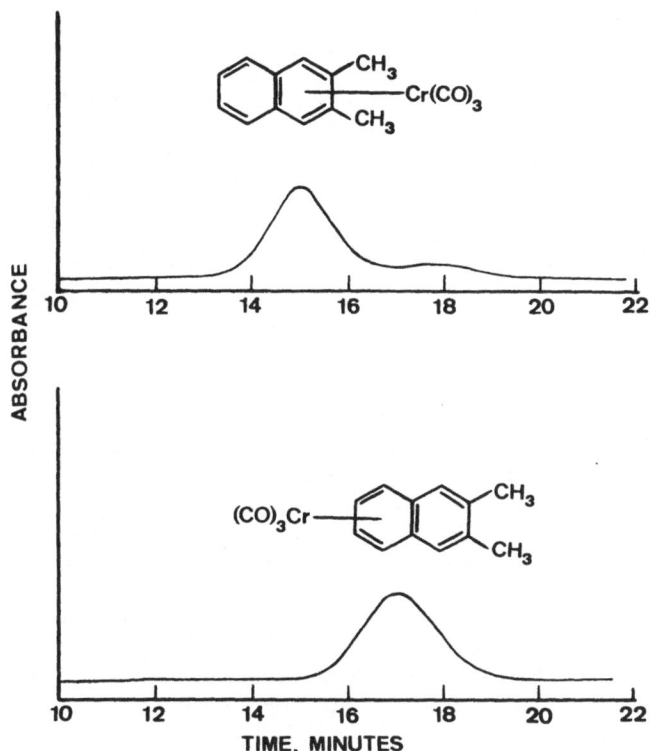

Figure 7.   Chromatograms of Separated DMNTC Isomers.   Same
             conditions as Figure 6.
             Reprinted from (9) with permission of the
             Journal of Organometallic Chemistry.

quantities of each isomer. The chromatograms of the separated
isomers are shown in Figure 7. They show single peaks of
retention times 15.0 and 17.1 minutes respectively, in good
agreement with the values obtained for the sample of mixed
isomers (Fig. 6).

     More recently, we have achieved the LC separation of
the isomeric forms of a newly synthesized complex, DTTC (11).
It was found that when 2,5-dimethoxytriptycene reacts with
hexacarbonylchromium, two isomeric complexes are formed.
These are shown in structures (III) and (IV). The Cr(CO)$_3$
group always bonds on one of the non-substituted rings, and
can be oriented toward (III), or away (IV) from the methoxyl-

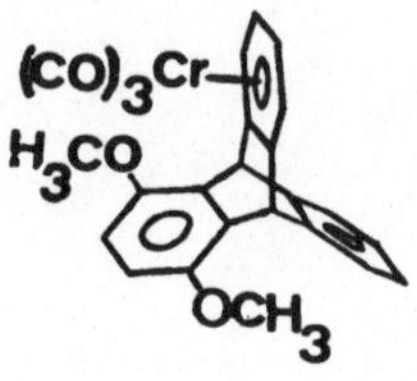

III 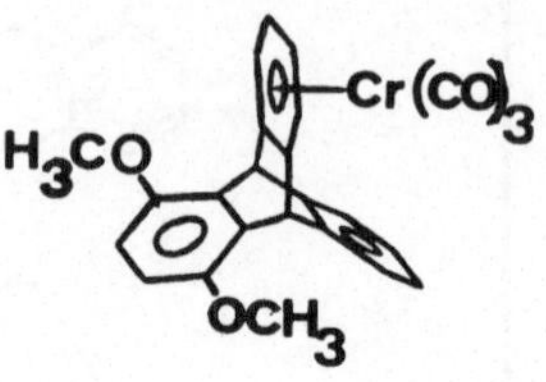   IV

DTTC

ated ring. Figure 8 shows the separation and resolution
achieved when a sample of the mixed isomers dissolved in
benzene was eluted with TMP from a 20 cm Carbowax-400/
Corasil I column at 28°C. The UV detector was operated at
325 nm. Figure 9 shows the same separation on a 20 cm uncoated Corasil I column at 28°C with the detector set at
235 nm. The capacity ratios were found to be 3.6 and 10.4
respectively for III and IV. The analysis is faster on
the Corasil I column (Fig. 9), however, the results shown in
Figure 8 indicate that preparative separations would be
easier on the Carbowax-400 column. This was indeed found to
be the case; the two isomers were separated by preparative
LC and the crystals isolated from the separated fractions
were utilized in subsequent UV and NMR analysis. From the
data obtained, it was shown that the first peak represented
isomer III and the second peak was isomer IV. A relative
isomer distribution of 63.0% to 37.0% was observed. Rein-
jection of the separated isomers into the column yielded
single peaks whose individual retentions corresponded to one
of the two peaks observed for the mixed isomers. Continuing
work on LC separations of additional arene metal carbonyls
is presently underway in these laboratories.

CONCLUSION

    The utility of high speed LC for the separation of
metal organic compounds has been demonstrated. It is felt
that the technique offers many possibilities for the continued study of separations and eventual quantitative
determinations of metals in their complexed form.

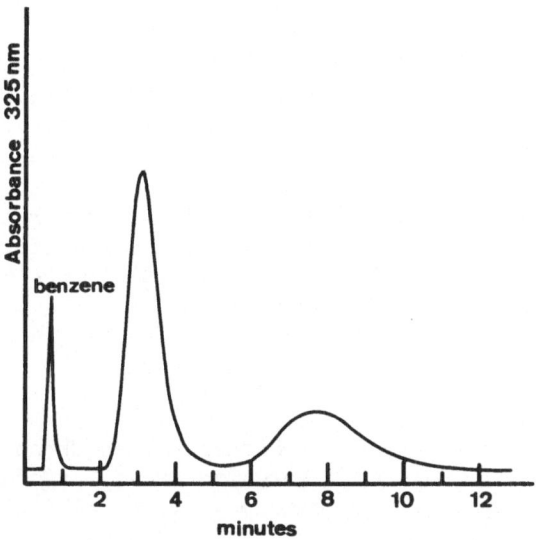

Figure 8.   Separation of DTTC Isomers.
             Column: 200 x 2.3 mm - Carbowax 400/Corasil I;
             Flow: 1.85 ml min^{-1}; temp: 28°C.

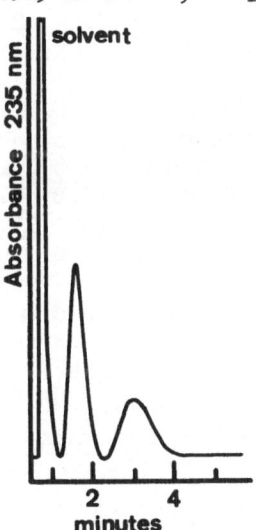

Figure 9.   Separation of DTTC Isomers.
             Column: 200 x 2.3 mm - Corasil I;
             Flow: 1.85 ml min^{-1} TMP; temp: 28°c.

ACKNOWLEDGMENTS

The author wishes to express his appreciation to Professor J. F. K. Huber of the University of Amsterdam and to Professor B. R. Willeford of Bucknell University. The work which was reviewed in this report, was previously published and coauthored with these investigators. Acknowledgment is also made to the donors of the Petroleum Research Fund administered by the American Chemical Society (PRF 3516-B), and to the National Science Foundation (GP-18755) for financial support of this work.

REFERENCES

(1) R. W. Moshier and R. E. Sievers, "Gas Chromatography of
       Metal Chelates," Pergamon Press, Oxford, 1965.
(2) H. Veening, W. E. Bachman and D. M. Wilkinson, J. Gas
       Chromatog., 5, 248 (1967).
(3) H. Veening and J. F. K. Huber, ibid., 6, 326 (1968).
(4) H. Veening, N. J. Graver, D. B. Clark and B. R. Willeford,
       Anal. Chem., 41, 1655 (1969).
(5) W. J. A. VandenHeuvel, J. S. Keller, H. Veening and B.
       R. Willeford, Anal. Lett., 3, 279 (1970).
(6) J. S. Keller, H. Veening and B. R. Willeford, Anal. Chem.,
       43, 1516 (1971).
(7) J. F. K. Huber, J. Chromatog. Sci., 7, 85 (1969).
(8) H. Veening, J. M. Greenwood, W. H. Shanks and B. R.
       Willeford, Chem. Commun. 1969, 1305.
(9) J. M. Greenwood, H. Veening and B. R. Willeford, J.
       Organometal. Chem., 38, 345 (1972).
(10) J. F. K. Huber, J. C. Kraak and H. Veening, Anal. Chem.
       44, 1554 (1972).
(11) S. A. Gardner, R. Seyler, H. Veening and B. R. Willeford,
       J. Organometal. Chem., in press, 1973.
(12) B. Deubzer, H. P. Fritz, C. G. Kreiter and K. Öfele,
       J. Organometal. Chem., 7, 289 (1967).
(13) J. F. K. Huber, J. Chromatog. Sci., 9, 72 (1971).
(14) E. O. Fischer and K. Öfele, Chem. Ber., 90, 2532 (1957).

LIQUID CHROMATOGRAPHY: ITS USE AS AN ANALYTICAL TOOL IN THE PHARMACEUTICAL LABORATORY

O. N. Hinsvark, W. Zazulak, P. Kraus, and
J. M. Marinello

Pharmaceutical Division
Pennwalt Corporation
Rochester, New York 14603

It may be implied by the title that the pharmaceutical laboratory is faced with measurement problems which differ from those of other analytical laboratories. This is not the case. Basically, we, as all others, are confronted with the problem: How is a component isolated from interferences for its ultimate measurement? The primary difference which exists between the pharmaceutical laboratory and that of a basic chemical industry, is the degree to which a product must be researched.

At one time, I had an excellent research director who later, by practicing this philosophy, became a very successful industrialist. He told me, "In 10% of the time, we can accomplish 90% of the research. We will be burned occasionally, but by practicing good scientific judgment, we can save the remaining 90% of the time for investment in other projects." In the pharmaceutical business, however, the market is people. In good conscience, our laboratories cannot afford that occasional burn; therefore, concerted efforts must be made at all times to minimize the possibility of error, and to maximize the safety and well-being of the patient. In addition, regulatory agencies, such as the FDA, also serve as powerful checks to insure that this research effort is continuous and the drugs are required, effective, and safe.

Common with all chemical laboratories is our need to establish identity, purity, potency, and stability of the

material which is to be employed as a drug. In addition,
however, it is necessary to ascertain how the drug is
handled by a biological system. Some questions which must
be asked concerning a drug are: Is it metabolized? What
are the metabolites? Does it or its metabolites selectively
accumulate in tissues? If so, does it develop toxicity?
Does the drug act directly, or does it, or its metabolites,
trigger a secondary mechanism? These and other questions
should be asked and attempts made to answer them. The
knowledge and technology has not advanced to the point where
it is possible to answer all these questions; however,
remarkable achievements have been accomplished. Even though
liquid chromatography is in its infancy of development, it
is proving to be a powerful tool, possessing the required
sensitivity and selectivity for measurement of drugs and
their action. Figure 1 is shown to help illustrate the
history of a drug and the steps in which liquid chromatography has proved useful. For a suspected drug action, a
structure is conceived and synthesized. The isolated product
is characterized both for purity and content. The compound
is then subjected to an initial screen. At this point,
interest in biological activity may be indicated. If so, a
more extensive biological characterization will be needed.
This includes, among other things, studies involving drug
action, preferred formulation, toxicity, and possible side
effects. At this point, it is hoped that a formulation
possessing maximum stability and biological availability can
be established, additional materials synthesized, and the
chosen formulation placed in the stability program. In the
meantime, if the results from the biological characterization
indicate continuing interest and need for a drug of this type
is indicated, an IND is prepared. On approval by the
regulatory agencies, this IND permits restricted testing in
humans. Ultimately, an NDA can be submitted; and if the drug
is needed, and if it is safe, and if it is effective, it will
be approved and placed in the hands of the prescribing
physicians. All of those steps marked with an asterisk
indicate points at which liquid chromatography can assist the
investigator in reaching his evaluation of the drug's
performance.

In this presentation, a new diuretic, metolazone, which
is the generic name for Zaroxolyn®, will serve as the primary
example to help illustrate how high-pressure liquid chromatography (LC) can assist in arriving at the solution of some of
the problems just mentioned. Throughout this study, a

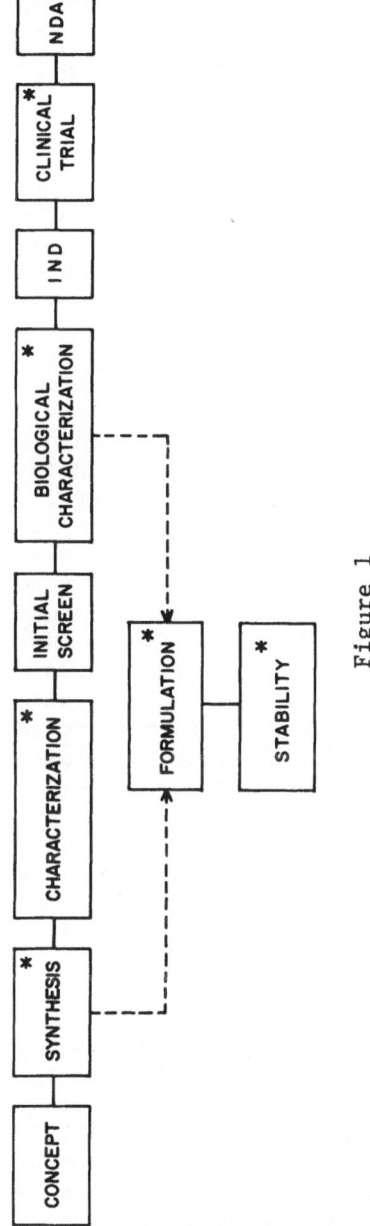

Figure 1

Waters ALC 100 liquid chromatograph equipped with an ultraviolet detector was employed. The low volatility and the
strong UV absorption of this compound at 254 mµ makes
metolazone an excellent example to demonstrate selectivity
and sensitivity of an LC technique.

The structure of metolazone is shown in Fig. 2.
Basically, it is a quinazolinone derivative to which active
groups have been added. Chemically, it acts as a weak acid
--the sulfamyl group reacting with strong bases to form water
soluble salts. The electro-negative nature of the chlorine
and sulfamyl groups tends to deactivate and neutralize the
amine in the 1- position. Physically, the drug is non-
volatile, and it exists in at least two stable polymorphic
crystalline modifications. At 37°C, approximately 60 µg of
the drug will dissolve in one ml of neutral or acidic
aqueous media, and it has a distribution coefficient between
octanol and distilled water of 42. Principle impurities
occurring during manufacture of the drug include hydrolysis
to form the "open ring benzamide", which results from
opening the ring between the 1- and 2-positions with a
subsequent loss of acetaldehyde or the "unreduced material"
which is the product of dehydrogenation between the 1- and
2- positions. It should be pointed out that the possibilities exist that isomers of the ortho-toludine are present.
In addition to the meta and para substitutions as possible
impurities, aniline, which is a known contaminant of
ortho-toluidine, could react with the final intermediate
and form the 3-phenyl derivative.

Fig. 3 shows a chromatogram of the "open ring benzamide",
the "unreduced material", and metolazone. By relating these
standard peak heights to those found in samples, we can
determine the concentration of individual components by a
simple proportionality.

These chromatograms were all obtained using an adsorption
column, Corasil®-II. Inevitably, however, improvements must
be expected, and this serves to illustrate that an indication
of purity must be defined by the explicit method of measurement before limits can be established. For example, while
exploring different column packings in an attempt to increase
the output, the Analytical Development Department, under the
direction of Dr. Robert Kieffer, was able to observe a second
peak. Using Zipax® as the packing material, not only was it

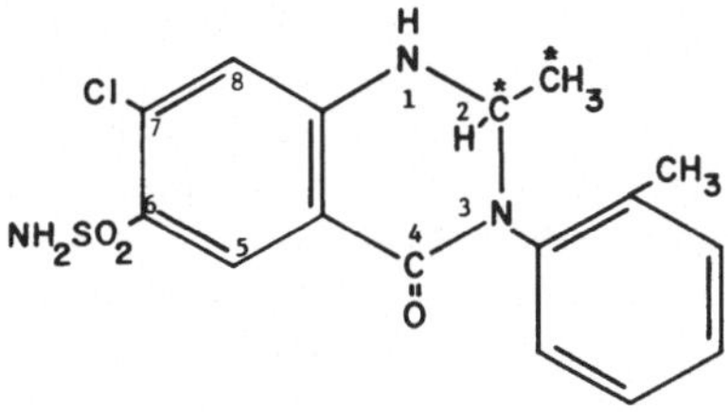

Figure 2

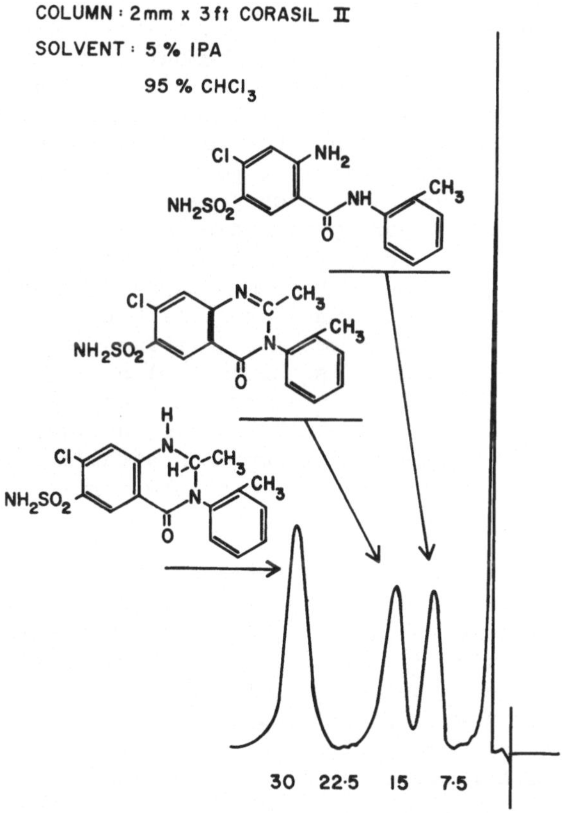

Figure 3

possible to increase the output of measurement of the
normal impurities, but in addition, the phenyl analog was
separated from the orthotolyl moiety. Fortunately, this
compound had been subjected to an acute biological screen,
both for toxicity and effectiveness. It was found to be
non-toxic, but less effective than the metolazone. The
finding of a new peak, however, did cause considerable
consternation, and a concerted effort on the part of the
development group had to be exerted to characterize this
phenyl analog.

During the development and synthesis of a drug such as
this, several intermediates must be prepared and isolated.
The organic chemist frequently desires information concerning the purity of the intermediate he has synthesized and
isolated. Fig. 4 illustrates some of the intermediates
which have been analyzed by liquid chromatographic techniques. These intermediates lend themselves to this method
of measurement, and valuable information can be provided to
the chemist or production personnel concerning the integrity
of the material being carried over to the subsequent step.
In addition to the intermediates, you will note that we have
included the "unreduced metolazone" mentioned previously.
This, indeed, is an impurity that can be introduced during
the production steps. However, with the exception of the
tolyl isomers, it is the only common impurity found in the
finished product. Metolazone is sensitive to light, and the
initial step of photolysis results in this product of
dehydrogenation. Since it is the initial product of photolysis, and since the final intermediate, the "open ring
benzamide", is the product of hydrolysis, a liquid chromatographic system keyed to these compounds affords a means for
following stability of the drug under the variety of
storage conditions employed in the program.

While liquid chromatographic techniques are valuable
for the evaluation of synthesis or stability programs, their
true value can best be illustrated in the measurement of
materials in complex systems -- such as biological fluids.
For example, a therapeutic dose of metolazone is 5 mgs. It
is necessary to have capabilities of measuring both the
intact drug and its metabolites at low concentration in the
complex biological fluids -- blood and urine.

Individually, biological fluids derived from different
subjects show considerable variation in themselves. The diet

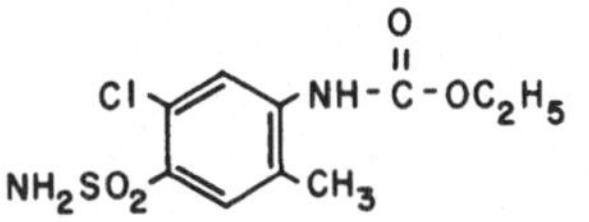

Figure 4

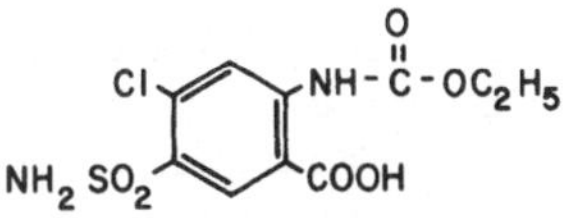

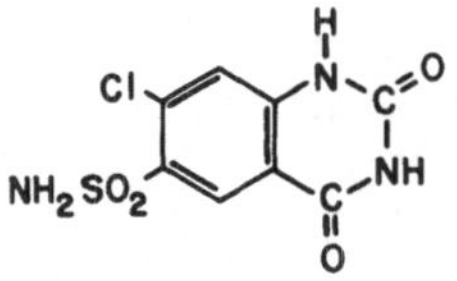

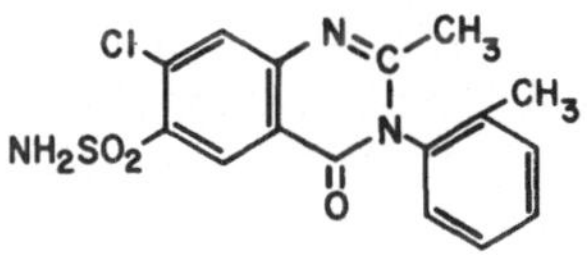

imposes still another variable, and the complexity of the
biological fluid matrix may be further increased. As a
consequence, it is difficult to describe a practical stepwise approach. Practical judgment on the part of the analyst
must be exercised in order to maximize the measurement
reliability and to minimize the number of erroneous results.
A combination of techniques, Carbon-14 counting with liquid
chromatography, has been used in our approach toward
providing an excellent measure of reliability to the study
of this drug's utilization by the human body, and to determine the distribution and extent of metabolism through
measurement of urinary metabolites.

Metolazone, doubly tagged with Carbon-14 in the 2- and
2-methyl positions, was administered to humans. Sequential
urine samples were collected, the volume read and recorded,
and frozen until ready for analysis. Two aliquots, one for
counting and one to be extracted for subsequent LC analysis,
were withdrawn. The number of Carbon-14 decompositions in
the aliquot is a measure of the total drug and its metabolites.
The aliquot for extraction was added to a 30-ml separatory
funnel, saturated with potassium dihydrogenphosphate, and
extracted into chloroform. The two layers were separated by
centrifugation, the organic layer (referred to later as the
chloroform soluble fraction) was drawn off and the exact
recovered volume recorded. An aliquot of the aqueous layer
was withdrawn for scintillation counting purposes. The
amount of drug found by conversion of the ^{14}C dpm in the
water layer is defined as that amount present in the aqueous
fraction. The organic layer is concentrated by evaporation
of the chloroform at $60°$ under a gentle stream of nitrogen.
The residue is reconstituted with a known amount of tetrahydrofuran, and this solution is then subjected to further
measurements, such as LC and/or TLC.

As can be seen in Fig. 5, some degree of separation can
be obtained during the extraction step. Approximately 95% of
the metolazone is extracted from water, at pH 4, by chloroform.
The remaining counts in the aqueous fraction are attributable
to metabolites. Complete extraction of the total counts can
be obtained using a more polar substance, such as ethylacetate,
for the organic phase. Selectivity, however, is lost in the
process. This is made obvious by referring to the extensive
front of the reconstituted solution (C) which completely
obscures peaks of interest.

TABLE I

Concentration of Metolazone in Blood and Urine
of Dog Following a 4.75 mg/kg Oral Dose

Metolazone μg/ml				
Time	Blood		Urine	
Hrs.	LC	^{14}C	LC	^{14}C
1	0.327*	0.061*	3.08	3.33
3	0.737	0.782	5.73	6.06
5	0.781	0.792	9.85	9.57
7	0.629	0.600		
22			7.03	8.48
48			2.18	1.71
72			0.62*	0.32

*Impurity under the metolazone peak

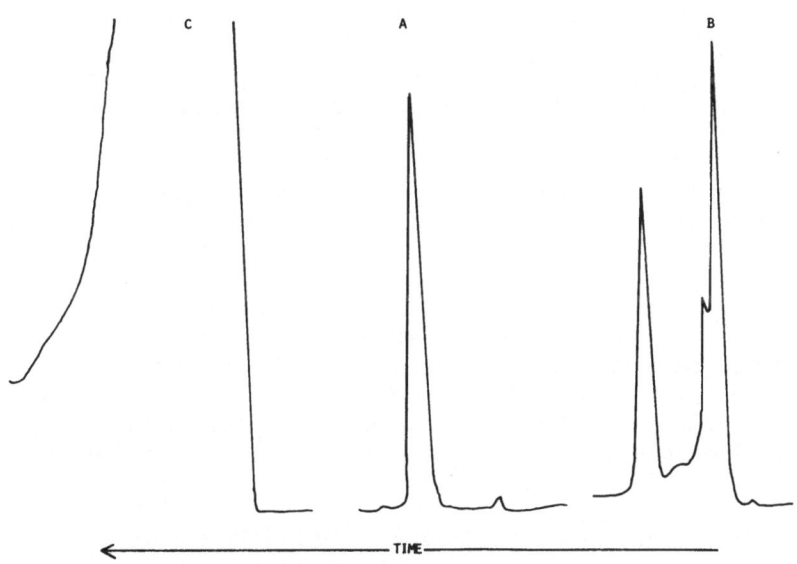

Figure 5

Fig. 6 shows the metabolites which have been found in
the chloroform-soluble fraction of dog urine. Hydroxylation,
oxidation, and possible conjugation appear to be the primary
routes of detoxification. Using TLC techniques, we had been
able to show that 40–60% of the total counts found in the
urine of dogs originated from metabolites. By contrast,
80–90% of the total counts appearing in the urine of humans
are attributable to unconverted metolazone. Approximately
95% of the drug found in the chloroform-extractable fraction
of the human urine is free and unmetabolized. This very
favorable distribution of drug to metabolites suggests that
keying to metolazone would provide a clinical measurement of
the drug as excreted by patients. This, of course, simplifies
the liquid chromatographic measurement. Initially, however,
it has to be demonstrated that a reliable measurement can be
attained. This is accomplished using the combined scintillation counting technique, TLC procedures, and LC measurements.

Fig. 5B shows a typical chromatogram of a urine extract.
The metolazone peak in the extract is compared against the
standard (Fig. 5A). Simple peak height measurements and
proportionalities are utilized to determine the amount of
drug present in that particular fraction.

By thin layer chromatography, metolazone can be isolated
from other Carbon-14 tagged metabolites. Scraping the spot,
extracting the drug, and counting provides a measure of the
amount of metolazone in that spot. Knowing the amount of
urine extracted and the amount of extract spotted, the drug
content of the urine sample can be calculated.

A comparison of these ^{14}C - TLC values to the LC
measurements of the extracted metolazone in chloroform is
shown by the data in Table I. The good agreement between
values indicates that, indeed, liquid chromatography will
provide a good, reliable, clinical measurement of a drug of
this type. This alternative method of measurement is
valuable since, not only is it capable of providing a
physician with potentially necessary information, but also
the research investigator who is studying drug action. It
is not feasible or economical to study a large population
with Carbon-14 tagged drug.

Even though it is possible to clinically measure cold
drug by keying to the unconverted metolazone, continuing
efforts and research made it necessary to measure metabolites

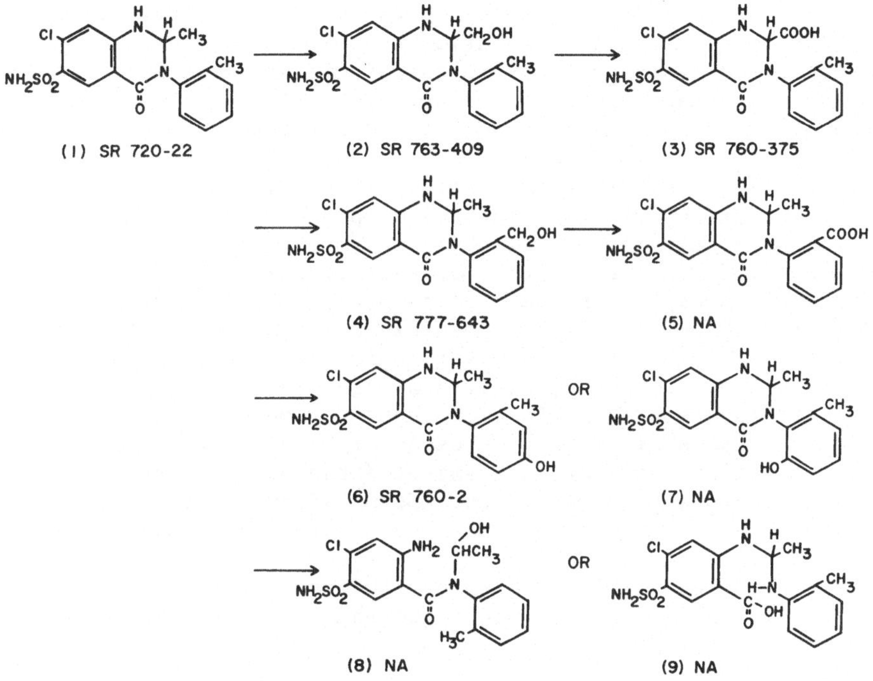

Figure 6

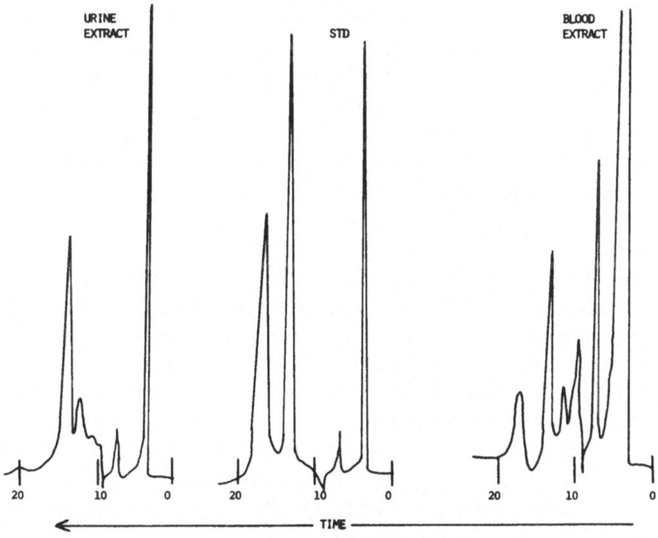

Figure 7

as well as drug -- not only in urine, but in blood. As
stated previously, hydroxylation and oxidation have been
established as the primary routes of metabolism. We
needed to establish the comparative distribution in blood
and urine of intact drug and metabolic products.

The low concentration of drug in circulation, coupled
with the strong protein binding, compounded the measurement
problems of drug in blood. Chloroform extraction of spiked
whole blood resulted in the recovery of only 10-20% of the
drug in the organic phase. Fortunately, complete distribution
away from blood protein is possible by using the more polar
solvent, ethylacetate. However, once the drug is separated
from the aqueous phase, additional cleaning is necessary.
The metolazone present in blood could be measured in a
manner quite similar to that used for urine. Its lower
concentration, however, necessitated the administration of
higher doses so that a greater amount of drug would be in
circulation at any given time. Dogs served as a model for
this experiment; and while this initial study shows results
from a single animal, subsequent studies seem to indicate
that this animal is not unique and that the distribution
is quite real.

To measure the more polar metabolites, the column-solvent
system required modifications. Gradient elution was necessary
to obtain a practical measurement of both the intact drug and
the metabolic fractions. The gradient elution system employed
was simply using a standby minipump to feed isopropanol at a
known rate to a vessel containing a measured volume of stirred
chloroform and cyclohexane. A continuing and repeatable
increase in the polarity of elution solvent is thus obtained.
Typical chromatographs of both blood and urine are shown in
Fig. 7. Unconverted metolazone elutes initially. It is
followed by the hydroxylated metabolites. Some displacement
in base line is noted after the metolazone peak; however,
the hydroxylated metabolites are separated and distinct
peaks are apparent. Finally, all counts are removed from the
column system by 1% acetic acid in isopropanol. Collection
of the individual fractions enables the use of counting procedures and subsequent detection of Carbon-14 fractions. The
measurement of specific activity in the individual fraction
establishes a secondary method of measuring concentration
and provides a means for detecting the presence of unknown
or unsuspected metabolic products.

TABLE II

Distribution of Drug in Blood and Urine of a Dog
Following a 4.75 mg/kg Oral Dose of Metolazone

Time Hrs.	Blood %			Urine %			
	Metolazone	Unr.	Metabolites	Metolazone	Unr.	Metabolites	Alb.
1	8.4	69.2	0.7	67.6	0.5	9.2	18.2
3	66.4	21.6	6.1	43.7	0.1	27.2	29.4
5	73.8	22.3	2.3	36.6	0.9	20.1	33.0
7	68.1	22.4	6.0				
22				40.9	1.0	15.1	39.8
48				31.0	1.5	16.6	46.1
72				31.8	1.8	25.4	46.8

Confirmation of liquid chromatographic measurements is obtained by comparing specific activity of the collected fractions to the LC measurements. This comparison is made in Table I. The comparative results are in good agreement and serve to demonstrate that the LC measurements are indeed valid.

The data in Table II are shown to illustrate the count distribution in the biological fluids. The individual fractions have been collected. Specific activities were converted to blood concentration of drug found in that particular fraction. The collected fraction 2 of blood contained about 20% of the total drug in circulation. The same fraction in urine, however, shows only negligible quantities. Further examination shows that these counts are attributable to the "unreduced material", mentioned earlier as a product of photolysis or mild oxidation. Its presence, therefore, should not be unexpected. What is puzzling, however, is the absence of this material in urine. Either the drug has been rehydrogenated prior to excretion via the urine, or it is preferentially metabolized and removed from circulation by way of the biliary system, and subsequently excreted in the feces. Or, this material is much more strongly bound to blood protein than the intact drug. In vitro studies, using albumin and hemoglobin, tend to indicate that this latter possibility is not the case -- to these proteins metolazone is more strongly bound than this particular moiety. This interesting mechanism is being further investigated; however, regardless of the ultimate explanation, the values serve to indicate that especially during the exploratory investigations, the researchers must resort to more than one system of measurement and detection; and they must not limit themselves to a single biological fluid.

During this presentation, we have tried to show how liquid chromatographic procedures can be employed by a pharmaceutical laboratory. Its intent has been to use a single material as the example, and to cite measurements which hopefully will serve as examples to scientists in chemical as well as pharmaceutical industries.

COLUMN PACKINGS AND EFFICIENCY
IN HIGH PRESSURE LIQUID CHROMATOGRAPHY

Karl J. Bombaugh

Chromatec, Inc.

Ashland, Mass.

Recent advances in high pressure liquid chromatography
have been brought about by a better understanding of the
column packing, particularly to the effect of particle size
on efficiency at high carrier velocity. It has now become
an accepted principle that the way to achieve high speed
in liquid chromatography is to use small particle diameter
packing (1, 2, 3). Since the principle was first reported
by such early workers as Huber (4), Snyder (5) and
Scott (6), the definition of the word small has changed,
and continues to change as column packing technology
develops. Each new development has a tendency to super-
cede previous work to such an extent that we often lose
perspective. It is, therefore, the purpose of this work to
attempt to review new developments in packing technology
in light of earlier developments, and, thereby, to
understand what it is that makes high pressure column
chromatography possible and to propose some considerations
for evaluating and optimizing LC performance.

A. HISTORICAL REVIEW

Liquid column chromatography was practiced for many
years as a slow and manual technique in which a comparatively
short, wide diameter column, packed with an active retardant,
was loaded with a sample mixture and then eluted with a
graduated sequence of solvents of increased strength. The
low efficiencies afforded by the large column diameter and

the wide range of particle diameter allowed only a few
components to be resolved at a given carrier strength. It
was a common practice to start with fully active packing
and then to strip the components off the column, one
activity level at a time, by an ever tightening grip of
the carrier. The problems encounted with this technique
included short column life, sample rearrangement on the
active surface, and poor resolution. At each increase in
carrier strength one could expect the "tail" of the
preceding peak to emerge on the leading edge of the new
peak. The technique as described, though useful, was
cumbersome and not easily automated. Because of the large
particle diameter ($\bar{d}p$) any increase in carrier velocity (U)
increased band width (W) which resulted in an increase in
plate height (H). Where:

$$H = \frac{L}{N} = \frac{L}{16V^2} \cdot W^2 = AW^2 \qquad (1)$$

when $\frac{L}{16V^2}$ is held constant $= (A)$. Further the limited
peak capacity (7) of the low efficiency columns made
solvent programming necessary even for relatively simple
mixtures. Several hours were required for the separations
followed by an equal amount of time to regenerate or repack
the column before introducing the next sample.

In contrast, by using current technology a separation
of many components, such as that shown in Figure 1 can be
made in 30-60 minutes at a single solvent strength, after
which the next sample can be introduced immediately. This
advance occurred through better understanding of the effect
of particle geometry on solvent band width in the moving
carrier stream. As a result of this advance, separation
speeds can be obtained by liquid chromatography comparable
to those afforded by gas chromatography as illustrated by
the data in Table I.

The systematic investigation of Huber and Hulsmen (4)
first demonstrated the implementation of a complete high
pressure liquid chromatography system using narrow bore
columns packed with small diameter closely sized support.
They attained Plate Heights of 0.6 mm at 0.4 cm/sec.
representing 7 theoretical plates per second. Their results
would be considered only intermediate speeds by today's
standards. The work is important, however, because it, in

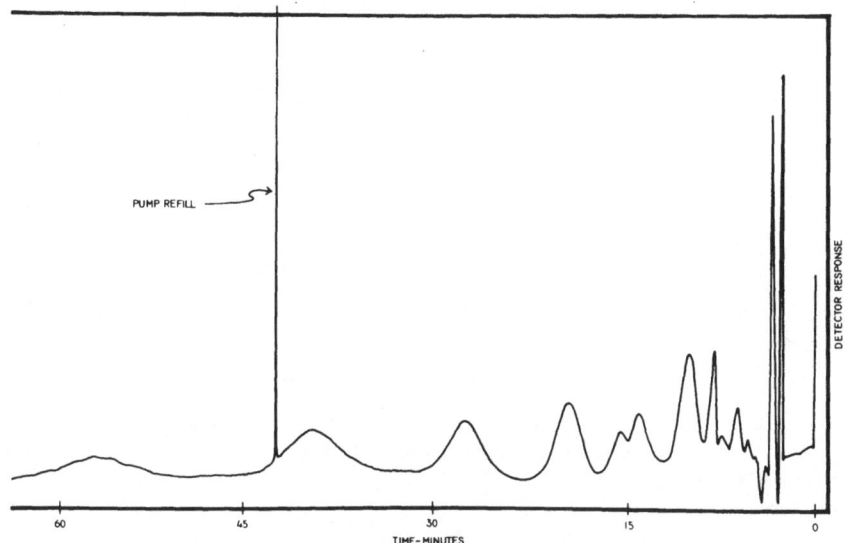

Figure 1. Typical Liquid Chromatogram
  Sample: Tergitol NPX (Ethylene oxide adduct of
                   octylphenol)
  Column: 4' x 2.1 mm ID ss pack with Chromasep S
  Carrier: Alcohol - Hexane - water
  Flow Rate: 1.8 ml/min; P 1800 psi
  Instrument: Chromatec Model LC 2100

Table I. Comparison of Column Efficiencies in Chromatography
Techniques

GC	Ref	$\bar{d}p$	N_{eff}/sec
Classical Packed	3	130	10
Open Tubular	3	250	25
Packed Capillary	3	10	40
LC			
Classical Packed	3	150	0.02
Porous Layer beads (silica)	1	43	0.81
Silica Gel (closely sized)	1	35	0.84
Silica Gel (closely sized)	1	20	2.5
Porous Layer beads (CSP)	2	27	10
Silica Gel (closely sized)	1	10	10
Silica Gel (closely sized)	1	6	23
Porous Microspheres	2	5-6	23

1. - R. Majors (15) 2. - J. J. Kirkland (16)
3. - General Literature

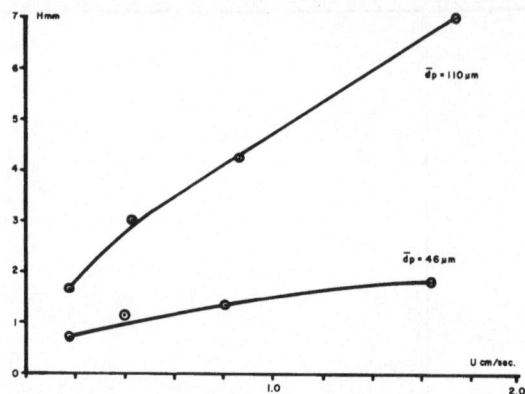

Figure 2. Effect of Particle Size (d̄p) on Plate Height (H)
           at Several Carrier Velocities
Neutral alumina with 6% water added
Carrier - n Hexane
Solute - Naphthalene K' - 0.5

d̄p	n	D	
110	0.63	0.47	
46	0.41	0.14	Ref (14)

effect, opened the doors to high speed liquid partition
chromatography.

   No review, however, would be complete without including
a most outstanding piece of work by Piel (8) reported in
1966. Using a column only 4 cm in length, packed with
sub micron porous particles, 6 components were separated
at 3500 psi into discrete zones. A band after 3 cm of
travel showed an H of 0.008 mm equivalent to 12,000 plates
per meter which is equal to the best efficiencies attain-
able with new packings reported in Table I. The work
clearly showed that micro particles can provide high speed
and high resolution in liquid chromatography. However,
the work was, to a large extent, unheeded. The belief was
generally held, and supported by published data from other
systematic investigations, that no benefits were gained
from a decrease in dp below 20-30 μm. (2, 9) Simultaneously,
favorable results were obtained using porous layer beads.
(10, 11) When the recent break through in the use of
porous micro particles occurred, a rivalry developed between
fully porous and pellicular packings. Therefore, a review
of the relative merits of each in the light of current
chromatography theory is also in order.

B. Basic Theory

1. Carrier Velocity Effect

It has been demonstrated in chromatography that H is proportional to $\bar{d}p$, in that at any given K' (retention level) speed of elution (t_r) depends solely on carrier velocity (U). H is related to U by the Van Deemter equation:

$$H = A + \frac{B}{U} + CU \qquad (2)$$

which defines three contributions to plate height: eddy diffusion, longitudinal diffusion, and mass transfer (12). In gas chromatography the curve showed a minimum, defining the optimum carrier velocity for maximum speed of separation. Any further increase in carrier velocity required an additional increase in column length to hold resolution constant which resulted in an undesirable increase in analysis time. However, the small diffusion coefficients encountered in liquid chromatography carriers minimizes the effect on the B term of the Van Deemter equation essentially eliminating the minimum (i.e. shifting the minimum toward the ordinate). Further, the slope of the ascending branch of the curve for liquids was less than the slope for gases. Thus, carrier velocity could be increased to gain speed in liquid chromatography. Synder (2) demonstrated that the relationship between H and U in LC could be defined by the emperical equation

$$H = DU^n \qquad (3)$$

where D is a constant and n an exponent with a value less than 1 at carrier velocities between 0.2 and 2.5 cm/sec. When the values of n are less than 1, carrier velocity may be increased without requiring an equivalent increase in column length to hold resolution constant. Consequently, an increase in speed is gained. Small n values are desirable since large increases in carrier velocity do not require a large increase in column length to maintain resolution. N values ranging between 0.7 to 0.3 have been obtained with various packings (13, 14). The value of the constant D (units $cm^{1-n} sec^{-1+n}$) in equation 2 also enters into the relationship between U and H. Since H is proportional to $\bar{d}p$, it appears that both D and n are affected by $\bar{d}p$. However, published data are somewhat in conflict and

inconclusive (2, 14, 15, 16). The H vs. U plot in Figure 2
reported earlier by Bombaugh, etal, showed an effect from
reducting particle size of alumina (14). Using tamp-tap
dry packed columns, they found that reducing d̄p from 110
to 46 μm reduced n from 0.63 to 0.41 while reducing D from
0.47 to 0.14. Synder in Figure 2 (2) drew parallel lines
(n = 0.4) through 5 sets of data for d̄p ranging from
150 to 20 μm. However, the data points indicate a steeper
slope could be drawn for the 150 μm d̄p (n > 0.4) and a
lower slope (n < 0.4) for the 44-56 μm particles. In a
more recent work Majors, using slurry packed silica columns,
showed no correlation between n and d̄p, but found a direct
correlation between D and d̄p. (15) The combined data are
plotted in Figure 3. The average d̄p values, for alumina (14)
are remarkably close to the line for slurry packed silica
reported by Majors (15). The values obtained by Randau
and Schnell (9), also reported in Major's plot, (obtained
with tamp-tap dry pack silica) did not follow Major's line,
and showed no decrease in H below dp = 25 μm. Thus it
appears that D and n are sensitive to both d̄p and packing
uniformity. Reliable measurement of these parameters in
terms of d̄p can be made only with a highly reproducible
packing procedure that produces a bed of optimum density
and uniformity.

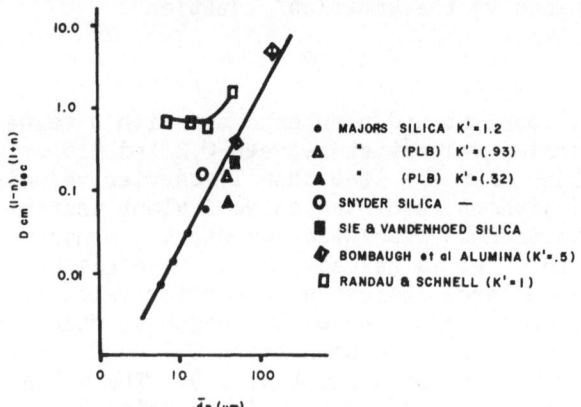

Figure 3.   Effect of d̄p on D for Several Adsorbents
            (Courtesy R. Majors (15))

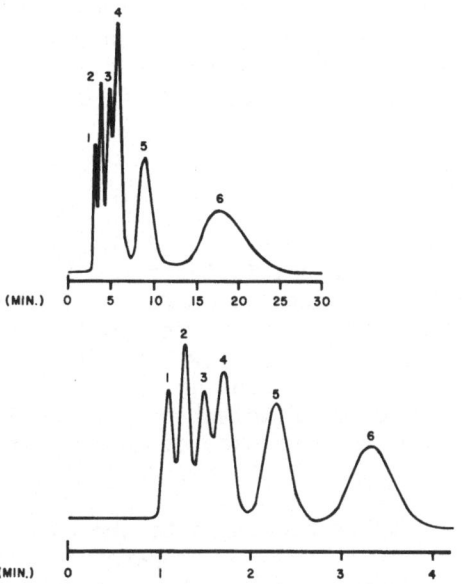

Figure 4. Illustrations of the Effect of d̄p on Resistanct to Mass Transfer (RMT) Using Alumina

Chromatogram	U	d̄p	H5	H6	H6/H5
top	0.5	110	7.45	16.67	2.23
bottom	1.6 cm/sec	46	3.63	4.87	1.34

Components: 1. Decalin, 2. Benzene, 3. Naphtalene, 4. Azulene, 5. O-Quaterphenyl, 6. M-Quaterphenyl

2. Resistance to Mass Transfer

The principle deterrent to high speed liquid chromatography has been resistance to mass transfer (RMT). The effect of RMT can be recognized as an increase in H with K' at increased U. By comparing W_6/W_5 in Figure 4, it is evident that W_6 relative to W_5 is greater with the larger d̄p packing even at one-third the carrier velocity. Thus, at low K' with low viscosity carriers and high diffusion coefficient solute relatively small H values may be obtained on a variety of packings at high carrier velocities. With such favorable test conditions the bad

effects of RMT are minimized. The true test of a high
performance packing is to test the same material with
representative samples using representative carriers at
K' greater than 10. If reasonable H values are obtained
at u = 1 cm/sec. over a wide range of K', then RMT has been
minimized and the packing is capable of providing a high
peak capacity.

Two approaches have been used to overcome RMT:
1, Reduce d̄p. 2, Use pellicular supports. Both approaches
have been used to reduce the diffusion path of the solute
in the stationary phase and in the pools of stationary
carrier. However, with but few exceptions the belief
prevailed that somewhere near d̄p = 25 μms, a limit was
encounted below which efficient columns could not be
produced reproducible for adsorption or partition
chromatography.

The most notable exception to the 25 micron barrier
was the work of Scott (6) who prepared 5 μm d̄p fractions
of porous ion exchange resins that afforded high performance separations, but which required comparatively high
column head pressures. A pellicular support was offered
as a compromise between RMT and column pressure drop.
Horvath etal (10) first applied porous layers of ion
exchange resin to soda-lime glass, thereby, providing a
uniform rigid 50 μm d̄p particle which could be dry packed
into a column uniformly offering minimal flow resistance.
The small particle provided relatively small inter-particle
diffusion paths and the thin coating of ion exchange resin
offered a relatively short intra-particle diffusion path.
Kirkland (11) developed a pellicular support for partition
chromatography in which the porosity of the layer was
controlled to further improve intra-particle diffusion.
A variety of commercial products followed offering a range
of porosities and surface activities for both direct and
reverse phase liquid chromatography. The pellicular
packings, for practical purposes, had reduced RMT to a
reasonable level. They were moderately efficient, easy to
dry pack into columns. However, the pellicular packing
had still not reached the performance level attained by
Piel (8) in 1966 nor that demonstrated by Scott (17) and
Hamilton (18) with 5 μm d̄p fully porous ion exchange beads.
Further the pellicular packings offered the disadvantage
of low sample capacity restricting their use for preparative
scale separations.

The break through occurred when Huber etal. (19)
reported the use of 5 μm silica supports for the separation
of surfactants by partition chromatograph.   The micro
particles were obtained by crushing 1750A Spherosil
(TM Pecheney St. Gobain, Paris, France) and fractionating
with an air separator. With this material Huber obtained
H values of 0.1 mm at carrier velocities of 0.16 cm/sec.
at K' = 1. This system afforded an estimated peak capacity
of 18 peaks in 25 minutes based on a resolution, R = 6 sigma.
While the material worked well as a support for partition
chromatography, severe tailing prevented its use in the
adsorption mode with the high polarity surfactants. The
packing procedure used by Huber consisted of adding small
increments of packing to a glass column and compressing the
increment to one-fifth its height with a teflon tipped rod.
The packing procedure from experience in our laboratories
was impossible to duplicate in 2 mm ID metal columns.

Majors (20) demonstrated that porous micro particles
could be packed into high efficiency columns reproducibly
using a slurry packing technique. The procedure used was
a modification of that previously described by Kirkland (21)

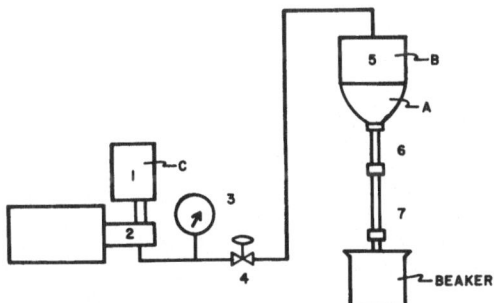

Figure 5. LC Slurry Packing Apparatus
1. Solvent reservoir,   2. High Pressure Pump (6000 psi),
3. Pressure gauge,   4. Shut off valve,   5. High pressure
slurry reservoir,   6. 5" precolumn,   7. Column.
A. Slurry   B.   Water   C.   Hexane

for use with 37 μm ether bonded Zipax (TM E.I. Dupont Co.)
having a particle diameter of 37 microns. By Major's
procedure a 1 gram quantity of dried micro particulate
silica was suspended in 10 ml of a balanced density solvent
consisting of 60.6 parts by weight of tetrabromoethane and
39.4 parts of tetrachlorethylene. An agglometate free,
degassed, suspension was added to the slurry chamber of the
apparatus pictured schematically in Figure 5. The slurry
was topped with water and driven into the column at 5000 psi
using a driving liquid such as n-Hexane. Kirkland (21)
reported that best results were obtained by forcing the
slurry into the column suddenly rather than by starting at
a lower pressure and increasing gradually to 5000 psi.
The slurry packing technique is applicable to any narrow
range particulate silica gel.

3. Reduced Plate Height

It has been a common practice in chromatography to
use reduced plate height $h = H/\overline{d}p$ to compare the performance
of LC packings by comparing h against reduced carrier
velocity (v) where $v = U\overline{d}p/D_m$. D_m is the diffusion
coefficient of the solute in mobile phase. Such comparisons
frequently show superior (lower) h values for larger particles.
A reduced plot is shown in Figure 6 using H and U values

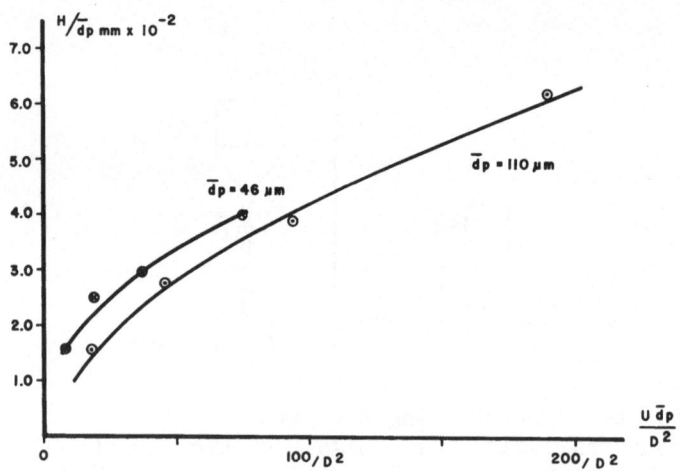

Figure 6.   Reduced Plate Height (h) Versus Reduced Velocity (v)
for $\overline{d}p$ - 46 μm, $\overline{d}p$ - 110 μm Alumina

from Figure 2. While the 46 micron material shows a
higher h vs. v plot, it is evident in the separation in
Figure 4 that the smaller particle provides the superior
separation. Reduced plots, therefore, do not show the
quality of packing material. Rather the reduced plot
enables one to assess how the performance of any packed
column compares with the ideal for that particular packing.
(2, 3) Highest efficiency and, therefore, greatest peak
capacity is attained with smaller d̄p which, to date, except
for the work of Piel, is limited to 5 - 10 um material.

The effect of d̄p on H over a wide range of carrier
velocities is demonstrated effectively by Figure 7 from
a systematic work by Majors (15). The investigation was
made with slurry packed LICHROSORB, an irregular silica
in narrow d̄p ranges, commercially available from E.M.
Laboratories, Elmsford, New York. The effect of d̄p on a
resolution is illustrated in Figure 8 which shows the
effect of 10, 20, and 30 μm Lichrosorb on the resolution
of vitamins. Ten μm packing provides higher efficiency
over the entire range of retention at increased U, thereby,
increasing both the peak capacity and the isocratic range
of operation.

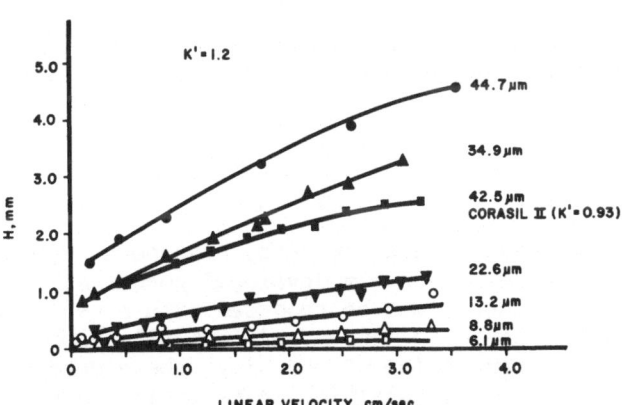

Figure 7.   Effect of Adsorbant Particle Size on the
            Slope of H vs. U Plots
          (Courtesy R. Majors (15))

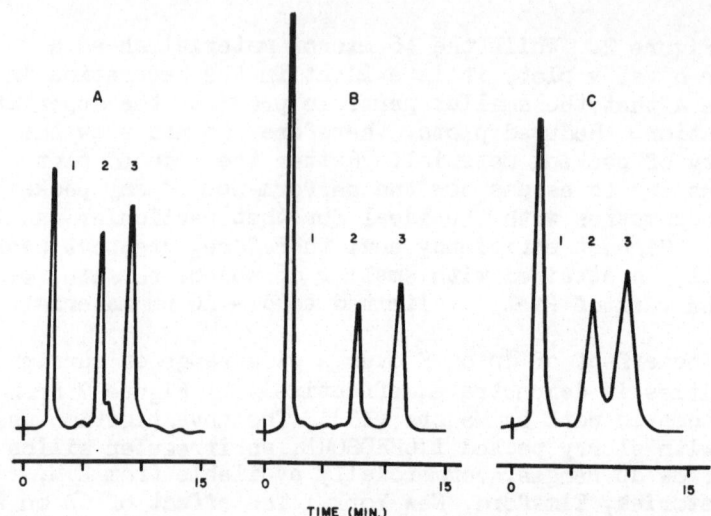

Figure 8.   Effect of Particle Size (d̄p) on Resolution of
            Vitamins.
Column: 30 cm x 2.9 mm, ID packed with Lichrosorb
Carrier: 10% CHC/₃ in n-Hexane
Instrument: Chromatec Model LC 2100
Sample: Vitamins 1 = E, 2 = D₂, 3 = A alcohol
Packing Particle Size: Left dp = 10 μm, Center = 20 μm
                        Right = 30 μm

4.   Speed of Separation

Perhaps the most meaningful measure of the speed of
separation in chromatography is "effective plates per second"
defined as:

$$N_{eff}/sec = 16 \left(\frac{V_r - V_m}{W} \right)^2 / t \qquad (4)$$

Among the highest speeds, to date, have been reported by
Kirkland (16) using a newly developed porous silica micro
sphere. The spheres can be produced with a very narrow
size distribution of 5 - 6 um and controlled pore size.
Thus, with proper choice of porosity, these spheres may be
used for adsorption, partition or exclusion chromatography.
Kirkland reported 23 effective plates per second at K' = 11
in partition separations and 14 effective plates per
second at K' = 10 in the adsorption mode. Majors (15),
however, reported 23 effective plates per second in the

adsorption mode using 6 μm $\bar{d}p$ Lichrosorb at $K' = 1.2$.

4. Pressure Drop Verses Particle Diameter

A decrease in particle diameter requires an increase in column head pressure to maintain flow velocity. The term which describes the pressure requirement of the column is permeability (K^O) defined as:

$$K^O = \frac{U n L f}{\Delta P} \qquad (5)$$

where U = carrier velocity, n = carrier viscosity, L = column length, f = total porosity, and Δ P = column pressure drop. A large K^O is desirable. Permeability is related to particle diameter by the Kozeny-Carman equation (23, 24).

$$K^O = \frac{\bar{d}p^2}{180} \frac{E^3}{(1-E)^2} \qquad (7)$$

where

$$E = \frac{V \text{ moving phase}}{V \text{ empty column}}$$

Combining the two equations:

$$\frac{U n L f}{\Delta P} = \frac{\bar{d}p}{180} \frac{E^3}{(1-E)^2} \qquad (8)$$

If the packing, carrier, carrier velocity and column length are all held constant, equation 8 becomes:

$$C_1 = \Delta P \, \bar{d}p^2 \frac{E^3}{(1-E)^2} \qquad (9)$$

A value of 0.42 has been given as a reasonable value for E at which equation 9 becomes (24):

$$C_2 = \Delta P \, \bar{d}p^2 \qquad (10)$$

which implies that Δ P depends only on $\bar{d}p^2$. The conclusion is misleading since E is a significant variable which makes permeability extremely sensitive to the E term in equation 7 as is shown by the data in Table II. Thus a very small change in packing density produces a large change in permeability. When permeability can be increased without a

Table II. Effect of E on K^o

E	$\dfrac{E^3}{(1-E)^2}$	$\dfrac{E^3}{180\,(1-E)^2}$	$K^o \times 10^8$
0.40	0.17	1040	0.09
0.42	0.22	820	0.12
0.44	0.27	665	0.15

proportionate loss in efficiency, then micro particles
can indeed be used at relatively low pressures. The
chromatograms in Figure 8 support this conclusion. The
30 cm column required only 250 psi to produce a flow of
1 ml/min. Since column efficiency relates directly to
packing uniformity, not packing density, it is evident that
pressure requirements for short columns packed with micro
particles might not be as large as was originally
anticipated. It is evident, however, that as column length
is increased to accomplish more difficult and more complex
separations, that column head pressure must be increased
proportionately.

5. Column Evaluation

Much emphasis has been placed on column efficiency
since band spreading caused by a poorly packed column can
prevent a separation. However, the plate number (N) is
merely an indication of the increase in band width relative
to the distance it traveled through the column. The
measure of the separation is resolution R defined as:

$$R = \frac{\Delta V}{W^r} = \frac{a-1}{a} \cdot \frac{K'}{1+K'} \cdot \frac{1}{4}\sqrt{N} \quad (11)$$

where a = net relative retention, K' = capacity factor, and
N = number of theoretical plates. The expanded equation
(25, 26) defines resolution in terms of a) selectivity
b) capacity, and c) plate number. Two peaks are considered
resolved when R = 1.5 peak widths, ie. when peak centers
are seprated by 1.5 peak widths. While the interactions
of the arms of the resolution equation have been discussed
previouly (26, 27, 28), a brief discussion is given here

as a basis for a simplified equation to follow.

R is extremely sensitive to a. At values of a = 1.2, the selectivity term becomes 0.02. Since R is the product of the combined terms, (R = A·B·C) a small selectivity term must be compensated by the other terms to maintain an R = 1.5 needed for base line separations.

The capacity term should also be used to gain needed resolution. The optimum value of K' = 2, as reported by Synder (27) pertains to speed of analysis, not to maximum resolution. By increasing K' from 2 to 8, the capacity term is increased from 0.67 to 0.89, at which resolution may be increased by 33%. The principle is illustrated by the diagrams in Figure 9. Peak widths were calculated for each retention volume assuming constant H. Thus, when

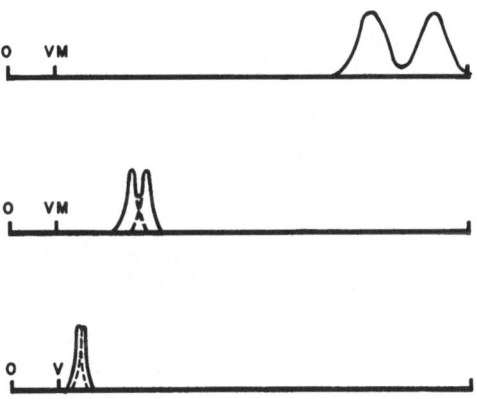

Figure 9.   Illustration of the Effect of K' on Resolution

Definitions: V_m = Volume of moving phase = column void volume.

$$K' = \frac{KV_r}{V_m} = \frac{V_r - V_m}{V_m} . \quad a = \frac{V_2 - V_m}{V_1 - V_m} . \quad N = 16 \left(\frac{V_r}{w}\right)^2$$

assumptions: a = 1.2, N = 1000

	bottom	center	top
K'	0.5	2	8
R	0.44	0.88	1.17

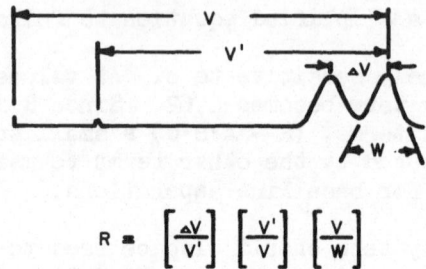

$$R = \left[\frac{\Delta V}{V'}\right]\left[\frac{V'}{V}\right]\left[\frac{V}{W}\right]$$

Figure 10. Expanded Resolution Equation Expressed in
Terms of Retention Volumes and Peak Width

the separation between a pair of compounds is inadequate,
an alternative to increasing column length may be to
decrease solvent strength, thereby, increasing the
capacity term's contribution to resolution, ie. making
greater use of the column's stationary phase.

6. Simplified Expression of the Expanded Resolution Equation

The expanded resolution equation may be defined by
three simple ratios which may be taken directly from the
chromatogram. The values are illustrated in Figure 10.

$$R \doteq \underbrace{\frac{\Delta V_r}{V'_r}}_{(A)} \cdot \underbrace{\frac{V'_r}{V_r}}_{(B)} \cdot \underbrace{\frac{V_r}{W}}_{(C)} \doteq \frac{\Delta V_r}{W} \quad (12)$$

Four simple distance measurements (V_r, V'_r, ΔV_r and W)
taken directly from the chromatogram and used in three
ratios, provide a numerical analysis of the separation.
Terms A and B approach 1 as the limit. The plate term is
simply the elution volume measured in peak widths. In

practice, the value of the selectivity term should be greater than 0.7 which would require a plate term of 12 (N = 2300 plates) to obtain base line resolution.

Expressing the resolution equation in terms of elution volume and peak width provides simple ratios which contribute directly to resolution. Table III shows the relationship between the commonly used fundamental units and the simplified term values.

Table III. Relationship Between Basic Units and Term Values of the Expanded Resolution Equation

a	A	K'	B	N	C
1.05	0.05	1	0.50	200	3.5
1.10	0.09	2	0.67	500	5.6
1.15	0.13	3	0.75	1000	7.9
1.20	0.17	4	0.80	2000	11.2
1.30	0.23	5	0.83	4000	15.9
1.50	0.33	6	0.86	6000	19.4
2.0	0.50	7	0.88	8000	22.4
5.0	0.80	8	0.89	10000	25.0
50.0	0.98	50	0.98	40000	50.0

As an illustration, with a selectivity term of 0.06 and the capacity term adjusted to 0.89, a C of 25 would be required to obtain base line resolution (ie. .06 x 0.89 x 25 = 1.5). The value of the plate term (C) increases as the square root of the increase of column length. Thus, to double resolution, it is necessary to increase column length for a factor of 4.

In summary, to maximize a difficult separation, select the carrier-retardant system to provide the best possible $\Delta V_r/V'_r$ then adjust retention to make V'_r = 0.8 or greater. Then increase column length until V_r/W is adequate to provide the needed resolution.

References

1. Huber, J. F. K., J. Chromatog. Sci. 85 (1969)
2. Synder, L. R., J. Chromatog. Sci. 7 352 (1969)

3. Knox, J. H., Saleem, M. J., J. Chromatog. Sci. _7_ 614
    (1969)
4. Huber, J. F. K., and Hulsman, J. A. R., Anal. Chim
    Acta _38_ 305 (1967)
5. Snyder, L. R., Anal. Chem. _39_ 698 **(1967)**
6. Scott, C. D., Anal. BioChem., _24_ 292 (1968)
7. Giddings, C. J., Anal. Chem. _39_ 1027 (1967)
8. Piel, E. V., Anal. Chem. _38_ 670 (1966)
9. Randau, D., and Schnell, W. J., J. Chromatog. _57_
    373 (1971)
10. Horvath, C. G., Preiss, B. A., and Lipsky, S. R.,
    Anal. Chem _39_ 1422 (1967)
11. Kirkland, J. J., Anal. Chem _41_ ·218 (1969)
12. VanDeemter, J. J., Zuiderweg, F. J. and Klinkenberg, A.,
    Chem. Eng. Sci. _5_ 271 (1956)
13. Waters, J. L., Little, J. N., Horgan, D. F.,
    J. Chromatog. Sci. _7_ 293 (1969)
14. Bombaugh, K. J., Levangie, R. F., King, R.M., and
    Abrahams, L., J. Chromatog. Sci. _8_ 657 (1970)
15. Majors, R. E., J. Chromatog. Sci. _11_ 88 (1973)
16. Kirkland, J. J., J. Chromatog. Sci. _10_ 593 (1972)
17. Scott, C. D., Jolley, R. L., P. H., W. W., and
    Johnson, W. F., Am. J. Clin. Pathol., _53_ 701 (1970)
18. Hamilton, P. B. "Handbook of Biochemistry: Selected
    Data for Molecular Biology", The Chemical Rubber
    Co., Cleveland, pp.43-55 (1968)
19. Huber, J. F. K., Kolder, F. F. M., and Miller, J.M.,
    J. Chromatog. Sci. _44_ 105 (1972)
20. Majors, R. E., Anal. Chem. _11_ 1722 (1972)
21. Kirkland, J. J., J. Chromatog. Sci. _9_ 206 (1971)
22. Giddings, J. C., J. Chromatog. _13_ 301 (1964)
23. Carman, P. C., "Flow of Gasses through Porous Media"
    Butterworths, London 1956, p.8
24. Karger, B. L., "Modern Practice of Liquid Chromatography"
    Ed., J. J. Kirkland, Wiley Interscience, New York
    1971, p.34
25. Purnell, J. H., J. Chem. Soc., 1268 (1960)
26. Snyder, L. R., Principles of Adsorption Chromatography
    Decker, New York 1968, p.18
27. Snyder, L. R., and Saunders, D. L., J. Chromatog. Sci.
    _7_ 195 (1969)
28. Bombaugh, K. J., Canadian Research and Development
    Sept./Oct. (1969)

QUANTITATIVE THIN LAYER CHROMATOGRAPHY OF AMINO ACIDS

Helen K. Berry

Children's Hospital Research Foundation
Institute for Developmental Research
Cincinnati, Ohio 45229

Chromatography on thin layers of loose adsorbent
was carried out as early as 1938. Ten years later
Meinhard and Hall used cohesive films which could be made
to stick to a glass plate for chromatographic separations
(1). With the introduction of paper chromatographic
methods the separation and identification of complex
mixtures of amino acids became possible. Following the
initial work of Consden, Gordon, and Martin, in 1944
(2), paper chromatography of amino acids developed
rapidly. Techniques were devised which permitted semiquantitative measurement of all amino acids in a complex
mixture, whether known or unknown. Development of thin-
layer chromatography of amino acids lagged somewhat
compared to techniques for separation of lipophilic
substances, partly because of the highly satisfactory
separations obtained by paper chromatography. However,
most of the procedures applicable to paper chromatography were found suitable for use with thin-layer chromatography (TLC).

TLC achieves sharpness of separation, high sensitivity, and speed by use of layers of adsorbents formed on
glass plates or other supports such as aluminum foil or
hard plastic. Solvent systems and reagents are similar to
those applied to paper chromatography of amino acids.
Limits of detection are lower than paper by a factor of
10 on one-dimensional plates and by a factor of 2 on
two-dimensional plates. Quantitative measurement of

thin-layer chromatograms, as with paper chromatograms, was achieved by measurement of spot size, photometry after elution or densitometry. In all quantitative procedures careful attention to thickness and quality of the layers and to reproducibility of the initial spot size were critical factors in obtaining reliable data.

Recently quantitative procedures for amino acid measurements have been described which employ automatic scanning devices taking into account both the size of the spot and degree of intensity of the reaction, whether with ninhydrin or another reagent.

In the present study commercially prepared cellulose plates (EM Laboratories) were used for TLC of amino acids. Quantitative measurements were made using a double-beam recording and integrating spectrodensitometer.

Preparation of Plates. Samples consisted of blood, urine, or protein hydrolysates. Blood or urine was neither deproteinized nor desalted prior to chromatography. For preparations of standards for use with blood, normal pooled serum was analyzed in triplicate on the amino acid analyzer. Known amounts of amino acids were then added to the serum to prepare standards containing 2.5, 5, 10 and 20 mg/100 ml of the particular amino acid. For use with urine or protein hydrolysates some standards were prepared in water.

Thin layer glass plates of precoated cellulose 0.10 mm thickness were used for all analyses. Plates, 10 x 20 cm, were marked lightly in pencil for 6 positions 1.5 cm from the bottom edge, leaving an empty lane of 1 cm beside each position. Samples or standards were applied in a streak using a 2 µl capillary. In most instances plates were prepared in duplicate, each containing at least two standards with 4 samples or duplicates of two samples (if a single plate were run). The same capillary was used for standard and sample on a set of plates. With practice each specimen could be applied in a uniform line of approximately 1.5 cm along the marked position. The same basic procedure was used for preparation and handling of all plates. After application of sample and standards, plates were run in the appropriate solvent. For certain amino acids a single run was sufficient for separation. For others, development a second time in solvent to which reagent had been added achieved better separations, and color development was more uniform and reproducible than when plates were sprayed with reagent. Solvents and reagents for determination of specific amino acids are shown in Table 1. Reagents were prepared as concentrated solutions using the major components of the solvent system.

For quantitative determination of phenylalanine and tyrosine the plates were placed in a small chromatography tank (22 x 12 x 6 cm) containing 40 ml of solvent (28 ml of a 1:1 mixture of butanol and acetone and 12 ml of acetic acid: water 1:2). Two plates were usually placed in each tank. When the solvent was within 1 cm of the top the plates were removed and dried. For development 3 ml of reagent (0.4 M ninhydrin in butanol: acetone) was added to the tank and mixed. The plate was then replaced in the mixture of solvent and reagent for a second development. When solvent again reached 1 cm from the top the plates were removed, dried in air, and heated 3 to 5 minutes at 80° C. Color on the ninhydrin-stained plates is stable for 24-48 hours at room temperature, for several days if preserved in a freezer.

Specific solvents can be used to achieve separations of particular amino acids. The need for a simple quantitative procedure for determination of branch chain amino acids in small volumes of blood from a child with maple syrup urine disease prompted us to modify a method described by Allen, et al. (3) as a micro procedure for measurement of leucine, isoleucine, and valine.

Blood was obtained from finger (or heel) puncture in heparinized microhematocrit tubes. The tubes, containing about 40 µl of blood were centrifuged to yield approximately 25 µl of serum. Samples and standards were applied to the plate in a streak as described. The plate was placed in a tank containing solvent as shown in Table 1. Following the first development the plate was removed, dried and then replaced in the solvent to which had been added 3 ml of 0.4 M ninhydrin in tert-amyl alcohol. Development in the second solvent, drying and heating were as described above.

Table 1.

Solvents, Reagents, and Procedures for Quantitative
        Determination of Specific Amino Acids

A. Phenylalanine-Tyrosine or Amino Acid Screening

        Development Solvent: n-Butanol   35
                               Acetone   35
                               Acetic acid   10
                               Distilled water   20

    Detection: 0.4 M ninhydrin in n-butanol/acetone(50/50)

    Directions: Add 3 ml ninhydrin reagent to solvent
       prior to second development. Dry plate; heat at
       80° for 3-5 minutes. Read at 555 mμ.

B. Leucine-Isoleucine-Valine

        Development Solvent: 2-Methyl-2-butanol   55
                               2-Butanone   19
                               Acetone   19
                               Distilled water   7

    Detection Reagent: 0.4 M ninhydrin in 2-methyl-2-
                        butanol

    Directions: Add 3 ml ninhydrin reagent to
       developing solvent during second run. Allow plate
       to dry before heating at 80° for 3-5 minutes.
       Plates may be stored in freezer after developing.
       Read at 555 mμ.

C. Proline

        Development Solvent: n-Butanol   35
                               Acetone   35
                               Acetic acid   10
                               Distilled water   20

    Detection: 3.5% isatin in n-butanol/acetone (50/50)

    Directions: Add 3 ml isatin reagent to solvent prior
       to second development. Dry plate; heat at 100°
       for 2-3 minutes. Read at 570 mμ.

D.   Histidine

Development Solvent: n-Butanol   80
                       2 N Hydrochloric acid   40
                       Ethanol (95%)   20

Detection Reagent: 0.9% sulfanilic acid in 1.2 N HCl
   diazotized with 4.5% sodium nitrite (in cold).   For
   spray reagent combine equal parts of diazotized
   sulfanilic acid and 10% potassium carbonate.

Directions: Spray lightly but uniformly.   Read plate
   when dry at 525 mμ.

E.   Tryptophan

Development Solvent: Isopropyl alcohol   80
                       Ammonium Hydroxide (15N) 10
                       Distilled water   10

Detection Reagent: Dissolve 0.25 g p-dimethylaminocinnamaldehyde in 10 ml conc. HCl.   Dilute to
   100 ml with water.

Directions: Spray lightly but uniformly.   Read plates
   within 30 minutes at 600 mμ.

Solvents and reagents developed for use in quantitative measurement of amino acids on paper chromatograms
(4) were adapted for determination of amino acids on
thin-layer plates (Table 1).   For measurement of proline
isatin reagent was added to the solvent prior to the
second run.   A blue color, characteristic of proline,
develops on heating the plates at 100° C for 2-3 minutes.
These plates if not scanned immediately may be preserved
by storage in the cold.

Details for measurement of histidine and tryptophan
are shown in Table 1.   For histidine, following a single
development the dry plate was sprayed with diazotized
sulfanilic acid combined immediately prior to use with
potassium carbonate.   Both reagents should be kept cold.
The red color characteristic of histidine is stable and
does not fade on standing at room temperature for several
days.   Tryptophan was detected following a single run by
spraying with p-dimethylaminocinnamaldehyde in 1.2 N

hydrochloric acid. The reagent is stable for several
weeks in a refrigerator. Tryptophan develops on drying
as a light blue band. The background gradually darkens
so quantitative measurements should be made within 30-45
minutes.

Quantitative Determinations. Measurements of spot
size or elution require many replicates and are tedious,
time-consuming procedures. Densitometry permits the
direct scanning of TLC plates. With the advent of
reliable spectrodensitometers in situ measurements of
thin layer chromatograms have become possible. A number
of instruments are available for densitometric scanning.
A double beam spectrodensitometer compensates for much
of the background produced by variables inherent in
thin-layer chromatography and is prerequisite for
quantitation of compounds at low concentration. A
Schoeffel Recording Spectrodensitometer Model 3000 with
Disc Integrator was used for quantitative measurements.
The densitometer was operated in the transmission mode,
its double beam permitting the adjacent blank area on
the plate to be scanned simultaneously with sample or
standard.

There are a number of techniques for quantitating
peak areas following densitometric scanning. These
include cutting out and weighing the peak, planimetry,
determination of area by triangulation, by using the
factor of height times width at half-height, or disc
integration. Procedures such as cutting and weighing
and planimetry are more time consuming than the others.
The disc integrator computes peak areas and displays the
computation as a second trace under the recorder curve
(see Fig. 2) so that quantitative data may be obtained
quickly and easily. Peak areas were determined using
disc counts from the integrator tracing in most instances.
For comparison some calculations were made using height
times width at half-height.

Reproducibility of repetitive scanning of bands
stained with the several reagents is shown in Table 2.
The standard deviations as percentage of total disc
counts ranged from 1-2% for high concentrations, intense
bands, and high disc integrator counts to approximately
10% for low concentrations, faint bands, and low disc
integrator counts. A series of 25 bands scanned twice

Table 2.

Multiple Scanning of Bands

Mean Disc Counts of 5 Scans	Std. Dev.	% Deviation
365	4	1.1
320	5	1.6
172	4	2.3
165	0	0
118	3	2.5
107	4	3.7
104	4	3.8
82	3	3.7
77	3	3.9
55	5	9.1
42	2	4.8
21	2	9.5

showed average difference between scans of 2.0% of total disc counts in the range from 50 to 400. The same tracings were calculated using peak height times width at half-height. Average difference between areas calculated from the same scans using peak height times width at half-height was 2.7%. Greatest difficulty in using the latter procedure came in determining the width of bands at half-height. Average difference between specimens on duplicate plates run the same day was 4.5%. Differences between duplicate determinations run on different days were approximately 6%. To minimize the effect of differences from plate to plate and from day to day, standards were always run on the same plates with samples.

Applications. Fig. 1 shows a TLC plate run in butanol-acetone-acetic acid solvent and stained with ninhydrin. This system may be used for purposes of screening of specimens for metabolic disorders which are characterized by increased concentrations of one or more amino acids. An important use has been in monitoring serum phenylalanine concentrations in children who are on phenylalanine-restricted diets for treatment of phenyl-ketonuria. In an earlier study (5) correlation between TLC and the fluorimetric procedure for phenylalanine was 0.96. We have since used TLC interchangeably with the

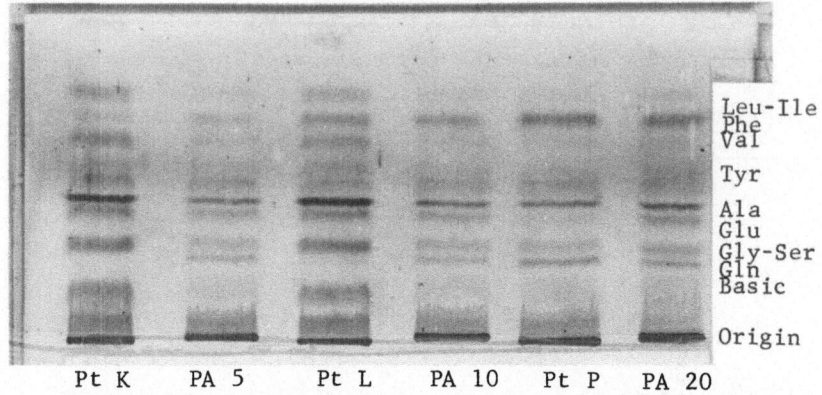

Fig. 1.   Thin-layer plate showing amino acids in sera
from patients treated for phenylketonuria (Pt. K, Pt L,
Pt P) and phenylalanine standards (5, 10, and 20 mg/100 ml)
prepared in normal sera.

fluorimetric procedure for purposes of monitoring serum
phenylalanine concentration.   Correlation between phenylalanine concentration measured by TLC and the amino acid
analyzer was also high.

The procedure described for separation of leucine,
isoleucine and valine has been used for monitoring blood
specimens from a patient with maple syrup urine disease
who is treated with a diet containing restricted amounts
of the three branch chain amino acids.   As in phenyl-
ketonuria, blood levels of the amino acids must be
followed to determine the effectiveness of dietary
restriction.   Fig. 2 shows a densitometric tracing
illustrating the separation obtained in the typical
chromatogram of blood or standards.   Differences between
concentrations of valine, leucine, and isoleucine/allo-
isoleucine measured by TLC and by the amino acid analyzer
were approximately 10% in the range of 5 to 20 mg/100 ml.

Urinary proline concentrations over 10 mg/100 ml are
abnormal, and the finding of concentrations in this range
is a reliable indicator of generalized aminoaciduria (4).
We have also used TLC determination of proline to measure
peptide-bound proline in urine specimens from patients

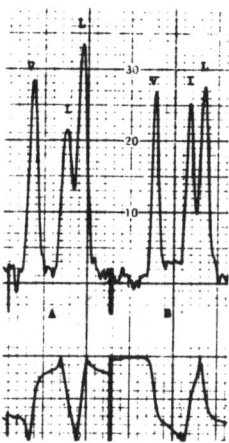

Fig. 2. Densitometric scanning and disc integrator trace
of a blood specimen (A) from a patient with maple syrup
urine disease and a standard solution containing
5 mg/100 ml each of leucine, isoleucine, and valine (B).

with skeletal disorders such as Paget's disease or poly-
ostotic fibrous dysplasia. The procedure for tryptophan
is useful in measuring the increased excretion of tryptophan by patients with Hartnup disease, a disorder in which
abnormal excretion of tryptophan results from a defect in
transport of the amino acid.

A typical standard curve for histidine is shown in
Fig. 3 comparing disc counts and absorbance as a means
of quantitation. There was little difference in the
results regardless of the method of calculation.

Sensitivity. Table 3 shows the sensitivity of some
of the quantitative procedures for individual amino acids
measured in blood. With the exception of valine, proline,
and leucine, values much below the normal mean cannot be
measured reliably, so the procedures are most useful for
detecting abnormal elevations. Recoveries of known
amounts of amino acids added to specimens ranged from 90
to 105%. Correlations of values obtained by TLC procedures for histidine, proline, tryptophan, valine,
leucine and isoleucine with those obtained on the amino
acid analyzer were high when concentrations were elevated,

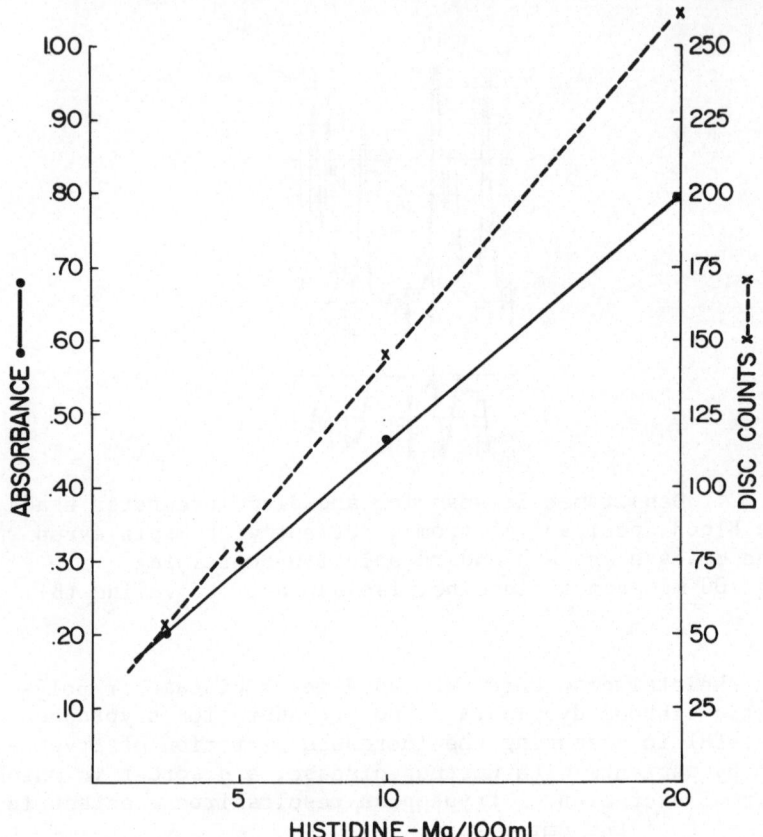

Fig. 3. Typical standard curve for histidine illustrating two methods for quantitating densitometric tracings.

less reliable in the low ranges, as expected from the
minimum concentrations shown.

Sources of Error. Obvious sources of error in
quantitative TLC of amino acids include lack of uniformity
in thickness of plate coating; errors in measurements of
the small volumes applied or non-uniform application;
variations in temperature and humidity during development
of ninhydrin color or irregular spraying of reagents not
included in the solvent; fading of stained bands prior to
densitometric scanning. To minimize errors as much as
possible the same commercially available precoated plates

Table 3.

Sensitivity

	Minimum Concentration Measureable by Quantitative TLC mg/100 ml	Normal Blood Concentration mg/100 ml
Phenylalanine	1.0	1.0
Tyrosine	1.0	1.0
Valine	1.0	2.5
Leucine	1.0	1.5
Isoleucine	1.0	0.8
Proline	1.5	2.2
Histidine	1.5	1.5
Tryptophan	1.5	1.0

were used for all determinations, and the same 2 µl capillary was used to apply both standard and sample to each plate or set of plates. Specimens were applied as uniformly as possible to reduce differences in band width. Conditions for color development were standardized as was the interval between development and scanning for substances which faded.

Conclusions. Quantitative TLC has been used for determination of amino acids in blood and urine using both a general reagent, ninhydrin, and specific reagents for specific amino acids. The procedures described are useful in screening large numbers of blood and urine specimens for a variety of biochemical abnormalities associated with amino acid metabolism. One of the chief advantages is in the small specimens required, particularly when repetitive testing of blood may be needed. Quantitative TLC procedures are not designed to compete with automatic ion exchange or gas chromatographic procedures for amino acids, but to complement them and to permit reliable quantitative data to be obtained routinely at a low cost.

Acknowledgement. This work was supported in part by Grants HD00324, HD05221, and HDAM06018.

References

1. J. E. Meinhard and N. F. Hall, _Anal. Chem._ 21, 185
    (1949).

2. R. Consden, A. H. Gordon and A. J. P. Martin,
    Biochem J. 38, 224 (1944).

3. R. J. Allen, H. J. Frey, L. M. Fleming and C. L.
    Owings, _Clin. Chem._ 18, 413 (1972).

4. H. K. Berry, C. Leonard, H. Peters, M. Granger and
    N. Chunekamrai, _Clin. Chem._ 14, 1033 (1968).

5. H. K. Berry, in "Quantitative Thin Layer Chromatog-
    raphy", J. Touchstone, Ed., John Wiley, New York,
    In Press.

SINGLE-COLUMN ANALYSIS OF AMINO ACIDS*

J. Robert Benson

Durrum Chemical Corporation
3950 Fabian Way
Palo Alto, California 94303

INTRODUCTION

Increased interest and improved technology in determining the primary structure of proteins have placed an emphasis on reducing the analysis time and increasing the sensitivity of routine amino acid analyses. The growing interest in the study and clinical diagnosis of metabolic disorders by analyzing ninhydrin-positive compounds in biological fluids has also dramatized the need for increased speed and sensitivity.

The two-column method of amino acid analysis as developed by Spackman, Stein and Moore (1) has been the method most used in automated systems. With this method, two identical sample aliquots are applied to two separate ion-exchange columns. Acidic and neutral amino acids are analyzed on one column and basic amino acids are analyzed on the other.

Associated with this method are inherent problems related to sensitivity, accuracy, and reliability. One obvious disadvantage is the requirement for two samples for a complete analysis. When the amount of available sample is small, the analysis itself can be the limiting factor in the experimental procedure. This requirement

*Portions of this paper originally appeared in AMERICAN LABORATORY, p. 51, October 1972.

also introduces the danger of pipetting error, thus dis-
torting the ratios of the amounts of basic to acidic and
neutral amino acids in the sample. Finally, automation of
the two-column method is complicated. Valves, relays, pumps
and electronics must always work synchronously if the analyzer is to be reliable.

The method of choice for rapid, automatic analysis of
amino acids utilizes a single column. Only one sample
aliquot is required. Sequential application of buffer
solutions of increasing pH and molarity is used to elute
all the amino acids from the single resin bed. With this
method, the correct ratios of the amounts of acidic and
neutral to basic amino acids in the sample are preserved.
With only one column, automation of the analyzer is greatly
simplified. However, several problems have prevented widespread acceptance of this technique, and only recently have
reliable, high-sensitivity single-column analyses been
possible.

THE SINGLE-COLUMN METHOD: PROBLEMS AND SOLUTIONS

Ammonia Contamination of Buffers

One of the chief problems with the single-column method
has been the change in baseline which appears during elution
of the high ionic strength buffer solution. This baseline
shift is due to ammonium ions which contaminate the acidic
buffers which are applied to the column during the early
part of the analysis. These ammonium ions are retained by
the resin and migrate slowly because of their strong basic
nature in the presence of low pH buffers. With the elution
of the high ionic strength buffer, however, this ammonia
band is displaced from the resin and its subsequent reaction with ninhydrin causes a baseline shift in the region
of the basic amino acids.

Ammonium ion contamination of buffer solutions is difficult to prevent. Most buffers are made with sodium citrate
which often contains ammonia as a contaminant. Ammonium
ions are frequently present in the hydrochloric acid used
for buffer pH adjustment. Unless water is deionized and
properly stored, it might also contain ammonia. Finally,

atmospheric ammonia can dissolve in low pH buffers. The
level of ammonia in these solutions can vary on a daily basis.

Several methods for preventing and removing ammonia
contamination have been devised. For example, prior to the
addition of hydrochloric acid, the freshly prepared alkaline
buffer solution can be heated. At this high pH, ammonia
(NH_3) is favored in the ammonium ion-ammonia equilibrium
and the gas is expelled from the heated solution. Subsequent addition of HCl for pH adjustment can add ammonium
ions, so care should be taken to obtain high-purity HCl.
To prevent contamination by atmospheric ammonia, concentrated sulfuric acid traps can be used on the air inlet of
buffer solution reservoirs.

The simplest and most efficient method for ammonia
removal, however, is to pass the buffer solution through
a strong cation-exchange resin ("ammonia filter"). This can
be accomplished in either of two ways. The first utilizes a
small column of resin on the high-pressure side of the pump.
Acidic buffer solutions flow through this filter; basic
buffer solutions bypass it. The filter is automatically
regenerated before the next analysis to elute the bound
ammonia. However, the resin in the filter column may not
contain sufficient capacity to remove all of the ammonia
when it is present in the buffers in large amounts.

In a second approach, resin columns are placed on
the inlet side of the pump. Each of the low pH buffer
solutions passes through one of these columns before it
enters the pump. (See Figure 1.) Regeneration of these
resin beds is required when the ammonium ions begin to
emerge. However, if the resin capacity is sufficient, regeneration may not be required for several months.

The choice of a resin for ammonium ion removal is
critical. It must have a high capacity for ammonia, must
permit unrestricted fluid flow and must not in any way interfere with the amino acid analysis. Most sulfonated poly-
styrene-divinyl benzene resins "leach" high molecular weight
polymeric material. Continued use of such resins results in
the deposition of this polymeric contaminant onto the analytical resin bed. This can produce anomalous baseline
behavior, spurious ninhydrin-positive peaks, loss of resolution, and increased column pressures. A commercially

available resin (2), used in this laboratory continuously during the past three years, has displayed none of these adverse effects.

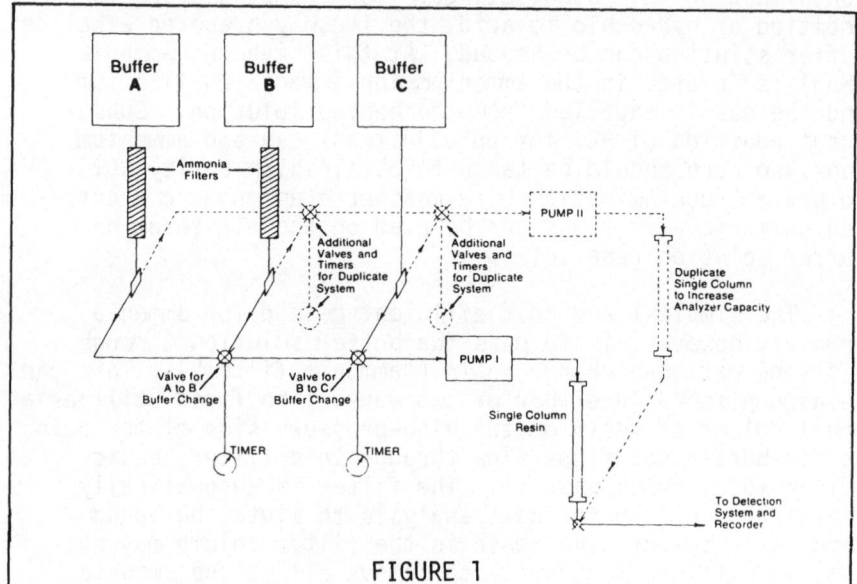

FIGURE 1

Schematic diagram of the single-column method showing placement of ammonia filters (see text). Most amino acid analyzers have an additional pump which can be utilized in a parallel system as shown above. Using dual columns allows optimum utilization of the analyzer since one column can be washed and equilibrated while the other is performing an analysis.

Baseline Anomalies

Baseline anomalies are often observed with the breakthrough of each buffer solution from the column. In the change from buffer "A" to buffer "B" for example, an artifact often referred to as "the buffer change peak" frequently appears. Even greater baseline changes are associated with the breakthrough of the final high ionic strength buffer solution. At sensitivities required for

detection of micromolar amounts of amino acids, these anomalies are usually not troublesome and peak areas can be integrated with reasonable accuracy. However, at sensitivities
needed for analysis in nanomolar quantities, baseline anomalies reduce the accuracy of peak area integrations.

These erratic changes in baseline are believed to be
caused by sulfonated polystyrene which elutes from the
crosslinked resin matrix. In some unknown manner, this
polymeric material reacts with ninhydrin, yielding a complex which absorbs at 570 nm, the wavelength used for detection of the ninhydrin-amino acid reaction product. A
recently-available commercial buffer system (3) eliminates
these baseline changes by controlling elution of this polymeric material from the resin matrix. Figure 2 illustrates
a single-column analysis of a 10 nanomole calibration mixture using these improved buffer solutions (4). No appreciable baseline change is observed throughout the analysis.

Another method of minimizing baseline changes requires
an alternating-wavelength photometer. Most amino acid analyzers use only a single-wavelength photometer operating at
the relative absorbance maximum of the ninhydrin complex.
If the signal from a reference wavelength is subtracted from
the signal at the absorbance maximum, a much quieter baseline results since background common to both wavelengths is
cancelled. The reference wavelength used must be transparent to the ninhydrin-amino acid reaction complex, but
show some absorbance for the ninhydrin-buffer mixture.

REDUCTION OF ANALYSIS TIME

Theoretical Considerations

With the way paved for high-sensitivity analyses of amino
acids on a single column, what remains is to reduce the total
analysis time without sacrificing sensitivity or resolution.
An examination of some of the results of Hamilton, et al (5),
who formulated a basis for a theoretical understanding of
liquid chromatography as it applies to amino acid analysis,
reveals how this can be accomplished.

Hamilton derived an expression for the resolution, R_{ab},
of two amino acid peaks, a and b:

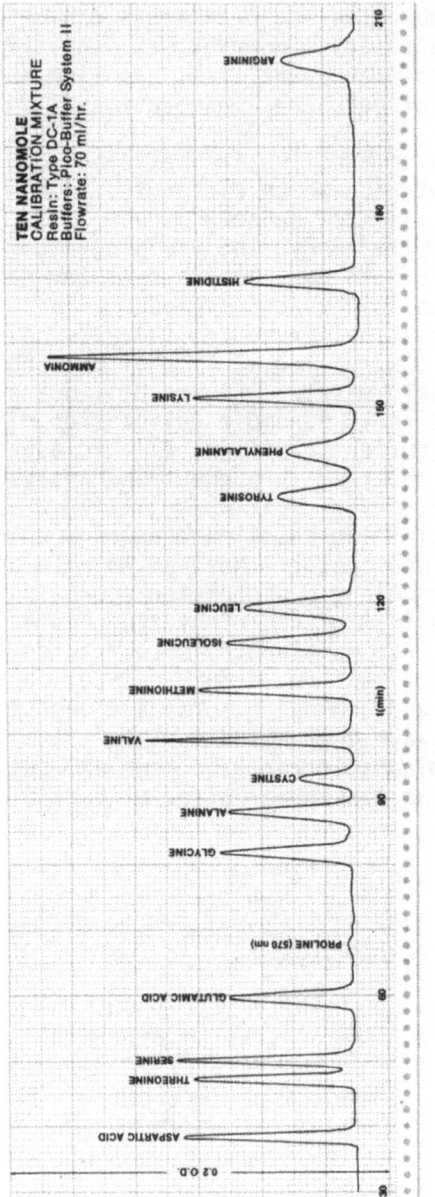

FIGURE 2

This chromatogram illustrates a high-sensitivity amino acid analysis on a single column. Three sequential buffers were used to elute all the amino acids in the 10-nanomole calibration standard. The first two buffers were each passed through an ammonia filtration column. (See Figure 1.) Note that the baseline remains flat throughout the run. Column pressure was about 300 lbs in^{-2}.

$$R_{ab} = \left(\frac{Z}{2U_o}\right)^{1/2} \frac{K_{d(a)} - K_{d(b)}}{(X_a)^{1/2} + (X_b)^{1/2}} \qquad \text{(Eq. 1)}$$

Z is the length of the resin bed and U_o is the linear velocity of the buffer that flows through the column. U_o is calculated from the volumetric flow rate and the cross-sectional resin bed area:

$$U_o = \frac{F(cm^3 min^{-1})}{A(cm^2)} = \frac{F}{\pi r^2} \qquad \text{(Eq. 2)}$$

K_d represents the distribution coefficient for each amino acid and is given by

$$K_{d(a)} = \frac{\bar{V}_a}{AZ} - F_I \qquad \text{(Eq. 3)}$$

where $\bar{V}_a$ is the effluent volume of peak a and F_I is the void fraction of the column. With d_p representing the mean resin particle diameter and $D_{s(a)}$ representing the solid phase diffusion coefficient of amino acid a,

$$X_a = \lambda \frac{d_p}{U_o} (K_{d(a)} + F_I)^2 + \frac{d_p^2 K_{d(a)}}{60 D_{s(a)}} \qquad \text{(Eq. 4)}$$

λ is a dimensionless constant related to eddy diffusion within the column.

The first term in Eq. 4 represents the axial diffusion of amino acid a in the column. This term decreases as resin particle diameter decreases and linear flow velocity increases. Either of these changes produces increased buffer solution flow resistance. Therefore, the axial diffusion component can be made small only if high column pressures can be tolerated. If this is the case, Eq. 4 becomes

$$X_a \simeq \frac{d_p^2 K_{d(a)}}{60 D_{s(a)}} = d_p^2 X_a' \qquad \text{(Eq. 5)}$$

Substituting Eq. 5 into Eq. 1 and factoring the resin particle diameter, d_p, yields

$$R_{ab} \approx \frac{1}{d_p} \left(\frac{Z}{2U_0} \right)^{1/2} \left(\frac{K_{d(a)} - K_{d(b)}}{(X_a')^{1/2} + (X_b')^{1/2}} \right) \quad \text{(Eq. 6)}$$

Thus, Eq. 6 gives the resolution of a pair of amino acids
when small diameter resin particles and high buffer solution linear flow velocities are employed. According to
Hamilton, complete resolution is accomplished when $R_{ab} \cong 2$.

The requirements for obtaining a second, faster analysis
after the conditions for a first analysis have been determined can be derived from Eq. 6. By demanding that the
resolution, R_{ab}, remain constant and by performing the second
analysis with the same buffers as were used with the first,

$$\frac{Z_1}{d_{p1}^2 \, U_{o1}} = \frac{Z_2}{d_{p2}^2 \, U_{o2}} \quad \text{(Eq. 7)}$$

is obtained. This expression gives the relationships among
particle diameter, column length and linear flow velocity
for the faster analysis in terms of those same quantities
in the original analysis.

If the mean resin particle diameter differs in the two
cases, the resin mass per unit column volume will also
differ. It is useful to introduce ϕ, the "packing density",
defined as

$$\phi = \frac{m}{V_c} = \frac{m}{\pi r^2 Z} \quad \text{(Eq. 8)}$$

where m is the mass of resin contained in the column and
V_c is the volume of the empty column. An amino acid inter-
acting with the greater mass of resin per unit volume of
column will have a longer elution time. Applying this
relationship, Eq. 7 becomes

$$d_{p2}^2 \, \frac{Z_1 \phi_1}{U_{o1}} = d_{p1}^2 \, \frac{Z_2 \phi_2}{U_{o2}} \quad \text{(Eq. 9)}$$

Eq. 9 is most often used to calculate the diameter of the resin required to maintain constant resolution when column length and linear flow velocity are changed to achieve a faster analysis.

The total analysis time is proportional to the product $Z\phi$ and inversely proportional to U_o. Thus, the time for the faster analysis (t_2) can be calculated from the data available from the initial analysis and the proposed modifications (Z_2, ϕ_2, U_{o2}):

$$\frac{t_2}{t_1} = \frac{Z_2 \phi_2 U_{o1}}{U_{o2} Z_1 \phi_1} \qquad\qquad (Eq.\ 10)$$

Results

To test the validity of Eqs. 9 and 10, three amino acid analyses were performed using resins of various mean particle diameters and columns of various dimensions. As used here, total analysis time (t) is defined as the time from sample injection to the zenith of the arginine peak. Table I summarizes the data from the three analyses.

TABLE 1

Experiment	1	2	3
Resin designation (6)	DC-1A	DC-6A	DC-4A-7
d_p (micrometers)	14 ± 2	11 ± 1	7 ± 1
Z (cm)	58.1	30.4	30.5
U_o (cm min^{-1})	1.89	1.89	5.75
ϕ (g cm-3)	1.08	1.30	1.6 ± 0.1
Analysis time (min)	196	120	42
Column pressure (lbs in^{-2})	300	400	2800

Experiments 1 and 2 were carried out under identical conditions except as noted in Table I. The three sodium citrate buffer solutions used to perform the two analyses were the same; in both cases, two column temperatures (45°C and 65°C) were used; both analyses were carried out using the same amino acid analyzer. In Experiment 3, the second and third sodium citrate buffers had slightly different ionic strengths and pH values than those used in Experiments 1 and 2. In addition, the value of ϕ was not determined exactly. Otherwise, the analysis conditions were similar to those in the other experiments.

In the first attempt to achieve a faster analysis than that obtained in Experiment 1, U_o was kept constant (1.89 cm min^{-1}) and the column was shortened. (An analysis performed under identical conditions as Experiment 1, except that only one column temperature was used (52°C), is shown in Fig. 2.) To compensate for the loss of resolution predicted by Eq. 6, a resin with a smaller mean particle diameter was required. Using Eq. 9, it was calculated that a resin 11 micrometers in diameter would fulfill the requirements. When the mean particle diameter of a resin is decreased, the packing density ϕ usually increases. In practice, ϕ is determined at the completion of the experiment by weighing the moist resin removed from the packed column and dividing this mass by the empty colume volume.

The analysis time which results from changing Z and ϕ is predicted by Eq. 10. This calculated value, 123 minutes, agrees closely with the actual analysis time of 120 minutes. The chromatogram resulting from using the 11 micrometer DC-6A resin in the shorter column is shown in Fig. 3.

An attempt was made to reduce the analysis time to less than one hour by increasing the linear flow velocity of the buffer solution, realizing that the smaller resin mean particle diameter coupled with the increased U_o would result in greatly increased column pressures. Consequently, this analysis was accomplished on the Durrum Model D-500 Amino Acid Analyzer, a computer-operated instrument capable of producing accurate and precise flow rates at elevated pressures. A description of this instrument can be found elsewhere (7).

As previously discussed, the analysis conditions for Experiment 3 were not identical to those in Experiments 1

and 2 so Eqs. 9 and 10 cannot literally apply. It is nevertheless interesting to use the experimental data from that
analysis and compare the results with the first two experiments. In this case, column length was held constant while
U_o was increased. (See Table 1.)

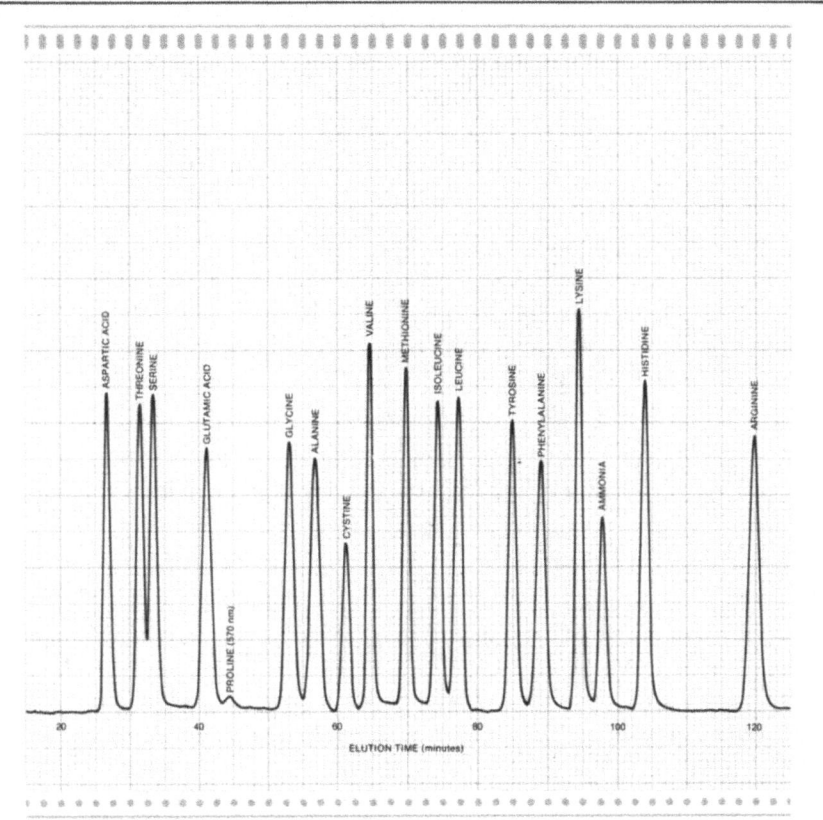

FIGURE 3

This amino acid analysis was completed in 120 minutes on a
single column. The column was about one-half the length of
that used in Figure 2, and the resin mean particle diameter
was reduced from fourteen to eleven micrometers. Column
pressure was about 400 lbs in^{-2}. Sample contained 25 nano-
moles of each amino acid.

Eq. 9 predicts that the mean particle diameter of the resin should be 7 micrometers. The DC-4A-7 resin was chosen to fulfill this requirement. The analysis time predicted by Eq. 10 is 48 minutes; the actual analysis time was 42 minutes. The chromatogram resulting from this analysis is shown in Figure 4. The difference in the calculated and experimental analysis time is most likely due to the variable changes previously discussed.

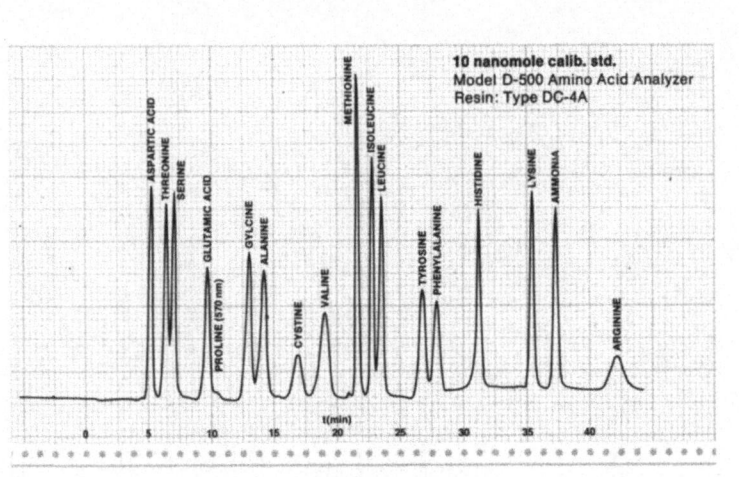

FIGURE 4

This chromatogram illustrates a high-speed amino acid analysis on a single column. The analysis was performed on a Durrum Model D-500 Amino Acid Analyzer at pressures of about 2800 lbs in^{-2}. Total analysis time is 42 minutes. Full scale absorbance is 2.0 O.D. The column length is the same as that shown in Figure 3. The reduced analysis time is due to the higher linear flow velocity of the buffer solutions.

DISCUSSION

It appears that Eqs. 9 and 10 can be used successfully to predict the changes in parameters required to reduce the time of an amino acid analysis without sacrificing resolution or sensitivity. These equations are an extension of the work of Hamilton et al. (5) and contain a new variable, the "packing density" ϕ. The packing density is simply the mass of resin in the column divided by the volume of the empty column. This quantity must be considered since the use of various packing techniques or resins of different mean particle diameter will yield columns with different masses of resin per unit volume of column. For a given column, a change in the resin mass will effect a change in the elution time of the amino acids. ϕ is dependent upon the method of column packing employed; consequently, its relationship with mean particle diameter is not obvious and it is usually determined empirically.

It is interesting to speculate about what changes in parameters could result in an analysis time of fifteen or twenty minutes. Whether the column length were reduced, the linear flow velocity of the buffer solution were increased, or whether some combination of both of these were employed, the mean particle diameter of the resin would have to be less than 5 micrometers if the resolution were not to be lost. With particles having this mean diameter, the column pressures would be very high, and problems such as resin bed supports and particulate-free buffer solutions become extremely important. Nevertheless, experiments in this laboratory are currently underway in an attempt to achieve this goal.

ACKNOWLEDGMENTS

The author wishes to thank Mr. John Reiland for reading the manuscript and for making helpful suggestions regarding the presentation, and Mr. Mike Stickney for his comments on the theory.

FOOTNOTES

1. Spackman, D.H., Stein, W.H. and Moore, S., Anal. Chem., 30, 1190 (1958).
2. Durrum DC-3 Resin, Durrum Chemical Corporation, Palo Alto, California.
3. Pico-Buffer® System, Durrum Chemical Corporation, Palo Alto, California. (reg. U.S. Patent Office, Washington, D.C.)

4. Details of experiments leading to some of the conclusions
    of this section, as well as the mechanism of action of
    the Pico-Buffer System will be discuss in a subsequent
    report, in preparation.
5. Hamilton, Paul B., Bogue, Donald C., and Anderson,
    Roberta A., Anal. Chem., 32, 1782 (1960).
6. All resins used were obtained from Durrum Chemical Cor-
    poration, Palo Alto, California.
7. Benson, J.R., "High-Speed, High-Sensitivity Single-Column
    Analysis of Amino Acids, AMERICAN LABORATORY, p. 51,
    October 1972.

FACTORS AFFECTING THE REPRODUCIBILITY
IN PYROLYSIS GAS CHROMATOGRAPHY (PGC)*

Clarence J. Wolf and Ram L. Levy

McDonnell Douglas Research Laboratories

McDonnell Douglas Corp., St. Louis, Mo. 63166

INTRODUCTION

The achievement of interlaboratory reproducibility in PGC, similar to that of IR, MS and NMR, will open new possibilities for analysis and greatly expand PGC's utility. Normalized pyrograms obtained under standard conditions could be compiled and used as reference spectra. The accumulation of compiled reference data will permit the deduction of modes of thermal fragmentation which, in turn, could be used for interpretation of new pyrograms not listed in the compiled data. To bring about the realization of this potential, the PGC subgroup of the British GC Discussion Group has undertaken a detailed study into the problems of interlaboratory reproducibility in PGC (1). It should be emphasized that excellent qualitative and quantitative intralaboratory reproducibility has been achieved by many workers (2) and therefore should not be confused with problems of interlaboratory reproducibility.

In PGC we should clearly differentiate between inter- and intralaboratory reproducibility. Nevertheless, the question of interlaboratory reproducibility remains unresolved and requires our attention. This paper reviews the general factors that affect PGC reproducibility with emphasis on similarities and differences in the actual pyrolysis conditions that exist in different pyrolyzers.

CLASSIFICATION AND DESCRIPTION OF PYROLYZERS

All the pyrolysis units described in the literature can be classified in three groups according to their mode of operation (3).

This research was conducted under the McDonnell Douglas Independent Research and Development Program.

237

1. Pulse mode units.
2. Continuous mode units.
3. Units which can be operated using both the pulse and the continuous modes.

The multiplicity of pyrolysis units complicates the problem of interlaboratory reproducibility primarily because of differences in the pyrolysis conditions existing in pulse and continuous mode units.

The pulse mode pyrolyzers are the most widely used devices in PGC (2). There are two principal types of pyrolyzers which belong to this group: the filament type and the Curie point type. In both types the sample is applied to a metal surface (as a thin film) and then the metal (ferromagnetic wire or Pt filament) is rapidly heated to a pyrolysis (450-750°C) temperature and kept at this temperature for a number of seconds; the energy source is then turned off allowing the metal to cool rapidly to subpyrolysis temperatures. Obviously the thermal energy is applied to the sample as a "pulse".

A basic characteristic of the filament and Curie point pyrolyzers is the fact that the sample is in direct contact with the heat source and that the primary pyrolysis products rapidly expand away from the heat source into the much cooler regions. The temperature profile in a typical pulse mode pyrolyzer is illustrated in Fig. 1. As soon as the primary products expand into the quench region, illustrated by the shaded area in Fig. 1, the probability of further reaction is greatly reduced. Secondary reaction occur only at the high temperatures in close proximity to the wire (3).

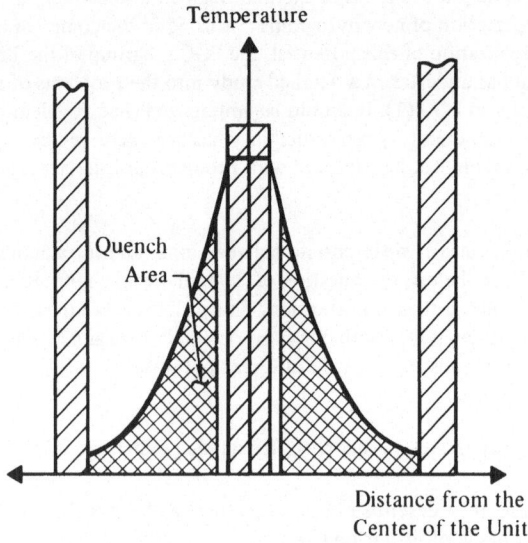

Figure 1   Momentary Temperature Gradient (Profile) in the Pulse Mode Pyrolyzer

Continuous mode pyrolyzers, or the so called furnace type, exhibit an entirely different set of conditions. With the most common variety of furnace microreactors, the sample is placed in a small reactor, usually a boat positioned in a cold zone of the furnace. Pyrolysis is performed by pushing the sample container into the continuously-heated hot zone of the pyrolyzer. The momentary temperature profile of the sample and its surroundings occurring under such conditions is schematically shown in Fig. 2. The temperature of the externally-heated walls of the microfurnace is higher than the temperature of the sample and heat transfer is relatively slow. The primary pyrolysis products released from the surface of the sample expand into the hotter zone of the microfurnace, thus increasing the probability for secondary reactions and secondary pyrolysis.

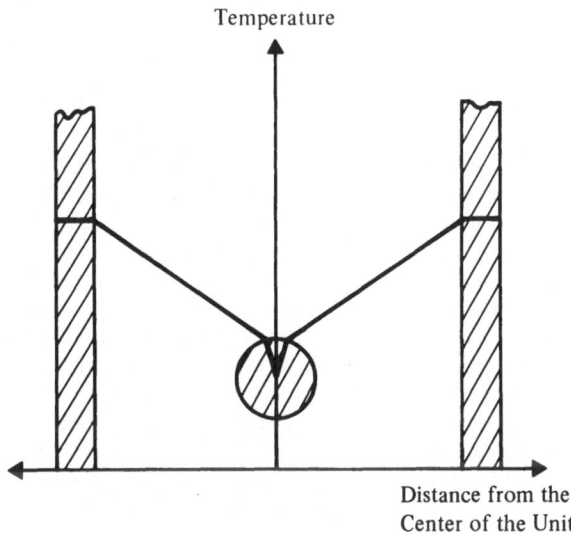

Temperature

Distance from the
Center of the Unit

**Figure 2   Momentary Temperature Gradient (Profile) in the
Continuous Mode Pyrolyzer**

Filament Pyrolyzers

Filament type pyrolyzers are the most widely used devices in PGC (2). The early models of filament type pryolyzers were extremely simple and easy to construct, which encouraged many workers to construct their own instruments. As a result, a wide variety of such pyrolyzers has been described in the literature (2). Recognizing the need to produce a well-defined pulse of thermal energy, Levy (3) formulated the "ideal" characteristics of a filament pyrolyzer. These characteristics include fast temperature rise time (TRT) to a well-defined equilibrium temperature T_{eq}. By securing fast TRT to relatively low pyrolysis temperatures, i.e., 450-650°C, one minimizes exposure of the sample to intermediate temperatures and makes the observed distribution of pyrolysis products representative of the final equilibrium temperature which is well defined and can be reproduced in different laboratories.

The second-generation filament pyrolyzers approach these characteristics. The so-called "capacitor discharge" filament pyrolyzer described by Levy (3) and Levy and Wolf (4), achieves a TRT of 10 msec and allows direct measurement of the final equilibrium temperature by a microthermocouple spot-welded onto the filament (5).

The filament type pyrolyzers, however, are best suited for work with soluble samples, and serious difficulties arise with attempts to apply nonsoluble samples as thin films. The commercial versions of the filament pyrolyzers, despite some improvements in the electronics, still lack the sophistication required for advanced PGC work and a high level of interlaboratory reproducibility.

Curie Point Pyrolyzers

Curie point pyrolyzers introduced by Simon and Giacobbo (6) are based on high-frequency induction heating of a ferromagnetic wire. The sample is coated on the surface of the wire as a thin film. The wire is positioned in the center of the induction coil; the temperature of the wire rapidly rises to the characteristic Curie point of the ferromagnetic material when rf power is supplied. The Curie point temperature of a ferromagnetic wire depends only on the composition of the alloy; therefore, different laboratories using wires of the same composition can easily reproduce the final pyrolysis temperature. This led many workers to believe that the Curie point devices would resolve the problem of interlaboratory reproducibility. Unfortunately, these expectations did not materialize because most commercial versions of the Curie point system did not meet the specifications of the original device. Both the frequency and power output of the rf power supply greatly affect the temperature rise time and thus the reproducibility of Curie point systems. This important point is often overlooked.

Microreactor Pyrolyzers

Microreactor or microfurnace pyrolyzers consist of tubes (quartz) which are connected to the gas chromatograph and which have provisions for external heating by small furnaces and for introduction of samples (loaded in microboats) into the hot zone of the tube. The temperature of the tube walls at the center of the hot zone is measured, and this temperature is frequently referred to as the "pyrolysis temperature". This, of course, is an error because the temperature of the microreactor walls and the true pyrolysis temperature of the sample are usually quite different.

The microreactors described in the literature differ mainly in dimensions, construction material, and the method of sample introduction. An early microreactor or "microboat" pyrolyzer which contains the essential features of all microreactors was described by Ettre and Varadi (7). The main and perhaps the only advantage of microreactor pyrolyzers is convenience in sampling which is important particularly when working with nonsoluble materials. The microreactor pyrolyzers are adequate for "fingerprint" type work. However, when a correlation between the structure of the original sample and the pyrogram is desired, microreactors are not suitable because the pyrolysis conditions promote secondary reactions which tend to mask correlations.

The ability to achieve interlaboratory reproducibility with microfurnace pyrolyzers obviously depends on very careful duplication of all the variables of the pyrolysis conditions including:
a) Reactor temperature
b) Heat transfer characteristics
c) Microboat dimensions
d) Sample size and form
e) Flow rate of the carrier gas
f) Temperature profile along the reactor.

It is unlikely that a high level of interlaboratory reproducibility can be achieved with microfurnace pyrolyzers due to the many variables that must be carefully controlled and reproduced.

Laser PGC

The use of pulsed or continuous laser energy to induce thermal fragmentation of solids introduces new possibilities in PGC analysis. A major problem in the use of lasers results from the fact that the output radiation from most lasers (with the exception of the CO_2 gas laser) is not readily efficiently abosrbed by organic materials. Consequently, it is necessary to add a radiation absorbing substance, such as powdered graphite, to the sample (8). The addition of graphite markedly affects intralaboratory reproducibility (9). To circumvent this problem, we developed a laser PGC system employing a blue glass sample holder which absorbs the radiant energy. (10) The interaction of the laser induced plume with the samples produces a characteristic fragmentation.

We conducted experiments to find the level of intralaboratory repeatability of laser PGC. The results of seven consecutive laser pyrograms are summarized in Table I. It is quite obvious from the results that the repeatability was less than desired. Unfortunately, there are insufficient data in the literature to determine the degree of intralaboratory reproducibility of laser PGC.

Table I   Repeatability for the Laser Pyrolysis of a Film of Polyethylene (0.5 mm) on a Blue Cobalt Glass Holder. *

Peak No.	Pyrogram Number					Av	2σ
	1	2	3	4	5	Av	2σ
1	100.0	100.0	100.0	100.0	100.0	–	–
2	25.7	28.3	26.2	30.4	31.9	28.5	2.4
3	8.0	8.6	8.3	10.3	10.8	9.2	1.1
4	2.9	2.9	3.0	3.0	2.9	2.9	0.6
5	3.0	2.7	2.7	1.3	3.2	2.6	0.6
6	1.6	0.8	1.0	2.5	0.5	1.3	0.7
7	1.2	1.8	1.4	1.9	0.5	1.3	0.4
Total Mass, μg	10.6	9.2	9.6	8.1	5.0	8.5	–

* The energy of the laser pulse was 0.76 joule and the products were separated on a 15-m by 0.3-cm diameter column packed with 20% DC200 on 60/80 mesh Chromasorb "P". The pattern is reported in terms of a distribution normalized with respect to the largest product.

Vapor Phase

Vapor phase PGC was developed as an aid for identifying GC peaks (11) and has exhibited excellent reproducibility. The pyrolysis techniques of vapor phase PGC, however, bare no similarity to the techniques utilized in pyrolysis of solids and represent a separate branch of PGC which developed independently. Therefore, a description and discussion of this method is beyond the scope of this article.

CONDITIONS AND FACTORS AFFECTING REPRODUCIBILITY

Equilibrium Temperature of the Heat Source (T_{eq})

Temperature has a pronounced effect on the pyrolysis process, and in many instances the product distribution varies significantly with temperature. For example, Barlow et al. (12) reported a different mechanism for the degradation of polymethylmethacrylate in different temperature regions. At $300^{o}C$ the depropagating chains are end-initiated and termination is primarily bimolecular. Above $500^{o}C$, initiation occurs at random, and the majority of the chains are terminated as the depropagation reaches the end of the molecule and the terminal radical distills out of the system. Thus, at $500^{o}C$ the primary product is monomer while at temperature greater than $1000^{o}C$, monomer is a minor product.

Temperature can influence the mechanism by either 1) affecting the initiation step or 2) controlling the rate of the subsequent reactions. The initiation reactions involved in the pyrolysis of most organic materials involve the generation of free radials by the cleavage of a simple bond (i.e., as C-C, C-H, OH, etc.) or the unimolecular elimination of a simple molecule such as H_2O, CO, etc. Both of these processes are highly temperature dependent. The subsequent reactions of these reactive species via diffusion, abstraction or combination reactions are also markedly sensitive to temperature.

It is well known to most workers in PGC that the final equilibrium temperature (T_{eq}) of the heat source is extremely important in determining the final product distribution, and most workers attempt to control this particular parameter carefully. However, what is not so well known and what may be equally important is that the heating rate, i.e., the time to reach the final equilibrium temperature, may also affect the product distribution.

Temperature Rise Time

Farre-Rius and Guichon (13) estimated that in most of the pyrolyzers used in PGC studies the actual pyrolysis temperature of the sample remains unknown and depends on the sample heating rate. Thus, the pyrolysis temperature cannot be accurately reproduced in different laboratories. They assumed a first order decomposition and with the aid of known thermodynamic data estimated the half-time for decomposition of polystyrene, polytetrofluoroethylene and polyethyleneglycoladipate at several different temperatures. Their calculations are summarized in Table II. Note that at $400^{o}C$, 97% of a polystyrene sample will decompose in less than 0.1 sec and even a relatively stable material such as Teflon (polytetrafluoroethylene) will be 97% decomposed in 0.13 sec at $600^{o}C$.

Table II   Half-Time for the Decomposition of Polymers as a Function
             of Temperature*

Temperature ($^{\circ}$C)	Polystyrene (sec)	Polytetrafluoro-ethylene (sec)	Polyethyleneglycol-adipate (sec)
300	18.0	2.8×10^8	4.5×10^3
350	0.42	1.3×10^6	180.0
400	0.015	1.3×10^4	11.0
450	0.95×10^{-3}	2.5×10^2	1.0
500	0.80×10^{-4}	8.0	0.14
550	0.90×10^{-5}	0.38	0.022
600	0.13×10^{-5}	0.026	0.0045

*Data taken from Reference 13

In addition, Farre-Rius and Guiochon (13) calculated the characteristic pyrolysis
temperature of polystyrene and polytetrafluoroethylene for different heating rates.
The characteristic temperature, T_s, is defined as that temperature at which a fraction
$1/e(37\%)$ of the sample remains unpyrolyzed. The temperature at which pyrolysis is
completed is defined as $T_s + \theta$. The characteristic temperature, T_s, and $T_s + \theta$ for
polystyrene as a function of heating rate, i.e., temperature rise time (TRT), are
shown in Fig. 3. This curve indicates that the characteristic pyrolysis temperature
of polystyrene varies by 300°C when the rise time varies from 5×10^{-4} to 10^3 sec.
It is important to note that the TRT's of many pyrolyzers encountered in normal
analytical PGC range between 15 and 5×10^{-3} sec. According to the data shown
in Fig. 3, within this range of TRT's, pyrolysis temperatures of polystyrene would
range between 350 and 570°C. Thus, the pyrolysis is complete at a temperature
well below the final equilibrium temperature (T_{eq}) of the heat source.

The direct measurement of pyrolysis temperature (or range of temperatures) is
highly desirable for the achievement of interlaboratory reproducibility. Levy, Fanter,
and Wolf (5) recently reported on a method to measure directly the "true pyrolysis
temperature". This method which is based on a determination of the temperature-
time profile of the heat source during pyrolysis is applicable to most systems utilizing a pulse mode pyrolyzer. The oscilloscopic tracings of the temperature-time
curves of a filament with and without sample are shown in Fig. 4. The appearance
of the plateau below T_{eq} in Fig. 4b indicates the temperature region where pyrolysis
actually occurs. The length of this plateau is indicative of the sample life-time on
the filament. This observation permits the direct measurement of the true pyrolysis
temperature.

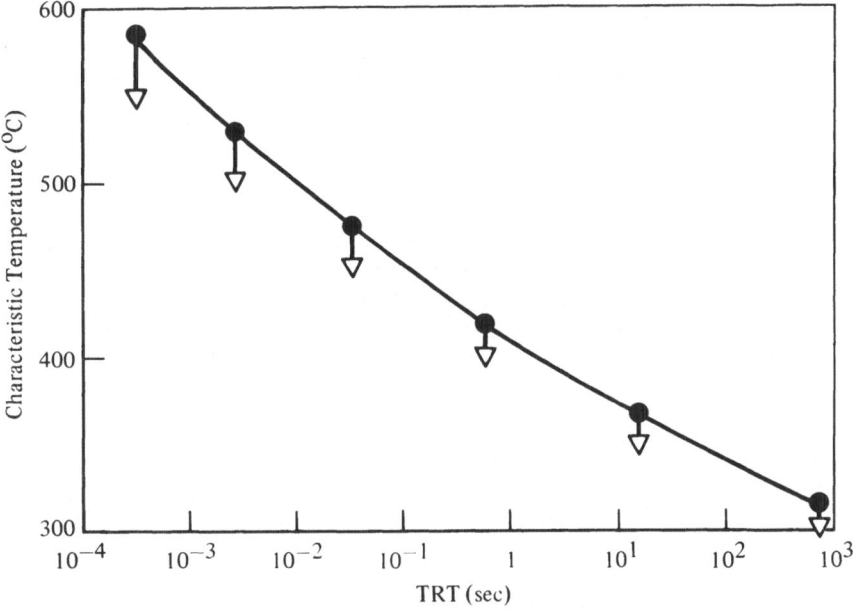

Figure 3   Characteristic Pyrolysis Temperature of Polystyrene T_s and $T_s + \Theta$ as a Function of the Temperature Rise Time (TRT). $\Delta \rightarrow T_s$ $\bullet \rightarrow T_s + \Theta$

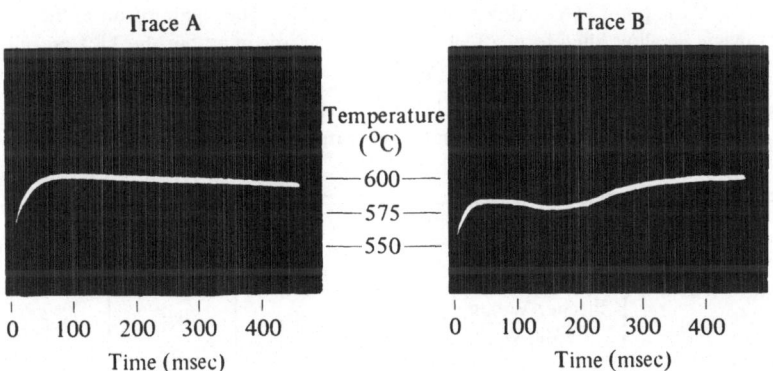

Figure 4   Expanded Scale Tracings of Temperature-Time Profiles of a Filament Obtained with and without Sample Showing the Time to Reach the Equilibrium Temperature, the True Pyrolysis Temperature, and the Decomposition Time on the Filament. Trace A was Obtained without Sample and Trace B was Observed with 150 μg of Acrylic Acid - Acrylonitrile Copolymer on the Filament.

In their calculations of T_s, Farre-Rius and Guiochon (13) tacitly assumed that the power consumed by the pyrolyzer-sample was not a limiting factor. In an actual system, however, pyrolysis of the sample consumes a considerable fraction of the power supplied to the heat source. When this occurs, the rate of temperature rise decreases resulting in a change in the temperature-time profile. In cases where the power needed to sustain the pyrolysis process equals the power applied, a plateau will be observed in the temperature-time curve. Therefore, the true pyrolysis temperature (TPT) is defined as that temperature at which the rate of energy (power) consumed by the sample is equal to the net power supplied to the system. The true pyrolysis temperature as defined here is related to the characteristic temperature, T_s, which was derived to estimate the actual pyrolysis temperature range as a function of heating rate (3). It is important to note that even with rapid rise time pyrolyzers, the pyrolysis process occurs at a fairly well-defined temperature which is, however, below the final equilibrium temperature of the heat source.

Measurements of the TRT of various ferromagnetic conductors carried out with two commercial Curie point pyrolyzers of different rf power output are summarized in Table III. The TRT's obtained with the 30 W Phillips unit are almost an order of magnitude slower than the TRT's obtained with the 1500 W Fischer unit.

The 600 and 610°C wires have very similar Curie points, but the 610°C wire exhibits a TRT almost twice that of the 600°C wire. The difference in TRT is a direct result of the composition of the wire. Correlation between the alloy composition and the corresponding TRT's indicates that alloys of higher nickel content show increased TRT's while higher cobalt content produces faster TRT's. Also, it should be noted that the 400, 600, and 700°C wires, which were specially fabricated for Curie point PGC, exhibit relatively fast TRT's. The composition of the ferromagnetic alloy, therefore, affects not only the Curie point but the TRT as well.

Table III   Temperature Rise Times (TRT) of Ferromagnetic Wires Obtained with rf Generators of Different Power Outputs. Composition of the Wires is Also Indicated

| Specified Curie Point of Wire, °C | TRT (msec) | | Ferromagnetic Alloy Composition | | |
| | Pyrolyzer | | | | |
	Fischer-Varian 1500 W	Phillips 30 W	Fe	Ni	Co
358	300	1300	0	100	0
400	40	500	61.7	0	38.3
510	150	700	50.6	49.4	0
600	70	500	42	41	16
610	130	1150	29.2	70.8	0
700	90	1350	33	33	33
770	110	2100	100	0	0

A number of other factors also affect the TRT of Curie point pyrolyzers (5). Measurement of the TRT at different points on ferromagnetic wires revealed that the TRT varies from point to point along the wire. For example, the top, center, and bottom thermocouples attached 5 mm apart to a 400°C wire exhibited TRT's of 70, 40, and 250 msec, respectively, with the 1500 W Fischer unit. A similar variation was noted with the 30 W Phillips unit. These measurements indicate that sample position with reference to the rf coil is extremely important. The sample must be deposited on a narrow region of the wire and positioned in the center (both axially and longitudinally) of the rf coil. Coiled ferromagnetic wires sometimes are used for increased sample capacity and convenience in application. However, the TRT's of these wires are approximately 10 times longer than those of the corresponding straight wires.

Sampling

Another major factor affecting PGC reproducibility is related to the method used to apply the sample to the heat source. When dealing with microgram quantities, great care must be used in selecting the sample, and the experimentor must choose a sample which is truly representative of the material under investigation.

Materials can be conveniently divided into two classes: those which are soluble in a volatile solvent such as water, benzene, acetone, etc., and those which are not. The soluble samples can be readily coated onto a wire (either a filament or Curie point wire), dried and pyrolyzed in a pulse mode instrument. Nonsoluble materials are usually placed in a microreactor and pyrolyzed in a continuous mode system. In some instances it is possible to grind a insoluble sample into a fine powder and apply it as a thin film on a ribbon filament. Unfortunately many materials such as rubbers cannot be ground into fine powder even when cooled to liquid nitrogen temperatures.

It is important to note that significant differences may occur in the pyrograms of the same material when pyrolyzed in pulsed or continuous mode systems. In fact, the use of the same method in different laboratories does not guarantee reproducibility. The influence of secondary reactions on the product distribution is extremely important and depends to a large extent upon the manner in which the sample is applied to the heat source. Barlow, Lehrle, and Robb (14) studied the rate of degradation of polymethylmethcralate with a filament pyrolyzer and found that rate of degradation was dependent on film thickness for samples thicker than 250 Å.

In thick or bulky samples three effects occur which are minimized with thin samples: 1) the products must diffuse through the hot interior prior to volatilization at the surface, 2) a large temperature gradient exists between the interior and exterior of the sample so that reaction temperatures cannot be specified, and 3) an unknown time interval is required to raise the entire sample to the desired temperature.

Therefore, it is imperative to use as thin a sample as is practical, and in addition, the approximate thickness of the sample should be reported along with other experimental parameters.

The thickness of the sample can also affect the true pyrolysis temperature. In a recent set of experiments, films of polyethylene weighing between 15 and 120 μg were deposited on a rapid rise time filament pyrolyzer. The T_{eq} of the filament was approximately 750°C and temperature-time profiles of the filament with and without sample were recorded. These results are summarized in Fig. 5. The empty filament reached 650°C in less than 20 μsec and then increased to 750°C in another 150 μsec. When the filament contained more than 15 μg, the pyrolysis temperature decreased in proportion to the weight of the sample. For example, with 120 μg of polyethylene the decomposition began at 450°C and the temperature slowly increased during the course of the experiment. Therefore, the true pyrolysis temperature varied between 450° and 750°C for samples ranging from 120 to 15 μg. It is immediately obvious that small samples are required to ensure rapid heating to a well-defined temperature.

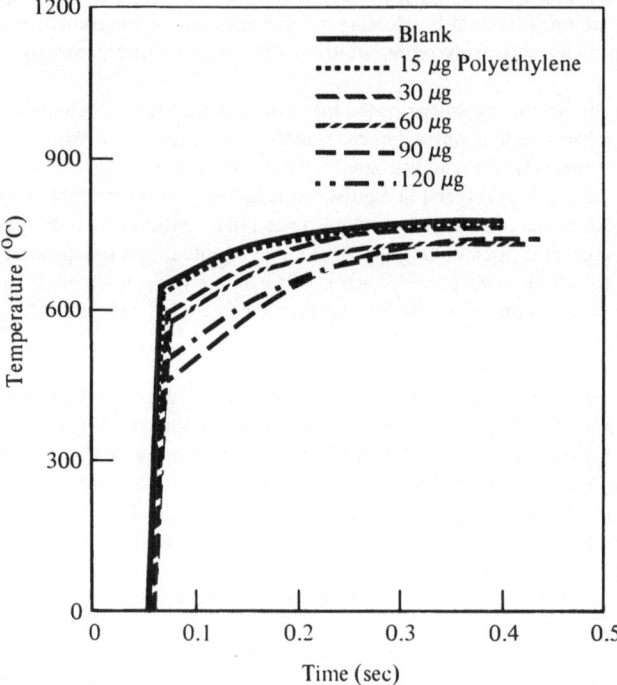

Figure 5   Temperature Time Profile of Filament Pyrolyzer as a Function of the Amount of Polyethylene Deposited on the Filament

Gas Chromatography

The separation of the products following pyrolysis cannot be overlooked. Reproducible PGC requires good gas chromatography. Problems of obvious concern include the stationary phase used for the separation, the carrier gas flow rate through the column, the temperature or temperature programming mode and the detection system. One important parameter often overlooked is the interface between the pyrolyzer and the GC column. A pyrolyzer can be considered a special inlet system for a chromatograph and all the normal problems associated with inlet systems must be considered. For example, the dead volume in the pyrolyzer must be small, heated transfer lines must be used, and it is necessary that the products be introduced as a sharp peak. This usually is not a major problem with pulse mode pyrolyzers but can introduce considerable band spreading with continuous mode systems unless special valves are used for product injection. Levy, Wolf, and Oro (15) recently described a special flow-through vaporizer-pyrolyzer coupled with subambient GC column operation to permit prolonged flow of carrier gas through the sample during the heating interval. The volatile products are trapped directly on the GC column, and when the source of heat is removed, the GC column is temperature programmed and the individual compounds eluted. Since the movement of products through the column is negligible until the temperature programming begins, a system of this type can be used with high resolution GC.

Conclusions

Pulse mode pyrolyzers appear to have higher potential for achievement of interlaboratory reproducibility. Regardless of the type of pyrolyzer used, we believe the optimum pyrolysis temperatures to be between 400-650°C which is considerably lower than the range of temperatures frequently reported in the literature. The TRT should be at least one order of magnitude faster than the half decomposition time of the sample at the chosen pyrolysis temperature. The sample should be applied as a thin film on the heat source. Special attention must be given to the interface between the pyrolyzer and the gas chromatograph.

REFERENCES

1. N.B. Coupe, C.E.R. Jones, and S.G. Perry, J. Chromatog. 47, 291 (1970).

2. R.L. Levy, Chromatog. Rev. 8, 48 (1966).

3. R.L. Levy, J. Gas Chromatog. 5, 107 (1967).

4. R.L. Levy and C.J. Wolf, 158th American Chemical Society Meeting, New York, N.Y., Sept.1969.

5. R.L. Levy, D.L. Fanter, and C.J. Wolf, Anal. Chem. 44, 38 (1972).

6. W. Simon and H. Giacobbo, Chem.-Ing.-Tech. 37, 709 (1965).

7. K. Ettre and P.F. Varadi, Anal. Chem. 35, 69 (1963).

8. O.F. Folmer and L.V. Azarraga, J. Chromatog. Sci. 7, 663 (1969).

9. W.T. Restau and N.E. Vanderborgh, Anal. Chem. 43, 702 (1971).

10. D.L. Fanter, R.L. Levy, and C.J. Wolf, Anal. Chem. 44, 43 (1972).

11. D.L. Fanter, J.Q. Walker, and C.J. Wolf, Anal. Chem. 40, 2168 (1968).

12. A. Barlow, R.S. Lehrle, J.C. Robb, and D. Sunderland, Polymer 8, 537 (1967).

13. F. Farre-Rius and G. Guiochon, Anal. Chem. 40, 998 (1968).

14. A Barlow, R.S. Lehrle, and J.C. Robb, Makromol. Chem. 54, 230 (1962).

15. R.L. Levy, C.J. Wolf, and J. Oro, J. Chromatog. Sci. 8, 524 (1970).

Laser Induced Pyrolysis Gas Chromatography

Nicholas E. Vanderborgh

Department of Chemistry
University of New Mexico
Albuquerque, New Mexico 87106

Although pulsed lasers have been commercially available
for only one decade, they have been utilized in chemical
analysis as highly monochromatic light sources, as intense,
short duration sources for photolysis, and during the last
few years, as thermal sources for degradation processes.
Pulsed laser energy offers unique properties for the
initiation of thermal degradation. Especially sensible is
the combination of laser induced thermal degradation with
gas chromatography. This paper reviews some applications
and experimental possibilities of this new pyrolysis method.
It should be noted that only a few papers have appeared on
this topic and certainly much still needs to be done to
fully characterize the potentialities of this technique.

Lasers have been described that operate in either
continuous or pulsed modes.[1] Although both types of
devices have utility for this type of study, here only
pulsed sources will be considered. Figure 1 shows
schematically the laser-induced degradation process, i.e.,
the steps occurring when pulsed laser energy impinges upon
a sample. Initially the sample is illuminated with pulsed
radiation (A). This radiation can be deposited into the
sample by a variety of energy absorbing processes. Many
of these are not the usual spectroscopic absorption
mechanisms. Thus, many compounds normally "transparent"
at 694.3 nm, the spectral output from a ruby laser, do
absorb ruby laser radiation, *vide infra*. Part of the
sample is then pumped into a plasma, the laser-induced

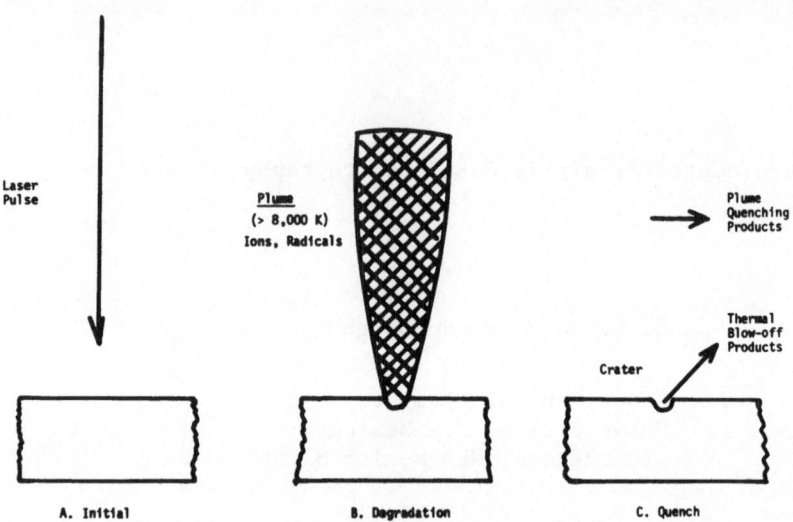

Figure 1:   Schematic Representation of Laser Pyrolysis

plume (B). A number of workers have attempted to estimate
temperatures of the plume; there is general agreement that
temperatures are reached in excess of 8,000 K and that the
plume consists of ionized atomic fragments and, perhaps, a
few radicals of unusual stability.[2] Upon termination of
the pulse, the plume quenches into an ensemble of products.
This is one mechanism, plume quenching, for product formation. The second results from thermal blowoff from the
material near the site of the laser zap. The sample, here,
undergoes less vigorous pyrolysis. One must conclude that
laser pyrolysis is the most vigorous of all the currently
used gas chromatographic pyrolysis techniques.

It is of interest to consider the time scale of this
process. Figure 2 shows this schematically. Normal pulsed
laser energy has a pulse width of some 500 microseconds;
the plume forms late in the pulse lifetime and rapidly
quenches with the termination of the pulse. The pulse
lifetime width is of the order of one millisecond. (These
time considerations refer to a normal burst lasing device;
Q-switching and mode locked lasers operate in much lesser
time periods.) Thus, the entire thermal event, at least
for the process that leads to the plume quenching products

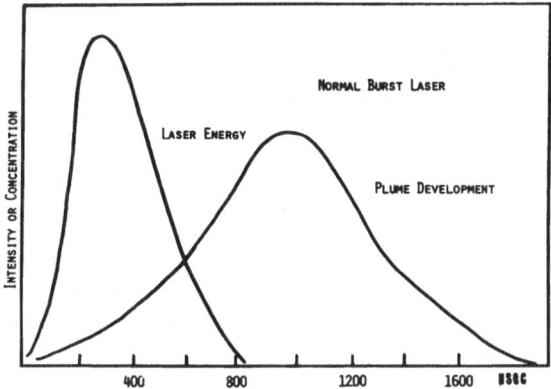

Figure 2:   Laser Plume Development-Time Relationships

is of shorter duration than the thermal rise time of the
fastest previous pyrolysis technique, the Curie point
pyrolyser.[3] It is difficult to estimate the time scale
for the formation of thermal blowoff products; there is
evidence that the rise time is shorter than the
cooling time and that probably this process occurs over a
longer time scale than the one which gives rise to plume
quenching products.[5]

EXPERIMENTAL CONSIDERATIONS

Laser pyrolysis should then result in two types of
products, those which come from thermal quenching of the
plume and those which come from thermal blowoff of the
solid material. Gas chromatography is an obvious technique
to observe these entities. The instrumentation for such
studies is shown in Figure 3. Here the laser, an Xenon
flash lamp and associated electronics along with the ruby
rod (the complete optics of the lasing system are deleted),
is shown in the upper section. From the ruby comes a

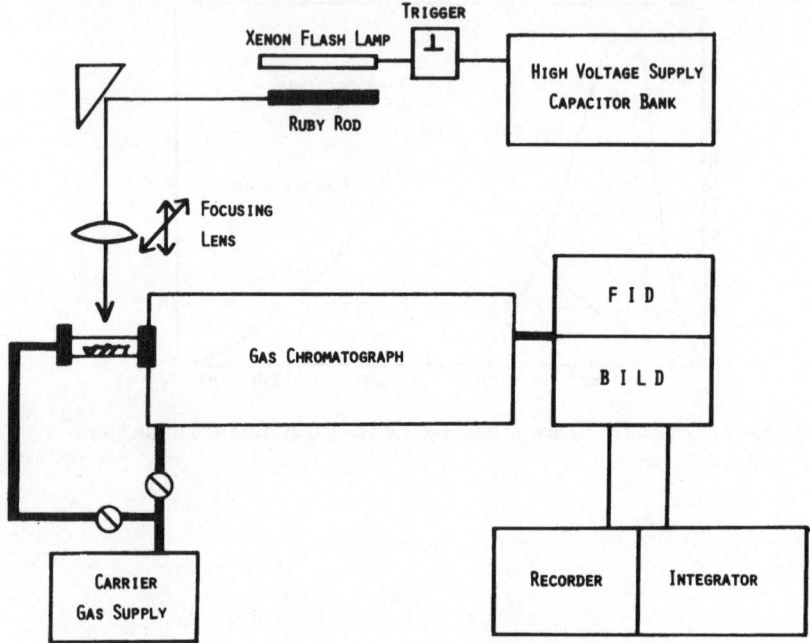

Figure 3:   Schematic of Laser Gas Chromatographic
Pyrolysis Instrumentation

laser pulse at 694.3 nm; this energy impinges upon a turn-
ing mirror and is focused onto the sample contained in a
small diameter quartz tube attached to the inlet port of a
gas chromatograph. During the experimental run, carrier
gas is fed through the inlet port; sample injection is
accomplished by firing the laser.

Commercially available lasers routinely deposit
single burst energies in the range of 0.1 to 5 joules.
This energy is not particularly impressive unless the
time scale for the process is included. Consideration of
this leads to the conclusion that a 5 joule pulse represents a power of 10^4 watts; should this be focused with a
beam diameter of 1 mm, the resulting power density is in
excess of 10^6 watts/cm^2. These rather impressive power
levels are available from "low power" lasing systems and
readily explain why this method results in extremely
vigorous pyrolysis.

Gas chromatographs used for laser pyrolysis do not have special requirements. For the studies presented here, a Perkin Elmer Model 800 equipped with two detectors, a FID and a BILD (a detector that works on the quenching of beta induced luminescence and is sensitive to low molecular weight gases), were utilized. Data is obtained in both digital and analog form.

Several workers have described pyrolysis assemblies.[4-7] Each consists of a chamber which permits firm positioning of the sample and allows transmission of the laser pulse through quartz windows. (Glass windows do not withstand the thermal shock.) Currently pyrolysis chambers in our laboratories are made from a short section (3-4 cm) of 5 mm OD quartz tubing. The intense thermal event does not couple with the flowing carrier gas; pyrolysis products are un-affected by the nature of the carrier: He, H_2 and N_2 all yield similar results.

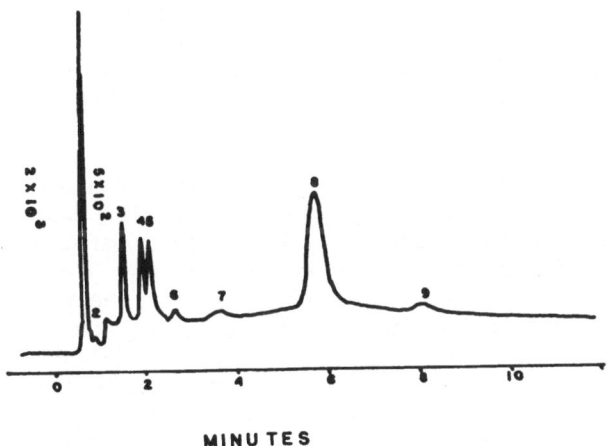

MINUTES

Figure 4:   Laser Pyrogram of Solid Polystyrene

Peak Identification: 1) unresolved methane, ethane, ethylene and acetylene; 2) methylacetylene; 3) butadiyne; 4) benzene; 5) C_5 or C_6 olefin; 6) toluene; 7) ethylbenzene; 8) styrene; 9) methyl-styrene (made by comparison of retention volumes with known compounds).

Figure 4 shows an actual recorder trace (FID) obtained during a laser pyrolysis of polystyrene. The first intense peak, attenuated by a factor of four, consists of unresolved low molecular weight gases, mainly C_1 and C_2 hydrocarbons. The rest of the chromatogram is identified on the figure; the predominate peak, 8, results from styrene. (Peak identification was done by comparison of retention volumes.) The two types of product are obvious, plume quenching products (low molecular weight gases) and thermal blowoff products (higher molecular weight fragments). Another degradation product is not shown here. Each shot results in a black deposit on the inside walls of the pyrolysis chamber which has the chemical properties of molecular carbon.

It is interesting to determine the affects upon the product distribution that result when the intensity of the laser is varied. Laser energies are quite easily varied by changing the total energy stored in the capacitor bank powering the Xenon flash lamp; however, varying this energy affects the pulse width of the laser pulse. For these experiments to be meaningful, both the pulse width and focus of the beam must be held constant while only the energy deposited in the sample should vary. This can be accomplished by attenuation of the beam after the exit optics of the laser.

We accomplished this by inserting a sandwich of microscope slides in the path of the laser light. It was possible to attenuate the laser power by a factor of five as the number of slides in the beam was increased. Results are shown in Figure 5; here the percentage of the various molecular products are shown as a function of input power. The upper trace (F) shows the styrene (thermal blowoff product) percentage of the product ensemble while trace D shows the similar results for acetylene (a plume quenching product). As the input intensity is decreased, moving from left to right, the styrene fraction increases while the acetylene fraction decreases. These data indicate that as the pyrolysis becomes more vigorous plume quenching products occupy a larger fraction of the total degradation ensemble and suggest that the two product forming routes have different energy thresholds. Gruen and coworkers have detected similar behavior when the power density was decreased; as the beam becomes less highly focused, the percentage of low molecular weight gases decreased.[4] These results dictate that considerations of reproducibility

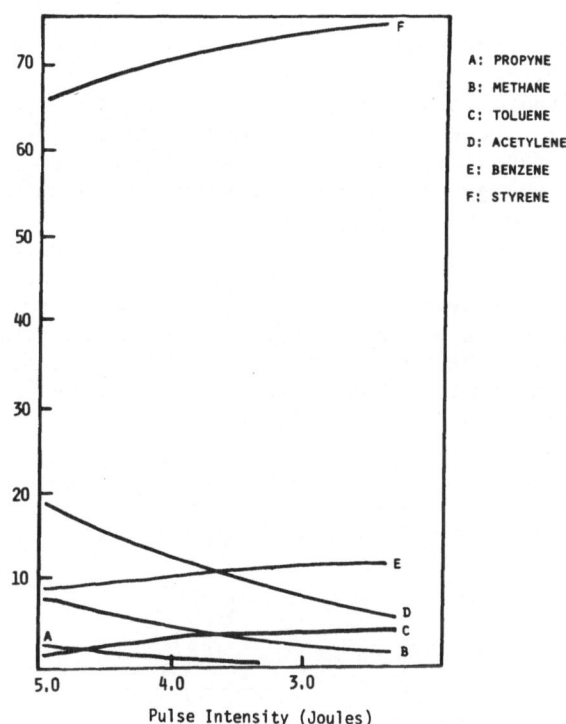

Figure 5: Variation of % of Product Distribution
with changes in input power; results
for polystyrene degradations.

for laser pyrolysis must include both the total power and
the resulting power density, watt/cm^2.

Another experimental difficulty is the problem of
samples which are "transparent" to energy at the laser frequency, i.e., material that exhibits a low absorbtivity to
the laser energy. (As mentioned previously, many samples
which do not "absorb" 694.3 nm radiation couple effectively
with ruby radiation. We need not go into the reasons for
this here.) There are two ways around this problem; either
the sample can be placed upon a sample support that will
accept the energy or absorbing centers can be mixed into
the sample that will accept the energy and then transfer
it to the compound of interest. Gruen and coworkers
report pyrolysis done by placing the compound on a colored
surface.[4] This technique is successful but must spread
the thermal pulse width and also decrease the temperature
reached during the thermal event. Several workers report
the utilization of carbon, graphite or coke as material
that can be mixed into the nonabsorbing sample. Figure 6

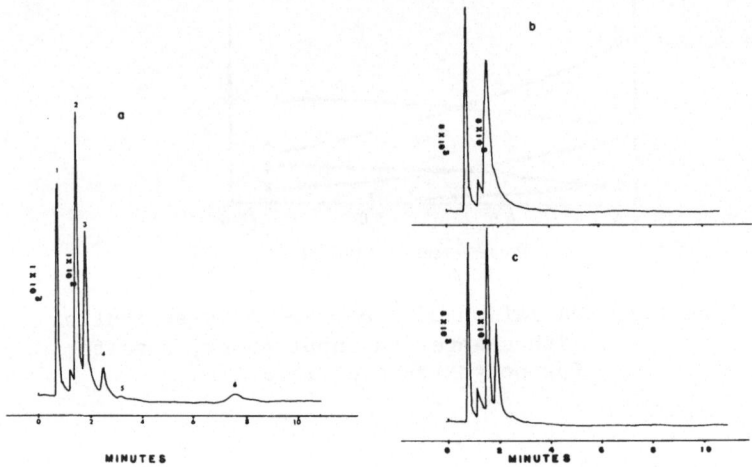

MINUTES

Figure 6:   Comparison of Degradation Pattern of p-Terphenyl
            with Different Additives
Chromatogram run isothermal at 150°C on the carbowax 20M
column: a) no additives; b) 5% ultrapure graphite added;
c) 5% Carbon Black added.   (Peak identification - a:
1) methane, ethane, ethylene, and acetylene; 2) butadiyne;
3) benzene; 4) toluene; 5) xylene or ethyl benzene;
6) methyl styrene.)   Analytical Chemistry, 43, 702 (1971).

shows the affects of carbon loading. Three different pyrograms of p-terphenyl are depicted. The left-hand trace
illustrates the degradation behavior of this compound with
no additives. (This compound was chosen so that this comparison could be made - it does effectively absorb ruby
radiation.) The upper trace on the right-hand side shows
results obtained with 5% graphite as an additive while the
bottom pyrogram shows results with 5% carbon black. As can
be readily seen, both the relative distribution of products
and the qualitative nature of the products change markedly,
both between the unloaded compound and the carbon loaded
compound and between the two shots with different types of
carbon loading. These results indicate that carbon loading
must be done with care and that these differences must be
expected.

PREDICTION OF PRODUCT ENSEMBLES

Degradation of organic molecules, such as polystyrene
and p-terphenyl has been shown to lead to products which
result by two pathways. It seemed desirable to explore
systems which would give either thermal blowoff products
or plume quenching ones. Consideration of the possibilities here led to a series of studies on inorganic salts,
compounds which should have no volatile thermal blowoff
products and which should yield gaseous products emanating
solely from the quenching route.[4,8]

Figure 7 shows, again, the stepwise pyrolysis event;
here we consider the degradation of silver sulfate. The
sample is heated by the laser pulse, part of the sample is
pumped into a laser induced plume; the stoichiometry of
the plume should parallel that of the original salt. With
the termination of the thermal event, the plume can quench
into several different possible product ensembles which
would be stable at ambient temperatures. We might find
silver oxide with sulfurtrioxide, or any of the other possible distributions listed 1 through 5 in Figure 7. (It
might also be possible to find mixtures of these.)

The most probable product distribution should be that
one which has the overall lowest free energy of formation
at the particular temperature of the chemical reaction.
Should that reaction be rapid, it can be thought to take
place at some elevated temperature, the quenching

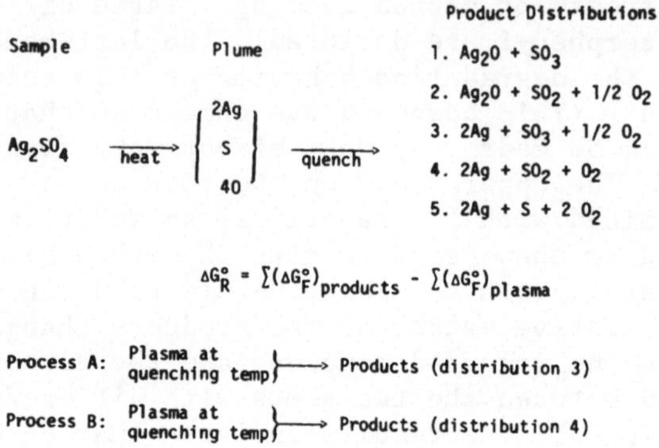

Product Distributions

Sample	Plume	

1. $Ag_2O + SO_3$

2. $Ag_2O + SO_2 + 1/2\ O_2$

Ag_2SO_4 $\xrightarrow{heat}$ $\left\{ \begin{array}{c} 2Ag \\ S \\ 4O \end{array} \right\}$ $\xrightarrow{quench}$

3. $2Ag + SO_3 + 1/2\ O_2$

4. $2Ag + SO_2 + O_2$

5. $2Ag + S + 2\ O_2$

$$\Delta G_R^\circ = \sum (\Delta G_F^\circ)_{products} - \sum (\Delta G_F^\circ)_{plasma}$$

Process A: Plasma at quenching temp $\}$ $\longrightarrow$ Products (distribution 3)

Process B: Plasma at quenching temp $\}$ $\longrightarrow$ Products (distribution 4)

Figure 7: Possible Distributions with Ag_2SO_4 Pyrolysis

temperature; this is not the uppermost temperature reached during the thermal excursion, but that temperature reached during the cooling cycle where stable chemical bonds first form. Then the free energy change is considered. This is given by the sum of the free energy of formation for a particular product ensemble minus the summation of the free energy of the various atomic components in the plume. Since we only seek for the most stable distribution, it is possible to assign the overall free energy of the plume some arbitrary value; this is most conveniently done by the assignment of zero. Thus, we need know only the free energy of formation for the various product ensembles as a function of temperature. For instance, we can compare energy differences (Figure 7) between process A, quenching the plasma to distribution 3 with process B, quenching the plasma to distribution 4. Even this might be difficult were it not for the experimental fact that each shot results in a silver mirror on the inside of the pyrolysis chamber and no trace of silver oxide was determined. Thus, product distributions 1 and 2 can be eliminated from these consider-ations.

Calculated free energy distributions for this sulfate system are shown in Figure 8.[4] The product distribution $(SO_3-1/2O_2)$ is thermodynamically unstable compared with the most favorable distribution (SO_2-O_2) which should predominate up to a temperature of 4,000 K. This (SO_2-O_2) is the distribution determined experimentally. In fact this type of prediction works well for a variety of compounds; if the free energy plot is examined to determine the most stable (lowest $\Delta G/T$) ensemble at 3,500 K, the temperature shown by the dashed line in Figure 8, that should be the ensemble found experimentally.

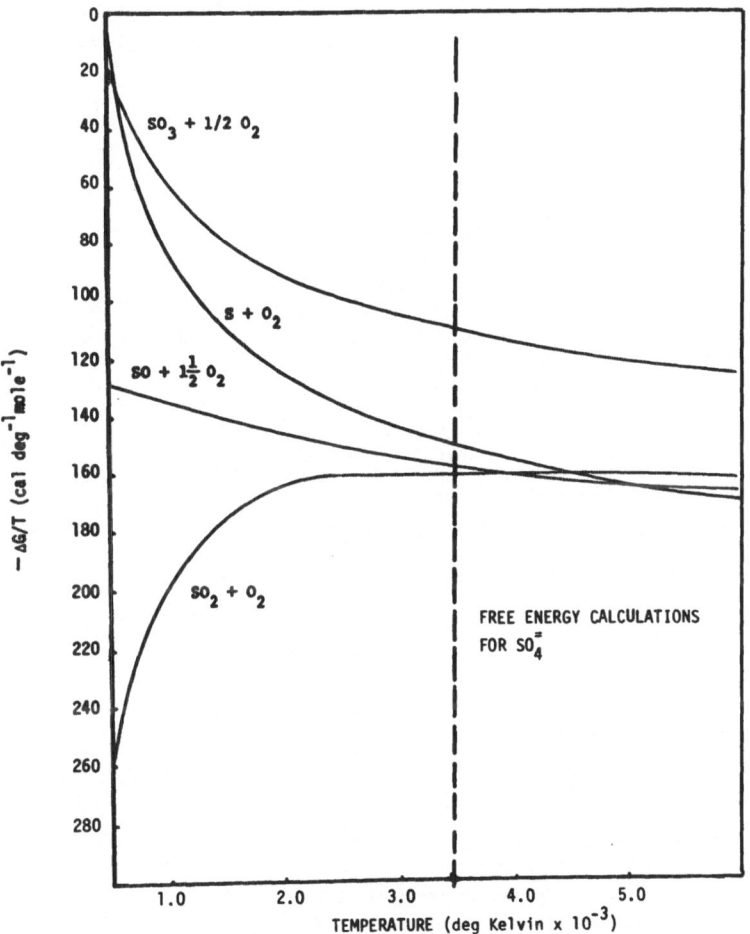

Figure 8: Calculated Free Energies for $SO_4^=$

Table 1 shows data for nitrates, sulfates, thiosulfates and oxalates. Consider the results for silver sulfate; the free energy considerations predict an equimolar mixture of (SO_2-O_2); This is exactly what is found experimentally. Now consider the same salt that has been mixed with carbon. The free energy considerations must now explore the differences in formation energies between sulfur and carbon oxides. Experimental and predicted results agree that the products should predominately include carbon oxides, not sulfur oxides.

The data in Table 1 point to several conclusions. First of all, the metal in the salt does little to affect

Table 1: Comparison of predicted and determined product distributions from thermally quenched plasma

System[a]	Predicted products[b]	T[c]	Observed products[d]
$AgNO_3$	$N_2(0.20)[e], O_2(0.70), NO(0.10)$	4000	$O_2+NO(0.99), NO_2(T), Ag$
$AgNO_3+C$	$N_2(0.20), O_2(0.30), CO_2(0.25),$ $CO(0.25), NO(T)$	3300	$CO+O_2(0.80), CO_2(0.20), Ag$
$AgNO_3+C$	$N_2(0.10), CO+O_2(0.80),$ $CO_2(0.10)$	3500	
$BaSO_4$	$O_2(0.50), SO_2(0.50)$	3500	$O_2(0.50), SO_2(0.50)$
Ag_2SO_4	$O_2(0.50), SO_2(0.50)$	4000	$O_2(0.50), SO_2(0.50), Ag$
Hg_2SO_4	$O_2(0.66), SO_2(0.33), S$	4500	$O_2(0.50), SO_2(0.50)$
Ag_2SO_4+C	$CO+O_2(0.50), CO_2(0.25),$ $SO_2(0.25)$	3800	$CO+O_2(0.85), CO_2(0.15),$ $SO_2(T), Ag, S$
	$CO+O_2(0.90), CO_2(0.05),$ $SO_2(0.05)$	4100	
BaS_2O_3	$O_2(0.05), SO_2(0.95)$	4000	$O_2(0.05), SO_2(0.95), S$
	$O_2(0.15), SO_2(0.85)$	4400	
BaS_2O_3+C	(see SO_4+C, above)		$CO+O_2(0.90), CO_2(0.10), S$
$Na_2C_2O_4$	$CO+O_2(0.50), CO_2(0.50)$	3500	$CO+O_2(0.92), CO_2(0.08)$
	$CO+O_2(0.80), CO_2(0.20)$	3700	
	$CO+O_2(0.99)$	4000	
$Na_2C_2O_4+C$			$CO+O_2(0.98), CO_2(0.02)$

(a) Carbon-loaded systems, 5% C, by weight.
(b) Numbers in parentheses show mole fractions with estimated accuracy of ±10%.
(c) Temperature (deg K) of predicted distribution.
(d) Numbers in parenthesis show mole fractions with relative standard deviation of 2.0%.
(e) Nitrogen cannot be detected with the experimental apparatus used.

the product distribution: barium, silver and mercury(I) sulfate all result in the same mixture (SO_2-O_2). Secondly, they point to the reason why carbon loading makes such a marked effect upon the product ensemble. The carbon does change the absorption mechanism, certainly, but its major contribution is to permit different quenching processes, processes that then lead to quite different ensembles of product. This type of study, inorganic compounds, certainly amplifies the extent of these loading parameters. However, carbon loading of organic systems leads to the same general phenomenon (*vide infra*).[9,10]

REPRODUCIBILITY

The reproducibility of laser induced pyrolysis was mentioned previously. Folmer has considered interpretation of degradation patterns and has shown that laser pyrolysis can lead quickly to the identification of a particular polymeric material. Other than this, there has been little done with interlaboratory comparisons, but we have considered those factors that control the precision on replicate pyrograms. Certainly the most important consideration is a reproducible thermal flux. Secondly, the sample must be carefully controlled since crystallites, grain boundaries, etc., will affect the processes of energy absorption.

Probably the most precise data will result from amorphous materials, materials such as polymers. Table 2 shows data taken from six replicate pyrograms, again from polystyrene chips. Seven separate peaks are shown; both the percentage of each peak and the number of integrator counts are indicated. In the extreme right-hand column, the total number of integrator counts, a measure of the total quantity of gaseous products detected, is listed. These data indicate that, under these conditions, a relative standard deviation of ± 6% is typical for this technique. Attempts were made to focus the laser beam on the surface of a particular chip; this cannot be done with certainty.

One possible way to improve this precision is to lase on thin sections of material. Should these be thin enough to ensure that the laser shot drilled a hole cleanly through the sample, each shot should be exposed to the

Table 2:   Reproducibility of Replicate Shots from Chips of Polystyrene

Shot Number	Chromatographic Peak													
	1		2		3		4		5		6		7	
	Count	%	Count	%	Count	%	Count	%	Count	%	Count	%	Count	%	Total
1	6404	20.60	559	1.79	1409	4.53	4232	13.62	1137	3.66	156	0.50	17181	55.28	31078
2	4598	17.58	401	1.53	1151	4.40	3896	14.89	1051	4.02	408	1.56	14643	56.00	26148
3	3897	13.47	344	1.19	1063	3.67	4059	14.04	1167	4.04	577	2.00	17797	61.58	28899
4	3585	21.40	218	1.30	730	4.36	2282	13.62	554	3.30	70	0.42	9313	55.59	16752
5	3247	14.93	223	1.03	659	3.03	3011	13.84	626	2.88	140	0.64	13843	63.65	21749
6	3057	15.30	111	0.55	644	3.22	2754	13.78	527	2.64	57	0.29	12830	64.21	19980
Average		17.20		1.23		3.87		13.97		3.42		0.90		59.39	24101
St. Dev.		±3.23		±0.43		±0.65		±0.48		±0.59		±0.70		±4.22	±2469

Peak Identification: 1: acetylene; 2:?:3:methane; 3:methane; 4:benzene; 5:toluene; 6:?:7:styrene
Under each shot and peak is the integrated number of counts followed by percentage for that peak.

same type of luminous flux; precise focusing would be less important in such a case. A series of experiments were run on thin sections (0.075 cm thick) of a polymeric material marketed as heat-shrinkable tubing (HST). The results from these replicate analyses are shown in Table 3. Only four significant peaks were observed; in each case a hole, approximately 1 mm in diameter, was drilled through the sample. As can be seen, the precision in this series is greatly improved at times approaching the stated precision of the integration instrument. These initial results certainly do not conclusively prove that this procedure,

Table 3:   Reproducibility of Replicate Shots from a Thin Section of Heat Shrinkable Tubing

Shot Number	Chromatographic Peak								Total Counts
	1		2		3		4		
	Counts	%	Counts	%	Counts	%	Counts	%	
1	347	3.29	6289	59.78	1383	13.14	2501	23.77	10,520
2	255	3.11	4985	60.94	1085	13.26	1854	22.66	8,179
3	391	3.50	6609	59.18	1518	13.59	2649	23.72	11,167
4	371	3.13	7059	59.73	1528	12.92	2860	24.20	11,818
5	385	3.14	7447	60.74	1477	12.04	2951	24.07	12,260
6	370	3.40	6657	61.27	1311	12.06	2527	23.25	10,865
		3.26 ± 0.17		60.27 ± .75		12.83 ± 0.58		23.61 ± 0.52	10,801 ± 3,200

lasing on thin sections of material, is the best strategy
for maximizing precision, but they do indicate that laser
pyrolysis can give excellent precision. They also point
to the type of considerations that must be explored to
obtain optimum precision.

APPLICATIONS OF LASER PYROLYSIS

If the chemical stoichiometry of the plume parallels
the similar atomic ratios in the sample, then the plume
quenching products should yield information about the
elemental composition of the sample. The simplest case to
consider is to deal with binary compounds and the most
interesting series of those is the hydrocarbons.

Consider again the processes that lead to the plume
quenching products. If the free energy of formation are
considered for a series of simple hydrocarbons, these calculations predict that at 3,500 K the most stable compound
is ethane, followed by methane, followed by acetylene. But
these computations do not consider the stoichiometry of the
reacting plume. Figure 9 shows, again, the degradation
process; the laser pumps hydrogen and carbon into the plume
environment; this then quenches into a variety of products.
The overall process leads to the conclusion that carbon rich
plumes result in predominately acetylene while hydrogen rich
ones should result in methane, even though acetylene is the
most thermodynamically stable at the temperature of interest.

Figure 10 shows some results for the quantification
of the low molecular weight gases from the laser pyrolysis
of a series of hydrocarbons. Here the percent methane
(percent of the low molecular weight gases) is plotted
versus a function termed the aromaticity function, A, the
number of aromatic hydrogen atoms divided by the total
number of hydrogens in the sample. (For example, 1,2-
diphenylethane has a total of fourteen hydrogen atoms per
molecule, ten of these are on the two phenyl groups: this
compound is plotted at 10/14 or 0.71.) Also shown are
data for paraffin, durene, polystyrene and a group of completely aromatic compounds like naphthalene. Figure 11
shows the same data plotted as a function of the hydrogen/
carbon ratio in the sample plotted versus percent methane
(Curve A) in the low molecular weight gases and versus

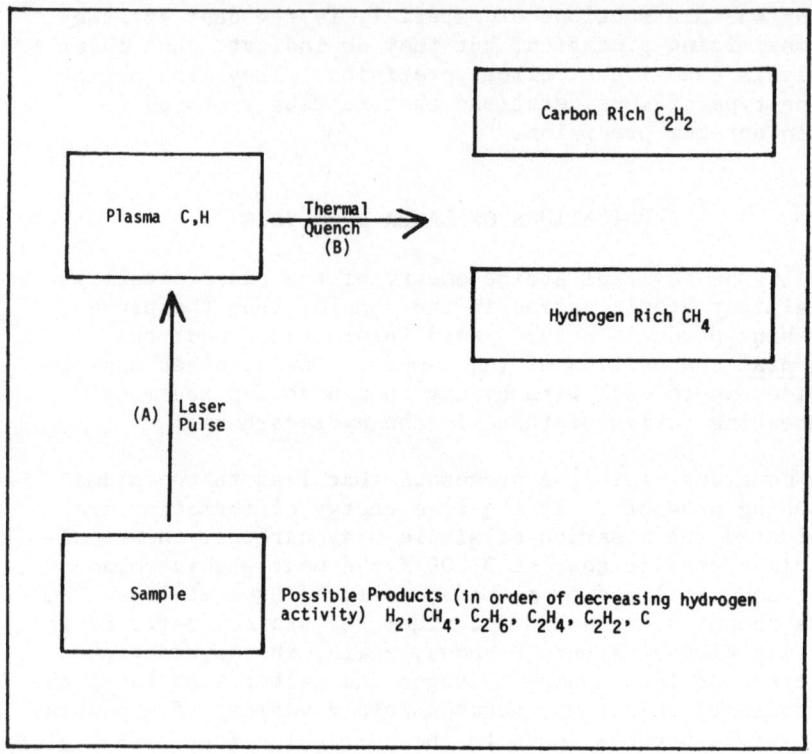

Figure 9:   Quenching Possibilities for Hydrocarbons

the total percentage of acetylene and ethane (Curve B).
As can be seen, as the compound becomes more saturated, the
methane percentage of these low molecular weight gases
steadily increases. Aromatic compounds yield a larger
percentage of acetylene and ethane. We term this technique plasma stoichiometric analysis.

The pyrolysis of polystyrene has already been mentioned.
One reason for striving for more vigorous pyrolysis methods
is to degrade compounds of good thermal stability, ones
that are difficult to analyze for using more conventional
pyrolysis methods. One such polymer is tetrafluoroethylene,
Teflon. Figure 12 shows the pyrolysis results with that
compound. Here two traces are shown, one for the FID and
one for the beta induced luminescence detector (BILD).
These simultaneous detector traces illustrate how the FID

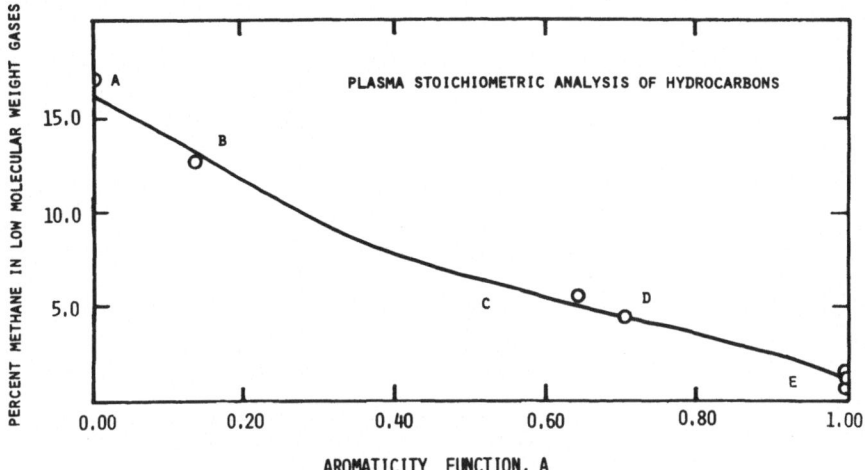

Figure 10: (A = number of aromatic hydrogens/total number of hydrogens per molecule); A) paraffin; B) durene; C) polystyrene; D)11,2-diphenylethane; E) naphthalene, terphenyl, phenanthrene.

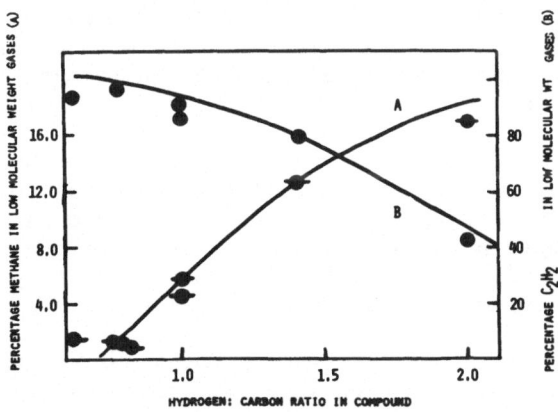

Figure 11: Plasma Stoichiometric Analysis of Hydrocarbons

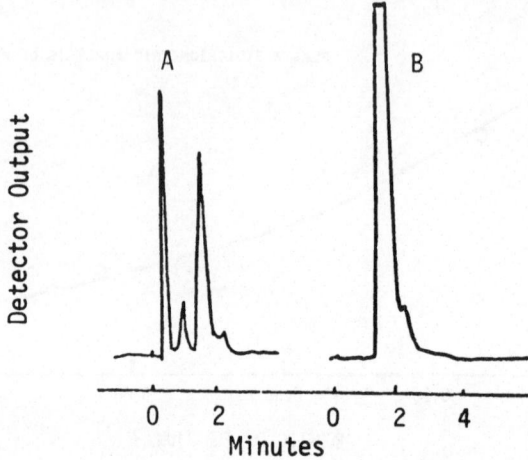

Figure 12: Laser Pyrogram of Teflon

Isothermal 20°C, 20 ml/min He
Porapak R
A.   BILD Detector
B.   FID Detector

completely misses two important products (one is hydrogen
and the other might be either fluorine or, possibly HF)
preceding the peak for tetrafluoroethylene. Laser pyrolysis
on teflon leaves the polymer surface cratered with a black
deposit, carbon. These results indicate that some of the
plume quenching products are gaseous while one, carbon,
at least, is not.

Figure 13 shows results for pyrolysis of Durette cloth,
another temperature stable polymeric material developed for
NASA. This result is included to illustrate again that
laser induced pyrolysis yields characteristic pyrograms
with little difficulty even with these high temperature
stability polymeric materials.

Figure 14 shows the results for a series of polycyclic
aromatic compounds. Each shows an unresolved low molecular
weight gas peak, predominately acetylene, followed by
characteristic pyrograms of thermal blowoff products. The

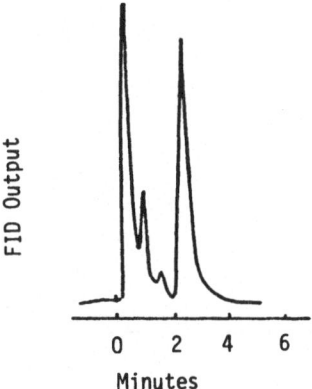

Figure 13: Laser Pyrograms of Durette,
Temperature Resistant Polymer,
10% Carbowax 20 M; 69-80 ABS
Isothermal at 150°C.

peak at 4 minutes is naphthalene, it appears in each pyrogram; the peak at 10 minutes retention time corresponds to
either anthracene or phenanthrene, the tricyclic compound.
Here again we see that sensible, easily identifiable pyrograms result from a series of very similar chemical compounds.

SUMMARY

It might be well to review some of the more obvious
possibilities for laser pyrolysis.

Plasma stoichiometric analysis has already been
described. The experiments done have only been on the most
simple chemical systen, that of hydrocarbons. They show

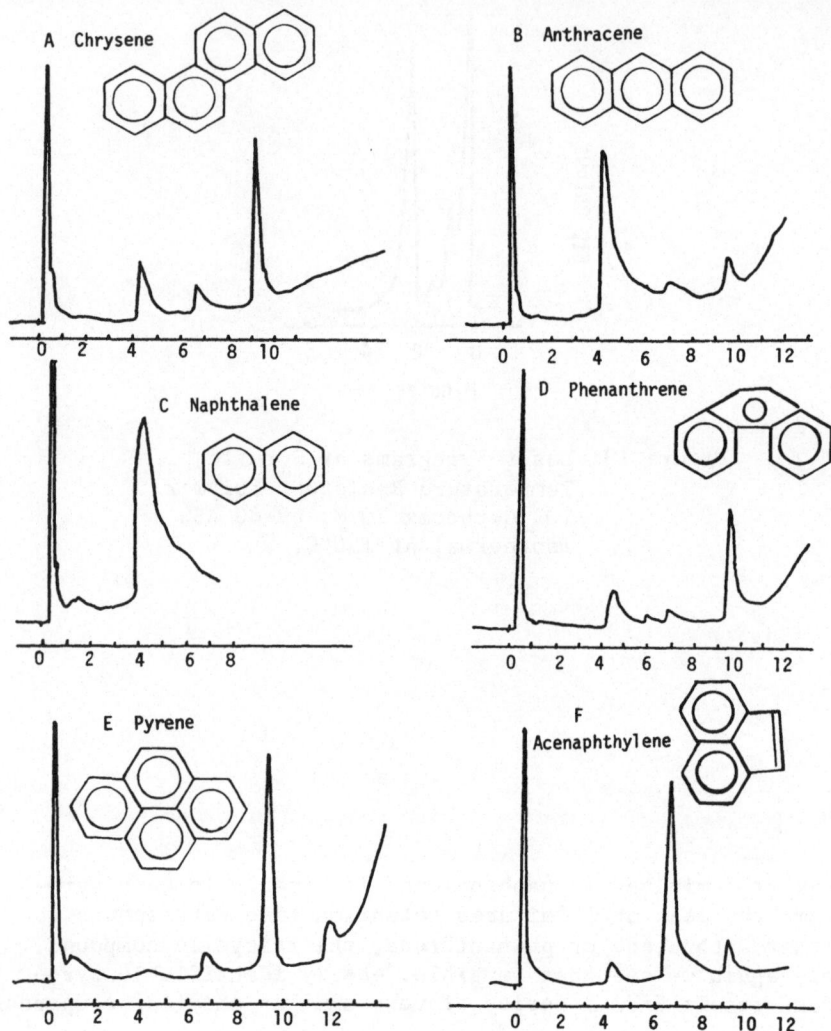

Figure 14: Laser Pyrograms of Polycyclic Aromatic Hydro-
            carbons. Column 10% Apiezon N-160-80 ABS;
            75 to 300C at 20° C/min

promise although certainly there are better methods to
determine the hydrogen/carbon ratio in a compound although,
perhaps, not more rapid ones. But the identification of
carbon oxides in the plume quenching products would clearly
point to an oxygen containing compound and quantification

of these oxides should allow a rapid estimation of the percentage of oxygen in the sample. Nitrogen containing compounds offer the same possibilities.

Laser pyrolysis offers rather unusual conditions for rapid power deposition. It is difficult to think of other techniques that will yield as rapid thermal rise times. Because of the very short TRT's, laser techniques might turn out to offer the highest precision in pyrolysis gas chromatography. It should also be stated that sample preparation, at least for the work described here, has been rather crude. Samples need not be dissolved in some suitable solvent; samples were not of controlled particle size; the orientation of samples to the incoming laser flux has not been well controlled. As this technique is developed, work needs to be done on sampling procedures. This means, of course, that methods must be found to make replicate samples highly reproducible as far as their ability to absorb laser radiation is concerned. There is no reason to doubt that significant progress will be made in this area.

Perhaps the most exciting possibilities for this technique, however, is the pyrolysis on compounds which have differing molecular geometries. For instance, optically active crystalline compounds should absorb light differently along their various crystallographic axes. Different tactic forms of structured polymers should absorb light in different ways.

In summary then, laser pyrolysis techniques offer promise for doing a variety of jobs in the characterization of materials. These pulsed thermal methods will probably not replace more conventional pyrolysis procedures but will certainly amplify the possibilities of pyrolysis followed by gas chromatography.

ACKNOWLEDGMENT

Much of the work described was taken from the Ph.D. dissertation of Dr. W. T. Ristau, UNM (1971).

REFERENCES FOR PYROLYSIS-GAS CHROMATOGRAPHY

1. B. A. Lengyel, Introduction to Laser Physics, John
    Wiley and Sons, Inc., New York, 1966.

2. D. Lichtmann and J. F. Ready, Phys. Rev. Lettrs, 10.
    342 (1963).

3. W. Simon, P. Kriemler, J. A. Voellmin and H. Steiner,
    J. Gas Chromatog., 5, 53 (1967).

4. B. T. Guran, R. J. O'Brien and D. H. Anderson, Anal.
    Chem., 42, 115 (1970).

5. W. T. Ristau and N. E. Vanderborgh, Anal. Chem., 42,
    1848 (1970); N. E. Vanderborgh, W. T. Ristau and
    S. Coloff, Record of 11th Symposium on Electron, Ion
    and Laser Beam Technology, R. F. M. Thornley, ed.,
    San Francisco Press, Inc., p. 403.

6. O. F. Folmer, Jr. and L. V. Azarraga, J. Chromatogr.
    Sci., 7, 665 (1969).

7. J. P. Biscar, Anal. Chem., 43, 982 (1971).

8. W. T. Ristau and N. E. Vanderborgh, Anal. Chem., 44,
    359 (1972).

9. W. T. Ristau and N. E. Vanderborgh, Anal. Chem., 43,
    701 (1971).

10. O. F. Folmer, Jr., Anal. Chem., 43, 1057 (1971).

APPLICATIONS OF COMBINED GAS CHROMATOGRAPHY-MASS SPECTROMETRY IN CLINICAL CHEMISTRY

John Roboz
Department of Chemistry
The Mount Sinai Hospital
11 East 100th Street
New York, New York 10029

The combination of gas-liquid chromatography and mass spectrometry, together with computerization, has created a highly efficient tool for the multicomponent analysis of very low concentrations of a variety of constituents in physiological fluids and tissues. Gas chromatography is used to separate and quantify the compounds of interest, and spectrometry is used to confirm the identity of known or suspected constituents and to determine the identity and structure of unexpected constituents when encountered. The computer provides rapid and accurate acquisition and reduction of data, and may eventually also be used for interpretation and translation of chemical information into biological information.

After a brief discussion of instrumentation and techniques, three areas of application are illustrated: screening, diagnosis, and searching for new diseases (inborn errors of metabolism); development of diagnostic procedures (bile acids); and determination of the concentration of drugs in body fluids (daunomycin in serum). These areas clearly overlap into medical biochemistry and clinical pharmacology, however, it is often the clinical chemist who develops the needed methodology, and subsequently provides routine analytical service.

A more detailed discussion of the use and potential of mass spectrometry in clinical chemistry is given in Reference 1.

Instrumentation

In most gas chromatograph-mass spectrometer combinations low resolution mass spectrometry has been used to
characterize the separated constituents. Here masses are
determined to unit mass number, and identification is
accomplished on the bases of mass spectral fragmentation
patterns. These patterns are compared, manually or by computer, to spectra compiled in data libraries. Most commercially available instrument combinations, including magnetic, quadrupole and to a limited extent time-of-flight mass
spectrometers, are adequate to confirm peak identity in
metabolic profile and analysis with a reasonable degree of
confidence. Where unknowns must be identified and their
structures determined, high resolution mass spectrometry
is usually indicated. Here the objective is to determine
masses to an accuracy of 4-5 mmu (1 millimass unit, mmu,
equals to 0.001 atomic mass unit, amu) so that probable
molecular compositions could be calculated. Both currently
popular ion optical designs for high resolution, the Nier-
Johnson geometry with electrical detection, and the
Mattauch-Herzog geometry with combination photoplate/elec-
trical detection, are well suited for combination with gas
chromatograph. The main problem is usually to provide adequate "plumbing" for the GC-MS interface because of geo-
metrical limitations. Since peak matching or manual evaluation of photoplates is very time consuming when a large
number of gas chromatographic peaks are involved, computer
facilities should be available for those planning to combine a gas chromatograph with a high resolution mass spectrometer.

The gas chromatograph-mass spectrometer-computer system (Figure 1) in the author's laboratory consists of a
Hewlett-Packard Model 6720 gas chromatograph interfaced via
a two-stage jet-type helium separator with a Jeol Model JMS
01SC double-focusing high resolution mass spectrometer of
the Mattauch-Herzog design which, in turn, is interfaced
with a data reduction system consisting of a Jeol Model
JMD-2C densitometer and a Jeol Model JEC-6 computer. Operation in the low resolution mode is on-line; photoplates
with high resolution data are analyzed with a densitometer
the output of which is fed into the computer (2, 3). With
photoplate detection high resolution data may be obtained

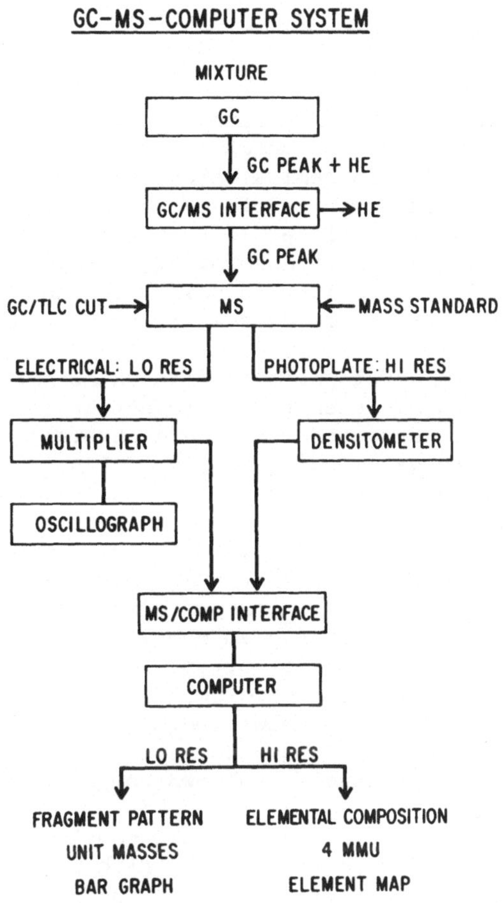

GC–MS–COMPUTER SYSTEM

MIXTURE

```
            ┌─────────────┐
            │     GC      │
            └─────────────┘
                   │ GC PEAK + HE
                   ▼
      ┌───────────────────┐
      │  GC/MS INTERFACE  │──→ HE
      └───────────────────┘
                   │ GC PEAK
                   ▼
GC/TLC CUT ──→ ┌──────────┐ ←── MASS STANDARD
               │    MS    │
               └──────────┘

ELECTRICAL: LO RES          PHOTOPLATE: HI RES
         │                          │
         ▼                          ▼
  ┌──────────────┐          ┌──────────────┐
  │  MULTIPLIER  │          │ DENSITOMETER │
  └──────────────┘          └──────────────┘
         │
  ┌──────────────┐
  │ OSCILLOGRAPH │
  └──────────────┘

              ┌──────────────────┐
              │  MS/COMP INTERFACE │
              └──────────────────┘
                       │
              ┌──────────────────┐
              │     COMPUTER     │
              └──────────────────┘
               LO RES │  HI RES
                  ▼         ▼
        FRAGMENT PATTERN   ELEMENTAL COMPOSITION
          UNIT MASSES            4 MMU
           BAR GRAPH          ELEMENT MAP
```

Figure 1

GAS CHROMATOGRAPH-MASS SPECTROMETER-COMPUTER SYSTEM

on as many as 25-30 gas chromatographic peaks in a single
analysis. Typical mass spectrometric conditions are summarized in Table 1.

Metabolic Profiles

An increasing number of diseases (more than 100 to-
date) are now recognized as metabolic disorders which result from defects in biochemical processes, often due to the
absence or decreased activity of an enzyme. This may result in a partial or complete block of a metabolic pathway
leading to the excretion of abnormal quantities of metabolites of low molecular weight in urine, or accumulation of
metabolites in other physiological fluids or tissues.

A significant development in clinical chemistry during
the last few years has been the introduction of the so-called
high resolution techniques. These are chromatographic
methods, including gas-liquid and liquid-liquid chromatography, in which a rather large number of low molecular
weight constituents of similar chemical nature are separated
and quantified in a single analysis. The diagnostic importance of such "profiles" is that they show the relative concentration of most or all members of a metabolic network.
For example, the gas chromatogram showing a "urinary acid
metabolic profile" may consist of some 40-50 peaks each
representing a different organic acid. In addition, the
profile will show a large number (perhaps 100) of additional
components, described as "compound noise" (4), which are
trace constituents of usually unknown composition.

When both pathological and normal samples of various
physiological fluids were analyzed by high resolution techniques, two kinds of differences were found: (a) increase
or decrease of one or more components in the pathological
sample with respect to the normal; (b) appearance of one or
more additional constituents in the abnormal sample. The
newly detected constituents must, of course, be identified
to establish their possible diagnostic significance.

The first step in a metabolic profile analysis (Figure
2) is the preparation of a "suitable sample." This refers
to a sample which contains only components of relatively low
molecular weight (<1,000), and from which interfering components have been removed. The physiologic fluid may be

CONDITIONS OF MASS-SPECTROMETRY

ION-SOURCE: TEMP: 280 °C   MASS RANGE: 12-400 OR 12-800
ION CURRENT: 200 uA
VOLTAGE: 23.5 OR 75 V
ACCEL: 7.6 KV

LOW RES: RESOLUTION: 2000   HIGH RES: RESOLUTION: 20,000
SCAN: 5 OR 10 SEC   EXPOSURE: $> 1 \times 10^{-10}$ COUL
MULTIPLIER: 2.0 KV   50 LINES ON PHOTOPLATE
OSCILL. PAPER: 10 CM/SEC   PFK MASS STANDARD

Table 1

TYPICAL CONDITIONS FOR MASS SPECTROMETRY

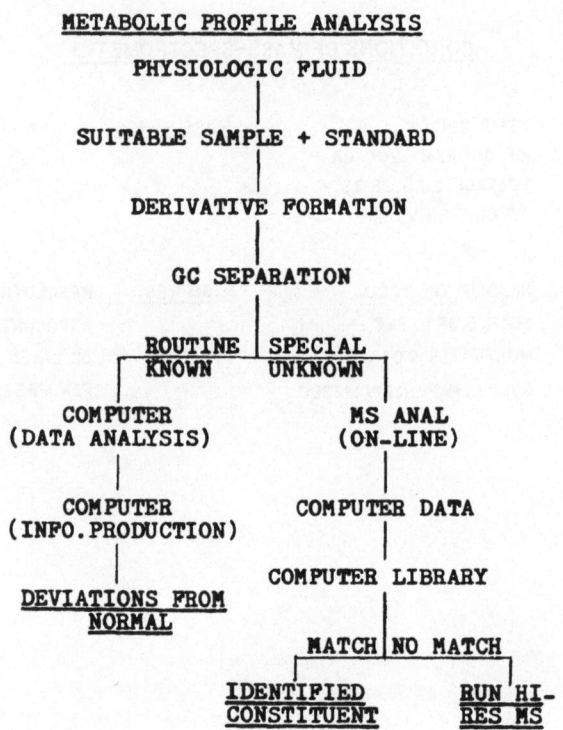

Figure 2

METABOLIC PROFILE ANALYSIS SYSTEM

urine, serum, amniotic fluid, cerebrospinal fluid, etc.
Sample preparation and derivatization are discussed in the
next section.

Once a proper sample preparation technique is selected,
a three-stage approach may be used for screening or diagnosis.
Initially a gas chromatographic survey is taken in a conventional manner using a flame ionization detector. The profile
thus obtained is compared to that established by analyzing
"normal" samples of the particular physiological fluid. The
term "normal" refers to pooled samples from healthy individuals who have not taken drugs for at least one week. The
computer printout of the gas chromatographic profile provide: (a) tentative identifications based on comparing relative retention times; (b) alert when the quantity of a normally occuring constituent is higher or lower than expected;
(c) indicate when a constituent not previously encountered
is detected.

When the quantity of a normal constituent is outside
the range established by normals, low resolution mass spectrometry is employed to confirm identity. Since the "ion
current monitograms" obtained in combined gas chromatography-
mass spectrometry are quite similar to those obtained in
conventional gas chromatography, the setting of the proper
mass spectrometric conditions is relatively easy. The low
resolution mass spectra are obtained on-line. The fragmentation patterns are compared, manually or by computer,
to a mass spectral data library. Information on a large
number of metabolites and drugs has become available during
the last few years (5, 6), and the possibility to search,
via telephone, the mass spectral data library of the National Institutes of Health (7, 8) should provide access to a
large quantity of useful data.

When the low resolution mass spectrum of a constituent
of increased quantity does not agree with that expected for
the constituent of the particular relative retention time,
the most likely reason is that chromatographic overlap
occurred due to the presence of an unknown. Gas chromatographic conditions must next be adjusted to resolve the two
constituents. The identity of the unknown is determined by
high resolution mass spectrometry. When a well-resolved
unknown is encountered, high resolution mass spectra are obtained without changing gas chromatographic conditions.

Sample Preparation: The procedure for the isolation
of a particular compound class in a physiological fluid involves some or all of the following steps: (a) removal of
proteins, inorganics, sediments; (b) pH-dependent extraction
with one or more solvents, or pre-separation according to
polarity, molecular size, or some other property of a group
of compounds; (c) hydrolysis, if necessary, by chemical or
enzymatic means; (d) suitable derivative formation for gas
chromatographic analysis.

Sample preparation and derivatization methods of
varying complexity have been published for virtually all
compound classes including "volatiles" (9), aromatic and
keto acids (10, 11, 12), free and peptide-bound amino acids
(13), carbohydrates (14), bile acids (15), and steroids (16).
Reference 1 lists a relatively large number of techniques
for sample preparation and derivative formation. Perhaps
the most important consideration in the selection of a
particular technique for sample preparation concerns specificity, i.e., the ability to fractionate as many components as possible of the desired class or group of compounds
with minimum interference from other classes. This is primarily important when an attempt is made to detect and subsequently identify additional, hitherto unsuspected, constituents associated with a particular disease. Required
sample quantity is important when screening infants.

Among techniques used for derivatization, methylation
(17), trifluoroacetylation (18), and trimethysilylation (19)
are the most popular, although many others are also in use
(20). Suitable derivatives should provide single peaks in
the gas chromatogram and yield molecular ions and high-mass
fragments with high abundance for mass spectrometric identification.

A "suitable sample" also includes one or more internal
standards for co-chromatography. Relative retention times
can best be established on the bases of "methylene units"
(21). To correct the vagaries of quantitative gas chromatography, an internal standard appropriate to the particular compound class in question should also be added.

In 1971 Jellum and co-workers at the Institute of
Clinical Biochemistry, Rikshospitalet, in Oslo, Norway developed a comprehensive sample preparation technique for

the routine gas chromatographic screening of urine for metabolic disorders (22, 23). Every urine sample is eventually
divided into 8 aliquots and analyzed with 8 gas chromatographs for a variety of constituents.

Starting with about 5 ml of urine (Figure 3), corresponding to 6 mg of creatinine, first three internal standards are added: n-eicosane, trehalose, and α-aminooctanic
acid. After acidifying with 6N hydrochloric acid, the urine
is extracted three times with 3 volumes of diethyl ether.
An aliquot of the ether extract is divided into two parts
and analyzed without any kind of derivatization. In system
A, on a Poropak column, very volatile constituents such as
lower alcohols, acetone, acetaldehyde, and the lower fatty
acids are separated. In system B, on a 10% OV-17 column,
higher alcohols and aldehydes, and possible hydrocarbons
are separated.

The second part of the ether extract is methylated
with diazomethane and divided into two portions. In system C, on an 8% butanediol succinate column, higher fatty
acids are separated. System D is the most important part
of the entire system. Here acidic metabolites are separated and detected. These include aliphatic and aromatic
mono-, di-, and tricarboxylic acids, including several Krebs-
cycle intermediates, all methylated, and also phenolic OH-
groups which are partially methylated yielding methoxy compounds.

The aqueous phase remaining after the initial ether
extraction is divided into three aliquots. One part is
evaporated to dryness, followed by treatment with n-butanol-
hydrochloric acid and trifluoroacetic anhydride. In this
fraction the free amino acids are converted into N-trifluoro-
acetyl-n-butyl esters and are analyzed in system E. Carbohydrates that may be present are converted to O-trifluoro-
acetyl derivatives and are analyzed in system F.

Finally, in systems G and H, the components originally
present in forms of peptides and conjugates are first hydrolyzed, then extracted and derivatized. Ether-soluble
acids which were conjugated with glucoronic acid and glycine are separated in system G. The total amino acid content, including amino acids hydrolyzed from peptides, is
obtained in system H. This should, of course, be compared
with system E where only free amino acids are separated.

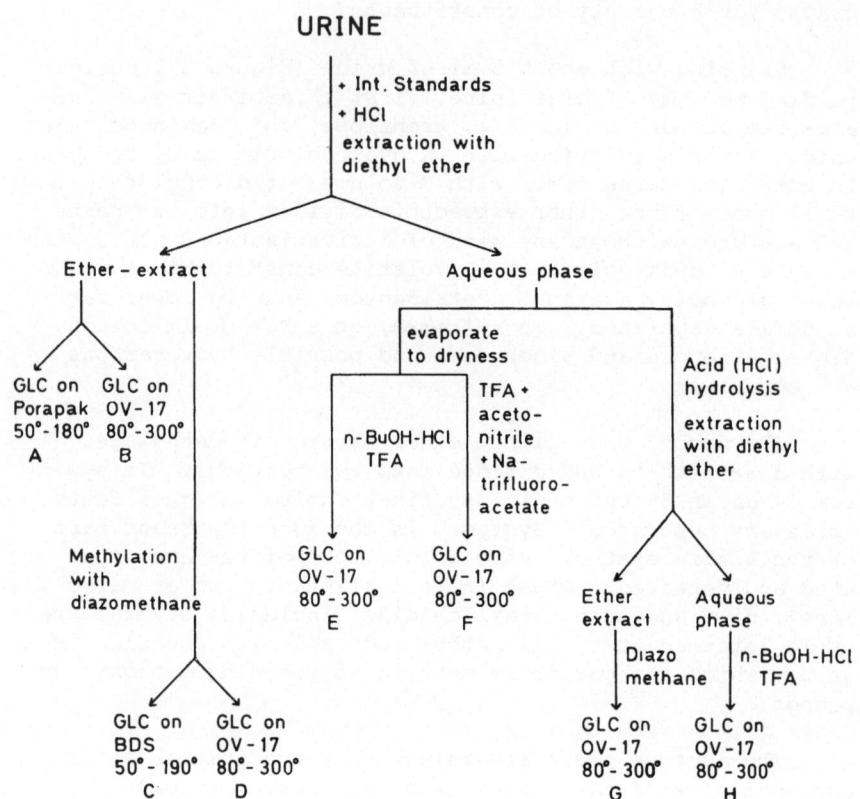

Figure 3

COMPREHENSIVE SAMPLE PREPARATION TECHNIQUE (URINE)
FOR SCREENING FOR METABOLIC DISORDERS (Reproduced
from Ref. 22 with permission.)

Of course, it takes considerable effort to perform all these analyses on a single urine sample. The authors recommend first to use only systems D and E which yield a great deal of information of acidic metabolites and amino acids. The complete system should only be used when the patient is particularly suspect as regard metabolic disease from the clinical point of view. Of course, mass spectral confirmation is offered on all peaks used for diagnosis.

Branched-chained keto acids, which are implicated in several metabolic disorders, cannot be analyzed conveniently by the technique described. Both keto acids and a variety of other acids can be analyzed in urine or serum when the keto acids are treated first with hydroxylamine to form oximines followed by trimethylsilylation; other components present are simply silylated (24). The total ion monitogram (Figure 4) of 9 important acids shows single, well-resolved peaks. As shown in Reference 24, the presence of other normally occurring acids does not interfere.

Applications: Table 2 (23) lists some 40 metabolic disorders where combined gas chromatography-mass spectrometry may be utilized for screening and diagnosis. In addition, of course, the technique is well suited to search for new diseases.

The concentration of α-keto isocaproic acid in normal urine is smaller than 10 mg/ml which is the detection limit of the technique mentioned above. The gas chromatogram (or ion current monitogram) of a patient with maple syrup urine disease (24) but under strict dietary control shows an elevated amount of α-keto isocaproic acid. When this child was given an oral load of leucine, a sharp increase occurs in the peak with retention time corresponding to α-keto isocaproic acid (Figure 5). The increase was rather dramatic, infact the peak was larger than that of hippuric acid (not shown in the figure because of much longer retention time) which is usually the largest peak in normal urine. A rather large increase was also noted in pyruvic acid. The identity of α-keto isocaproic acid was confirmed by both low resolution (Figure 6) and high resolution mass spectra (Table 3). The molecular ion of the silyl derivative of the oxime of α-keto isocaproic acid was readily detectable at 23 V ionization energy.

Ketone Acid Oxime - TMS Mixture

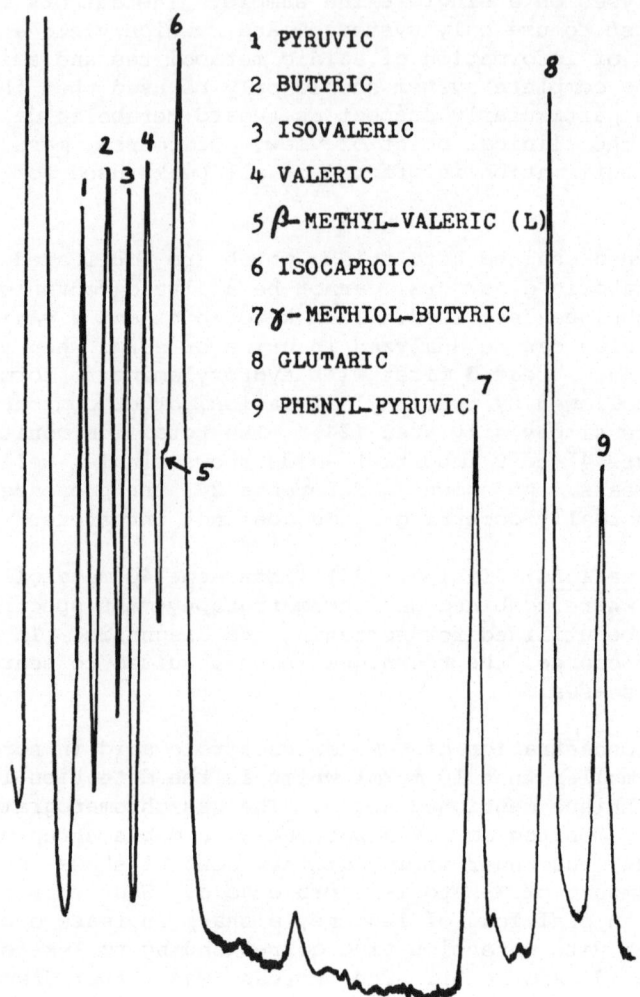

1 PYRUVIC

2 BUTYRIC

3 ISOVALERIC

4 VALERIC

5 β-METHYL-VALERIC (L)

6 ISOCAPROIC

7 γ-METHIOL-BUTYRIC

8 GLUTARIC

9 PHENYL-PYRUVIC

Figure 4
TOTAL ION MONITOGRAM OF KETO ACID MIXTURE USING OXIME-
TRIMETHYLSILYL DERIVATIVES (24)

Name of disease	Compounds detectable by the GLC–MS methods	Name of disease	Compounds detectable by the GLC–MS methods
Alcaptonuria	homogentisic acid	Hypertryptophanemia	tryptophan
Carnosinemia	carnosine	Hypervalinemia	valine
Congenital lactacidosis	lactic acid	Isovaleric acidemia	isovaleric acid, β-hydroxy-isovaleric acid, iso-valerylglycine
Cystathioninuria	cystathionine		
Cystinuria	cystine		
Diabetes mellitus	glucose, β-hydroxybutyric acid, acetoacetic acid	Maple syrup urine disease	valine, leucine, isoleucine, α-ketoisovaleric acid, α-ketoisocaproic acid, α-keto-β-methylvaleric acid
Essential fructosuria	fructose		
Essential pentosuria	L-xylulose		
Galactosemia	galactose, amino acids		
L-Glyceric aciduria	L-glyceric acid, oxalic acid	Methylmalonic acidemia	methylmalonic acid
Hartnup disease	neutral amino acids	Non-ketotic hypergly-cinemia	glycine
Histidinemia	histidine, imidazoleacetic acid	Oast-House disease	α-hydroxybutyric acid
Homocystinuria	homocystine, methionine	Ornithinemia	ornithine
β-Hydroxyisovaleric aciduria and β-methyl-crotonylglycinuria	β-hydroxyisovaleric acid, β-methylcrotonylglycine	Orotic aciduria	orotic acid
		Phenylketonuria	phenylalanine, phenyl-pyruvic acid, phenyllac-tic acid, o-hydroxy-phenylacetic acid
Hydroxylysinuria	hydroxylysine		
Hydroxyprolinemia	hydroxyproline		
Hyper-β-alaninemia	β-alanine, β-aminoisobu-tyric acid, γ-aminobu-tyric acid	Propionic acidemia	propionic acid
		Pyroglutamic aciduria	pyroglutamic acid (pyr-rolidone-2-carboxylic acid)
Hyperlysinemia	lysine		
Hypermethioninemia	methionine, α-keto-γ-methiolbutyric acid	Refsum's disease	phytanic acid
		Renal glycosuria	glucose
Hyperoxaluria	oxalic acid, glycolic acid, glyoxylic acid	Short chain fatty aci-demia	butyric acid, caproic acid
Hyperprolinemia	proline	Tyrosinosis	tyrosine, p-hydroxyphen-ylpyruvic acid, p-hy-droxyphenyllactic acid
Hypersarcosinemia	sarcosine		

Table 2

INBORN ERRORS OF METABOLISM DETECTABLE BY THE SCREENING SYSTEM. (Reproduced from Ref. 23 with permission)

MSUD Leucine-Loaded

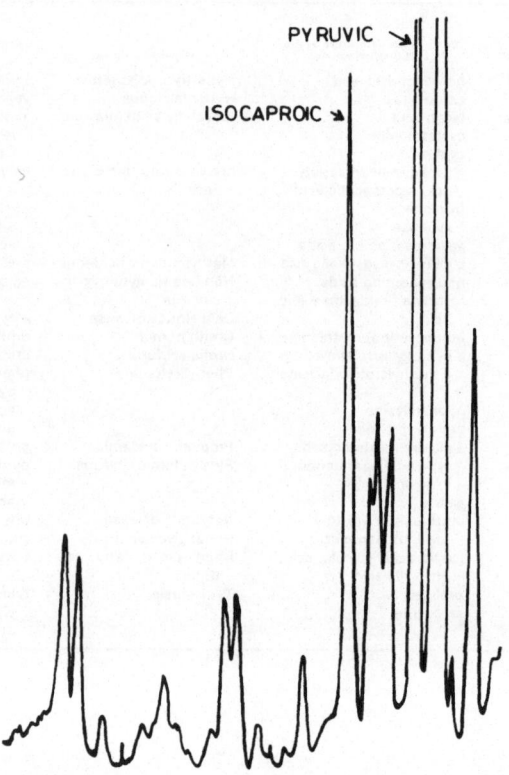

Figure 5

TOTAL ION MONITOGRAM OF EXTRACT FROM URINE OF PATIENT
WITH MAPLE SYRUP DISEASE AFTER AN ORAL LOAD WITH l-
LEUCINE, 100 mg/kg BODY WEIGHT; EXCRETION OF α-ISO-
CAPROIC ACID IS INCREASED ABOUT TENFOLD (24)

RUN 72-46/11   ALPHA-KETO-ISOCAPROIC-OXIME-TMS

START 56

```
MASS   INT%  DIFF
56.0   258    37  ----------
59.0   145    36  -----
73.0  1000    32  -----------------------------------------BS
74.0   105    32  ----
75.0   311    32  ------------
89.0    94    26  ---
133.0  104    10  ----
143.0   92     6  ---
147.0  735     6  ----------------------------
148.0  148     6  -----
172.0  311    -3  ------------
200.0  298   -15  ------------
201.0   91   -14  ---
246.0  131   -33  ------
247.0  304   -33  ------------
261.0  135   -42  -----
274.0  298   -48  ----------
275.0  131   -48  -----
289.0   94   -55  --- TABLE END
```

Figure 6

COMPUTER OUTPUT OF THE LOW RESOLUTION MASS SPECTRUM OF α-
KETO ISOCAPROIC ACID-OXIME-TMS DERIVATIVE. Starting mass,
m/e = 56; decimal point is not shown for intensity values,
e.g., 258 means 25.8; only peaks with intensity larger than
9.0 are shown; the column "diff" refers to mass measuring
accuracy and need not be considered here.

RUN 72-39/30 ∝ KETO ISOCAPROIC-OXIME-TMS

INT	OBSVD	ERROR	C12	H	N	O	SI8	ASSIGNMENT
67	289.1540	+1.0	12	27	1	3	2	M
79	274.1306	+1.2	11	24	1	3	2	M-CH$_3$
83	247.1086	+2.6	9	21	1	3	2	M-3CH$_3$
85	200.1101	-0.5	9	18	1	2	1	M-OTMSI
83	172.1134	-2.3	8	18	1	1	1	M-COOTMSI

Table 3

HIGH RESOLUTION DATA FOR ∝-KETO ISOCAPROIC ACID OXIME-
TRIMETHYLSILYL DERIVATIVE

The exciting potential of discovering new diseases is illustrated by a case (25) when an unknown peak was found in the urine of a 19-year old mentally retarded patient who was admitted to the hospital for unrelated surgery but was discovered to have chronic metabolic acidosis. The mass spectrum of the unknown (Figure 7) revealed the molecular ion at m/e 143 and the base peak at m/e 84, clearly indicating the loss of a $COOCH_3$ group. Eventually the peak was identified as the methyl ester of pyroglutamic acid (pyrrolidone-2-carboxylic acid). Further studies revealed that pyroglutamic acid is formed in the kidneys and brain, and the metabolic defect is in the conversion of pyroglutamic acid to glutamic acid in the δ-glutamyl cycle.

In addition to pyroglutamic aciduria, several more metabolic diseases have been discovered on the basis of identifying unknown peaks by mass spectrometry. Examples are methylmalonic acidemia (26), β-hydroxy-isovaleric aciduria (27), β-mercaptolactate cysteine disulphiduria (28), and the discovery that adipic and suberic acids are greatly elevated in ketoaciduria (29).

Bile Acids

Combined gas chromatography-mass spectrometry has been used extensively by Sjovall and co-workers in the Clinical Chemistry Department of the Karolinska Institutet, Stockholm, Sweden both to study the metabolism of chloresterol and the bile acids, and also to develop routine procedures for the determination of bile acids in urine, serum, bile and feces (30). Several bile acids, hitherto unknown to occur in the human, have been reported to be present in connection with various liver diseases. For example, the decrease of the content of deoxycholic acid in the urinary bile acid fraction appears to be of diagnostic value in recognition of obstructive jaundice (31).

Studies that might lead to diagnostic procedures usually start with rat livers and human liver biopsies that are left after histologic examination. After an appropriate biochemical procedure designed to study a particular type of metabolic pathway, the bile acids are extracted, a derivative is formed, and the mixture is analyzed by gas chromatography. Commonly occurring bile acids are identified on the basis of retention times; however,

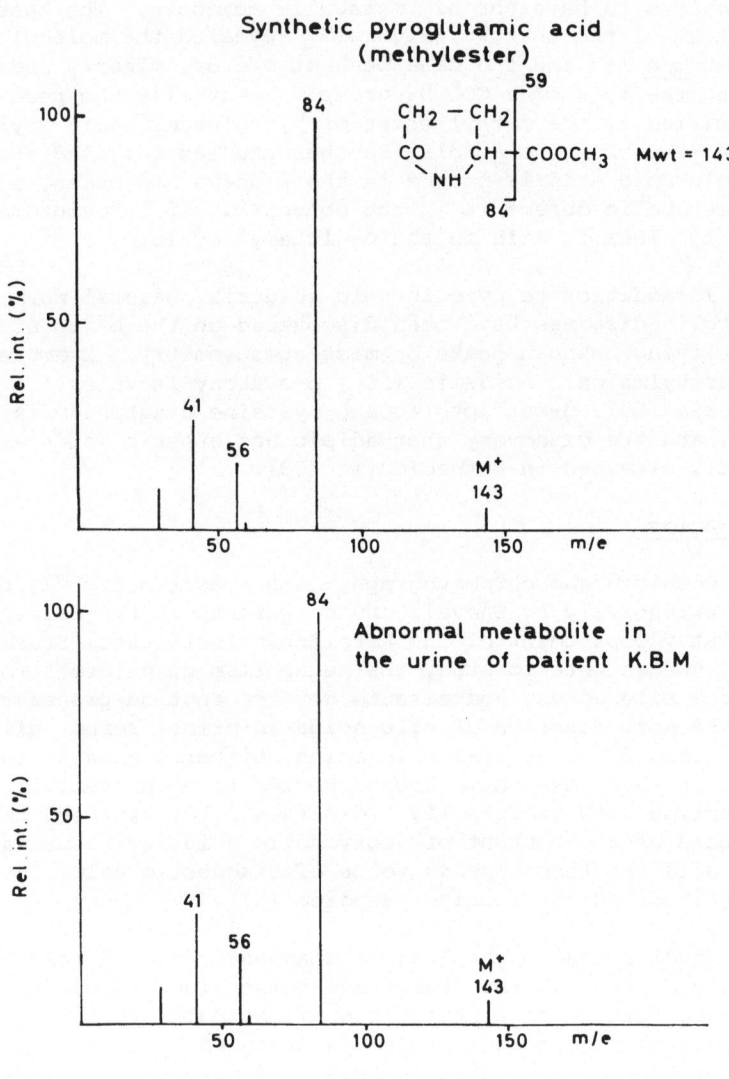

Figure 7

MASS SPECTRA OF SYNTHETIC PYROGLUTAMIC ACID (METHYLESTER)
AND OF THE UNKNOWN GAS CHROMATOGRAPHIC PEAK. (Reproduced
from Ref. 25 with permission.)

when unusual peaks are found, or large concentration changes occur, mass spectrometric identification is warranted.

The gas chromatographic profiles of trifluoroacetylated bile acid methyl ester derivatives from a normal rat liver and a liver three days after bile duct ligation show considerable differences (Figure 8). Here peak #5 appears as a new constituent and peak #7 shows a rather large increase in a previously unidentified small peak. The high resolution mass spectra of these peaks (Table 4) reveal these constituents to be a dihydroxy and a trihydroxy bile acid, respectively. With the aid of gas chromatographic retention times of different derivatives on different columns, and thin layer chromatographic separations, the compounds were finally identified as ursodeoxycholic acid and β-muricholic acid, respectively (32).

In another study concerning the toxic effects of lithocholic acid upon the liver, it was observed (33) that isolated liver microsomes from surgical biopsy speciments were capable of transforming taurolithocholic acid into another compound. The compound was characterized by high resolution mass spectrometry as a dihydroxy bile acid. With additional information from gas chromatographic and thin layer chromatographic analyses, the compound was identified as hyodeoxycholic acid.

This area of research is expected to benefit greatly with the introduction of stable isotope labeled bile acids. Indeed, deuterium labeled bile acids have already been used for the study of bile acid kinetics in man (34). A combined gas chromatograph-mass spectrometer system specifically modified for stable isotope ratio measurements has been described (35, 36).

Daunomycin in Serum

Daunomycin is an experimental antileukemia drug. It consists of a pigmented aglycone called daunomycinone in glycoside linkage with an amino sugar called daunosamine. A gas chromatographic method was developed (37) for the determination of daunomycin in serum in the silylated form. The low resolution mass spectrum is shown in Figure 9. Once an adequate gas chromatographic method is developed

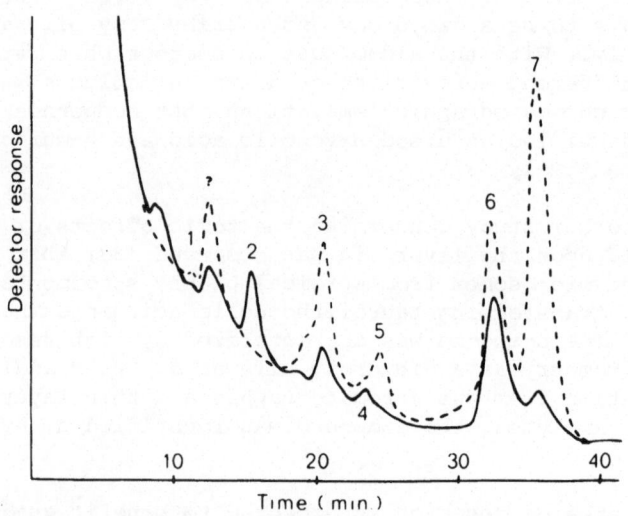

Figure 8

GAS CHROMATOGRAMS OF TRIFLUOROACETYLATED BILE ACID METHYL
ESTER DERIVATIVES IN A NORMAL RATE LIVER (SOLID LINES) AND
IN A LIVER THREE DAYS AFTER BILE DUCT LIGATION (DOTTED
LINE). 2% OV 210 column (Gas Chrom Q 100/120 mesh), 196°C
isothermal. The numbers represent: (1) lithocholic acid,
(2) deoxycholic acid, (3) chenodeoxycholic acid, (4) hy-
deoxycholic acid, (5) ursodeoxycholic acid, (6) cholic
acid, (7) β-muricholic acid. The question mark represents
an unidentified peak. (Reproduced from Ref. 32 with permission.)

Mass	Error[a]	Elements			Fragment[b]
		C	H	O	
370.2830	−4.1	25	38	2	M − (2 × TFA)[c]
355.2592	−4.5	24	35	2	M − [(2 × TFA) + CH₃]
339.2625	−5.2	24	35	1	M − [(2 × TFA) + OCH₃]
316.2408	+0.6	21	32	2	M − [(2 × TFA) + A ring]
262.1910	−2.2	17	26	2	M − [(2 × TFA) + A + B rings − CH]
255.2132	+1.9	19	27		M − [(2 × TFA) + side chain]
249.1877	+2.2	16	25	2	M − [(2 × TFA) + A + B rings]
228.1912	+3.4	17	24		M − [(2 × TFA) + side chain + C_{16} + C_{17}]
213.1655	+1.2	16	21		M − [(2 × TFA) + side chain + C_{16} + C_{17} + CH₃]

[a] Deviation from calculated value in millimass units.
[b] M, molecular ion; the minus sign indicates loss of fragments from the molecules. The lost fragments are shown in parentheses. Only fragments of relatively high abundance were selected.
[c] Trifluoroacetate residue.

Mass	Error[a]	Elements			Fragment[b]
		C	H	O	
368.2716	0.0	25	36	2	M − (3 × TFA)[c]
253.1902	−5.3	19	25		M − [(3 × TFA) − side chain]
249.1870	+2.9	16	25	2	M − [(3 × TFA) + A + B rings]
211.1441	−4.5	16	19		M − [(3 × TFA) + side chain + C_{16} + C_{17} + CH₃]

[a] Deviation from calculated value in millimass units.
[b] M, molecular ion; the minus sign indicates loss of fragments from the molecules. The lost fragments are shown in parentheses. Only fragments of relatively high abundance were selected.
[c] Trifluoroacetate residue.

Table 4

HIGH RESOLUTION MASS SPECTRAL DATA FOR METHYLATED-TRI-FLUOROACETYLATED (Me-TFA) DERIVATIVES OF TWO BILE ACIDS ISOLATED BY THIN-LAYER CHROMATOGRAPHY AND ANALYZED BY COMBINED GAS CHROMATOGRAPHY-MASS SPECTROMETRY. (Reproduced from Ref. 32 with permission.)

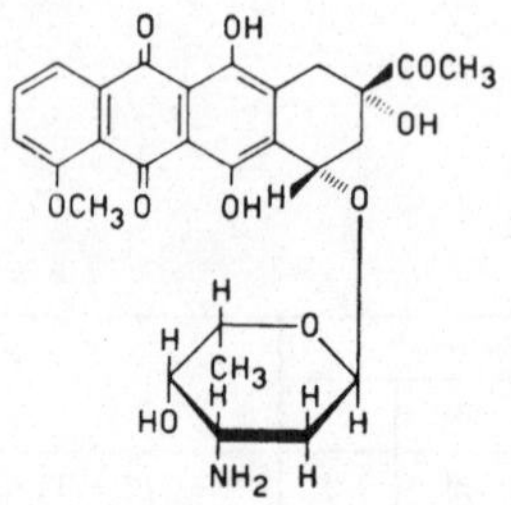

RUN 72-9 PEAK #7 DAUNOMYCIN-TMS IN SERUM

JOB= 5   START 200

```
NO.    MASS    INTX    DIFF
119    204.0   1080    -14    ------------------------------------------*OVER
120    205.0    546    -16    ----------------------
121    206.0    207    -15    --------
129    216.0   1080    -12    ------------------------------------------*OVER
130    220.0     98    -12    ---
138    228.0    131    -13    -----
139    230.0    122    -13    ----
159    254.0     48    -14    -
183    277.0     96      5    ---
190    291.0    139      3    -----
191    292.0     47      2    -
192    293.0     32      1    -
200    307.0    927      1    -------------------------------------
201    309.0    144      1    -----
206    319.0   1080      2    ------------------------------------------*OVER
207    320.0   1080      0    ------------------------------------------*OVER
208    322.0    148      0    -----
213    331.0    203     -2    --------
214    333.0     30     -1    -
220    345.0    266     -2    ----------
221    347.0     43     -2    -
248    409.0    216    -10    --------
249    411.0     43    -12    -
253    421.0    254    -13    ----------
254    423.0     52    -14    --
274    524.0     36    -36    -
276    599.0     31    -55    -
```

Figure 9

FORMULA AND LOW RESOLUTION MASS SPECTRUM OF THE TRIMETHYL-
SILYL DERIVATIVE OF DAUNOMYCINE (37). Starting mass, m/e
= 200; decimal points are not shown for intensity values,
e.g., 546 means 54.6; the term "over" refers to saturated
amplifier; the first and third columns (No. and Diff) need
not be considered here.

and the derivatives are properly identified by mass spectrometry, the technique can be used routinely to follow
the course of treatment. Of course, when gas chromatographic survey analyses reveal the presence of metabolites,
mass spectrometry should again be employed for identification. It is noted in passing that work is currently in
progress to apply this methodology in clinical studies and
to extend it to other chemotherapeutic agents.

Problems and Limitations

     Table 5 lists a number of problems in no particular
order. Most of them have nothing to do with gas chromatography or mass spectrometry, however, they must be considered when extra work leading only to disappointment is to
be avoided.

     Proper sampling is a major problem. Drugs must be
avoided for at least one week before a representative
sample is taken. A single aspirin will result at least two
very large peaks in the urinary aromatic acid profile.
The major metabolite of aspirin, salicyluric acid, appears
in a quantity even higher than hippuric acid which is
normally the largest peak in urinary aromatic acid profiles.
Another large peak that appears is also a metabolite of
aspirin, o-hydroxybenzoic acid. Much time may be wasted
to identify drugs unrelated to the disease studied.

     To learn about the so-called normal values, this
author made several analysis of the urinary acid content
of the urine of a healthy person under various dietary
conditions. As expected, major differences were found in
a variety of constituents. Instead of the customary 24-
hour urine collection, it might be better to take fresh
morning urine only, preferably after a day of fasting.
Genetic variations probably also contribute to variations
observed in healthy individuals. The ideal situation would
be to establish normal values for individuals; for example,
compare profiles of a gouty patient under normal conditions
and during an attack.

     Unknown peaks should first be treated with scepticism.
Cases were reported where innocent skin ointments showed up
as unknown peaks.

<u>Normal Values</u>:	. Different for different components
	. Values are usually not known
	. Significance of increase/decrease with respect to normal is not known and is variable
<u>Samples</u>:	. Usually small quantity (pediatrics)
	. Variations: diurnal, diet, genetic, drugs taken
<u>Identification</u>:	. It takes just as long for unimportant constituents
	. Unknown could be anything
	. Speed is vital

Table 5

SOME PROBLEMS IN CLINICAL CHEMICAL APPLICATIONS

Measured mass[a] (av[b])	Range,[c] mmu	Empirical formula	Observed metastable losses[d]	Difference from theoretical, mmu
142.0269 (3)	+0.1, −0.1	$C_6C_6O_4$		+0.4
141.0201 (1)		$C_6H_5O_4$		+1.4
125.0223 (3)	+6.6, −7.2	$C_6H_5O_3$	−HCO	−0.6
123.0086 (1)		$C_6H_3O_3$	−COOH	+0.4
113.0236 (3)	+1.0, −1.7	$C_5H_5O_3$	−CO	−0.2
98.0317 (3)	+1.3, −3.8	$^{13}C_1C_4H_5O_2$		−0.6
97.0276 (3)	+1.6, −1.2	$C_5H_5O_2$		−1.3
79.0180 (3)	+1.0, −1.0	C_5H_3O	−CO	−0.3
69.0334 (2)	±0.05	C_4H_5O		−0.6
68.0251 (2)	±1.6	C_4H_4O	−CO	−1.0
51.0239 (1)		C_4H_3		+0.5

(Metastable loss of $-H_2O$ shown between the 142/141 and 125/123 rows.)

[a] Number shown in parentheses refers to number of independent laboratory measurements.
[b] Results are average values if more than one measurement is indicated.
[c] Range is shown in + and − millimass units from average.
[d] Determined from the low-resolution mass spectrum.

Figure 10

HIGH RESOLUTION MASS SPECTROMETER DATA FOR UNKNOWN
FOUND IN URINE; MOLECULAR WEIGHT = 142. (Reproduced
from Ref. 38 with permission.)

Considerable effort might be needed to identify unknowns. Mrochek and Rainey of Oak Ridge Laboratory reported (38) that 3 peaks found in urine by liquid chromatography could not be identified by low resolution mass spectrometry alone after isolation by preparative chromatography. Since one of the peaks appeared in great excess in a patient with lymphocytic leukemia, further studies appeared warranted. Elemental composition was determined by high resolution mass spectrometry (Figure 10) as $C_6H_6O_4$ with two replaceable hydrogen, based on studies of silylation. From data using metastabile peaks it was determined that the compound must contain a carboxylic acid group. Ultraviolet data revealed the presence of conjugated double bonds. Finally, with a great deal of additional work the unknown was identified as 5-hydroxymethyl-2-furoic acid. Identification of the other two peaks was easier on the basis of the experience gained; the compounds were 2-furoylglycine and 2,5-furandicarboxylic acid. Although these compounds turned out to be known urinary metabolites, the work provides considerable insight into the metabolic pathways of uronic acids to substituted furans in humans.

Summary

Considerable progress has been made on the application of combined gas chromatography-mass spectrometry in clinical chemistry. Sample preparation techniques are now available for virtually all compound classes. The major constituents have already been identified in several metabolic profiles, including aliphatic, aromatic keto, and amino acids, steroids, bile acids, and carbohydrates in urine and serum. There are some 50 diseases where the technique may be of practical help in screening and diagnosis. Although most work reported to-date concerns methodology, the technique is clearly beyond the development stage as illustrated by a number of applications to pathologic cases.

Current methodology developments involving stable isotopes, and also the widespread of recent ionization techniques such as chemical ionization and field ion desorption, will open new areas of applications. Methodology already available for urine and serum is beginning to be applied to other physiological fluids, such as spinal fluid and amniotic fluid, and rapid progress is expected in these areas of application.

REFERENCES

1) Roboz, J., "Mass Spectrometry in Clinical Chemistry,"
    in Advances in Clinical Chemistry, Vol. 17, Bodansky,
    O., and Latner, A., editors, Academic Press, 1973.

2) Hutterer, F., Roboz, J., Sarkozi, L., Ruhig, A., and
    Bacchin, P., Clin. Chem., 17, 789 (1971).

3) Roboz, J., Sarkozi, L., Ruhig, A., and Hutterer, F.,
    Proc. 19th Ann. Conf. Mass Spectrom., 1971, p. 106.

4) Horning, E., and Horning, M., in Recent Advances in Gas
    Chromatography Domsky, I., and Perry, J., editors, Mar-
    cel Dekker, 1971, p. 341.

5) Markey, S., Thobhani, H., and Hammond, K., Identifica-
    tion of Endogenous Urinary Metabolites by Gas Chroma-
    tography-Mass Spectrometry: A collection of mass
    spectral data, Dept. Pediatrics, Univ. Colorado, Med-
    ical Center, Denver, 1972.

6) Mass Spectra of Drugs: A collection of mass spectral
    data, Biemann, K., editor, Dept. Chem., Massachusetts
    Inst. Technology, Cambridge, Mass., 1972.

7) Heller, S., Anal. Chem., 44, 1951 (1972).

8) Heller, S., Fales, H., and Milne, G., Org. Mass
    Spectrom., 6, 816 (1973)

9) Zlatkis, A., Mon, A., Robinson, A., Cary, P., and
    Pauling, L., Anal. Chem., 44, 18 (1972); also Anal.
    Chem., 44, 18 (1972); also Anal. Chem., 45, 763 (1973).

10) Mamer, O., Crawhall, J., and San Tjoa, S., Clin. Chim.
    Acta, 32, 171 (1971).

11) Chalmers, R., and Watts, R., Analyst, 97, 951 (1972).

12) Chalmers, R., and Watts, R., Analyst, 97, 958 (1971).

13) Zumwalt, R., Kuo, K., and Gerhke, C., J. Chromatog.
    57, 193 (1971).

14) Horning, E., and Horning, M., in Methods in Medical
    Research, Vol. 12, Olson, R., editor, Year Book Med-
    ical Publishers, Chicago, 1970, p. 369.

15) Horning, E., and Horning, M., ibid, p. 407; also Sand-
    berg, D., Sjovall, J., Sjovall, K., and Turner, D.,
    Lipid Research, 6, 182 (1965).

16) Frensdorf, E., Vankampen, E., and Hoek, W., Clin. Chim.
    Acta, 28, 77 (1971).

17) Schlenk, H., and Gellerman, J., Anal. Chem., 32, 1412
    (1960).

18) Lamkin, W., and Gerhke, Anal. Chem., 37 383 (1965).

19) Pierce, A., Silylation of Organic Compounds, Pierce
    Chemical Co., Rockford, 1968.

20) Hammarstrand, K., and Bonelli, E., Derivative Formation
    in Gas Chromatography, Varian Aerograph, 1968.

21) Dalgliesch, C., Horning, E., Horning, M., Knox, K., and
    Yarger, K., Biol. Chem. J., 101, 792 (1966).

22) Jellum, E., Stokke, O., and Eldjarn, L., Scand. J.
    Clin, Lab. Invest., 27, 273 (1971).

23) Jellum, E. Stokke, O., and Eldjarn, L., Clin. Chem.,
    18, 800 (1972).

24) Sternowsky, H., and Roboz, J., Ped. Res., 7, 58 (1973);
    (full article in press).

25) Jellum, E., Kluge, T., Borresen, H., Stokke, O., and
    Eldjarn, L., Scand. J. Clin. Lab Invest., 26, 327 (1970).

26) Stokke, E., Eldjarn, L., Norum, K., Steen Johnsen, J.,
    and Halborsen, S., Scand. J. Clin. Lab. Invest., 20,
    313 (1967).

27) Eldjarn, L., Jellum, E., Stokke, O., Pande, H., and
    Waaler, P., Lancet, II, 521 (1970).

28) Ampola, M., Efron, M., Bixby, E., and Meshorer, E., Amer. J. Dis. Child, 117, 71 (1969); also 117, 66 (1969).

29) Pettersen, J., Jellum, E., and Eldjarn, L., Clin. Chim. Acta, 38, 17 (1972).

30) Sjovall, J., Eneroth, P., and Ryhage, R., in The Bile Acids, Chemestry, Physiology, and Metabolism, Nair, P., and Kritchevsky, editors, Plenum Press, 1971, p. 209.

31) Back. P., Clin. Chim. Acta, 44, 199 (1973).

32) Greim, H., Trulzsch, D., Roboz, J., Dressler, K., Czygan, P., Hutterer, F., Schaffner, F., and Popper, H., Gastroenterology, 63, 837 (1972).

33) Trulzsch, D., Roboz,. J., Greim, H., Czygan, P., Rudick, J., Hutterer, F., Schaffner, F., and Popper, H., in Bile Acids in Human Diseases, Back, P., and Gerok, W., editors, Schattaner Verlag, 1972, p. 71; also Biochemical Medicine (in press).

34) Danzinger, R., Hofmann, A., Schoenfield, O., Berngruber, P., Szczepanik, P., and Klein, P., Gastroenterology, 60, 192 (1971).

35) Klein, P., Haumann, J., and Eisler, Wm., Clin. Chem., 17, 735 (1971).

36) Klein, P., Haumann, J., and Eisler, Wm., Anal. Chem., 44, 490 (1972).

37) Roboz, J., Kruman, D., and Hutterer, F., (abstract), 20th Ann. Conf. Mass Spectrom., Dallas, 1972, p. 76.

38) Mrochek, J., and Rainey, W., Clin. Chem., 18, 821 (1972).

APPLICATIONS OF MASS SPECTROMETRY IN DRUG METABOLISM AND RELATED FIELDS

W. J. A. VandenHeuvel

Merck Sharp & Dohme Research
Laboratories
Rahway, New Jersey   07065

The role of mass spectrometry (MS) in drug metabolism studies is to supply qualitative and possibly quantitative information from micrograms of partially purified isolates. The earlier in an isolation procedure the great sesitivity and structure elucidating power of MS can be focused upon a sample from a small quantity of biological material the greater the value of the method. Manipulation of grams of tissues or milliliters of fluid is preferred to handling kilogram or liter quantities. The ability to identify metabolites in a moderately complex matrix from small amounts of tissue, urine, etc., facilitates work with individual animals and offers the possibility of carrying out dynamic experiments (e.g., following changes in metabolite pattern with time). Direct probe (DP)-MS is useful in such studies but often it is combined gas-liquid chromatography (GLC)-MS, independently or together with DP-MS, which is especially effective in dealing with mixtures of this type. Successful application of combined GLC-MS often requires derivatization techniques for the conversion of non-volatile compounds to substances with suitable GLC properties. Derivative formation should not be considered strictly as GLC-oriented, however, for this approach can prove highly illuminating with respect to the MS aspects of the analysis. Metabolism studies

usually employ radioactively-labeled drugs,
offering the opportunity of characterizing a
component in a mixture as drug-related. This is
very important, as it reduces the possibility of

participation in a structure elucidation "wild
goose chase". The use of heavy atom labeled
drugs in MS structure determination studies and

as internal standards for quantitative analysis
by GLC-MS techniques is increasing in frequency

(1). Readily recognized doublets in the mass
spectra of isolates are an attractive means for
characterizing a component in a mixture as drug-

or substrate-related, and provide an alternate to
the use of radioactivity. Application of stable
isotopes will probably become more common, but at
present radio-labeling (especially with carbon-
14) remains the approach of choice in drug meta-

bolism studies. Unfortunately, at the specific
activities normally employed with radioactive

drugs the excess atom percent is too low to permit
recognition of the label by MS. The ideally
labeled drug probably would possess both radioactive and stable isotopes for isolation and

structure determination purposes,
This article will concentrate on approaches
found successful in our laboratory for sample

preparation and application of DP- and GLC-MS in
drug metabolism (and also several areas of bio-

chemistry) and describe procedures for overcoming
experimental difficulties which may be encountered.

        An understanding of the GLC and MS behavior
of a drug should be obtained before structure

determination work is undertaken on its metabolites.
Cambendazole yields two peaks when chromatographed
at temperatures > 200°, the result of a thermal

transformation (loss of isopropanol) to the
corresponding isocyanate (2). The component of

shorter retention time possesses a molecular
weight (by combined GLC-MS) of 242, whereas the
slower moving component exhibits an MS entirely

$$
\begin{array}{c}
CH_3 \\
CH_3
\end{array}
> \quad
\underset{\underset{\text{Cambendazole}}{}}{
\overset{\overset{\text{H} \quad \text{O} \quad \text{H}}{|\quad\quad ||\quad |}}{
C-O-C-N}}
$$

Cambendazole

in accord with the DP-MS of the parent drug.   An
intense M-42 fragment ion is found in the spectra
of compounds containing the isopropoxycarbonylamino moiety, and results from loss of C_3H_6.   This
fragment is, of course, not found in the isocyanate, for which the most intense fragment ion is
M-27, loss of HCN from the thiazole ring.   When
subjected to trimethylsilylation conditions
cambendazole forms a di-TMSi derivative (molecular
ion of 446 by DP-MS).   GLC of the derivative
yields only one peak, however, and a molecular
ion of 314 is found for this compound.   The IR
spectrum of the collected component indicates
the presence of an isocyanate group, and this
compound is clearly the mono-TMSi derivative of
the isocyanate from cambendazole.

An unknown porcine metabolite of cambendazole
was found to produce a single peak when subjected
to GLC-MS; its retention time and mass spectrum
were identical to the early GLC peak from
cambendazole.   GLC-MS of the metabolite following
exposure to trimethylsilylation conditions
resulted in elution of a single peak; the
retention time and mass spectrum were identical
to these found with GLC-MS for the TMSi derivative of cambendazole.   DP-MS of the metabolite,
on the other hand, indicated a molecular ion of
318, with an intense fragment ion at M-58, not
M-42 as found with the parent drug.   The m/e
values of the fragment ions from the metabolite
were identical to those from cambendazole.   When
the TMSi derivative of the metabolite was
subjected to DP-MS, the molecular ion was found

at m/e 534, whereas that for the TMSi-d9 derivative was found at m/e 561. The metabolite thus
forms a tri-TMSi derivative and possesses a
molecular weight of 318. The metabolite and its
TMSi derivative are thermally transformed to the
same product as cambendazole - namely, the
isocyante. The structural difference between the
parent drug and the metabolite must thus reside
in the isopropyl group. This is also deduced
from the DP-MS results; the only differences are
in the m/e values for the molecular ions - the
fragmentation patterns are identical. The
structure presented below was proposed for the
metabolite; comparison with authentic synthetic
alcohol confirmed this proposal.

A second porcine urinary metabolite of
cambendazole was found, in contradistinction to
the first, to exhibit the same molecular ion (m/e
258) both by DP- and GLC-MS. An intense M-42 ion
suggested the presence of the isopropoxycarbonylamino group. An intense signal was also present
at m/e 231, M-27, possibly loss of HCN from the
thiazole ring. On molecular weight consideration
both of these moieties cannot be present if the
metabolite retains the entire benzimidazole ring
intact. The lack of difference between the DP-
and GLC-MS results suggests the absence of the
isopropoxycarbonylamino side chain, a proposal
strengthened by the fact that the DP-and GLC-MS
results with the trimethylsilylated metabolite
are identical (molecular ion of m/e 402; 420 for
for the TMSi-d9 derivative). The metabolite
possesses two reactive functional groups, one of
which appears to undergo loss of 42 mass units
and also forms a TMSi derivative. The data point

to 5-N-acetylaminothiabendazole (below) as a
logical structure for the metabolite, and an exact
mass measurement on the molecular ion of the

metabolite did yield an empirical formula of
$C_{12}H_{10}ON_4S$. The amide was synthesized and found
to possess spectral and chromatographic properties
identical to those of the metabolite. It is clear
from these two examples that both DP- and GLC-MS
analysis of metabolites and their derivatives
should be attempted. Complimentary data may result
which facilitate in the determination of structure.

One approach to the characterization of
compounds in a mixture as drug-related (radioactive) is to carry out a two dimensional TLC
analysis of an isolate and then contact the
surface of the plate with X-ray film for an
appropriate period of time. Experience in our
laboratory (3) indicates that one week's exposure
to a tight zone containing 125 dpm of carbon-14
produces a suitably dark spot on the developed
film. If the parent drug possesses a UV-
absorbing system, the plate should also be
examined under UV light. Each radioactive zone
might be expected to also show UV-absorption, but
there will probably be UV-absorbing components
for which there is no evidence of radioactivity.
The metabolite zones can then be eluted from the
silica gel and examined by MS techniques.

A urinary metabolite of cambendazole from
the sheep was isolated by solvent extraction, two
dimensional TLC and radioautography, elution from
the silica gel being effected with methanol. An

aliquot equivalent to 2 μg (based on radioactivity)
was placed in a probe tube and subjected to DP-
MS.   Considerable organic material was vaporized
from the tube at low temperatures, but even at

>250° no spectra were obtained which possessed
fragmentation patterns suggesting they arose from
drug-related compound(s).   The probe tube was
removed from the instrument and its radioactivity
content determined.   Nearly all of the initially
introduced radioactivity was still present in the
tube.   It thus appeared that the cambendazole
metabolite had not been volatilized.   As this TLC
zone resulted from an ethyl acetate extraction of
urine, it would not be expected to contain highly
polar (non-volatile) compounds.   One possible
explanation was that the metabolite was strongly
adsorbed by silica gel present in the methanol
eluate.   A portion of the eluate was taken to
dryness and partitioned between ethyl acetate and
water; nearly all of the radioactivity went into
the organic phase.   Two μg of this solvent
soluble material subjected to DP-MS.   This time a
spectrum was obtained with an apparent molecular
ion of m/e 279 and a fragmentation pattern which
could be accounted for by a cambendazole transformation product with an intact isopropoxycarbonylamino side chain.   The base peak in the
metabolite spectrum, m/e 206, can be considered
M-(42 + 31), and an M-31 ion of 20% relative
intensity is noted.   Loss of water is also observed.   The low intensity of the signal at M + 2
($\sim$ 2%) suggests that the sulfur atom is not present in the metabolite, and a possible structure is
given below:

$$CH_3 \diagdown C-O-C-N \longrightarrow \bigcirc \diagdown \diagup N \diagup CH-CH_2OH$$

the loss of 31 mass units can be explained by
carbon-carbon scission of the glycol side chain.

Some qualitative organic chemistry on the μg
scale using MS to follow functional group changes
was then employed (4).   An aliquot of the metabolite equivalent to 10 μg was exposed to periodate
oxidation conditions, and the resulting radioactive compound isolated and 2 μg subjected to DP-
MS.   A change in molecular weight from 279 to 247
was indicated, the required shift for
R-CH-CH$_2$OH ⟶ R-CHO.   The remaining
   |
  OH

oxidation product was treated with methoxylamine·
HCl in pyridine; a derivative was formed with a
molecular ion of m/e 276.   Both the oxidation
product and the methoxime exhibited intense M-42
fragment ions, and the proposed metabolite
structure appeared very reasonable.   In another
experiment, an aliquot (2 μg equivalent) of the
original methanol TLC plate eluate was taken to
dryness in a probe tube in the vacuum lock.   The
tube was removed from the instrument, a few μl
of BSA added to the residue, and the tube returned
to the vacuum lock within 30 sec.   DP-MS of this
sample resulted in a spectrum with a molecular
ion of 423, 144 amu higher than that of the parent
compound, and reflecting the introduction of two
TMSi groups; virtually none of the radioactivity
remained in the tube.   Evidently the glycol side
chain was trimethylsilylated, and the derivatized
metabolite, much reduced in polarity, volatilized
readily even in the presence of silica gel.
Trimethylsilylation with a large excess of reagent
at 70° for 1/2 hr resulted in production of a
derivative of molecular weight 567, in accord with
the proposed structure.   Comparison of the metabolite with synthetic glycol confirmed its structure.

An example of the use of both DP- and GLC-MS,
and also radiochromatographic techniques, is found
in a study of the metabolism of ^{14}C-bis(chloromethyl)sulfone in sheep and cattle (5).   An
acidic metabolite was obtained by a combination of

techniques including selective solvent extraction,
ion-exchange, countercurrent distribution and TLC
with radioscanning. An intense signal at m/e 65
and pairs of signals at m/e 49/51 and 114/116
exhibiting the chlorine isotope cluster were
observed in the DP mass spectrum. One major peak
was observed when the metabolite was subjected to
trimethylsilylation conditions and then analyzed
by GLC; more than 80% of the injected radioactivity was found to be associated with this peak. A
newly developed GL radiochromatographic technique
was employed to give this result (6). The column
effluent is split between a mass detector and a
combustion system, and the CO_2 from each
chromatographic component is collected in vials
containing Hyamine. The vials are then subjected
to radioassay by liquid scintillation counting.
The drug-related component thus characterized,
the sample is subjected to combined GLC-MS. The
monochloro ion of m/e 49 was again observed with
the TMSi derivative, but the molecular ion
(monochloro) was found at m/e 186; other pertinent
ions include m/e 171 (chlorine-containing), 122
and 137 (see Figure 1). Data from the two spectra
clearly identify the metabolite as chloromethane-
sulfinic acid.

An acidic canine urinary metabolite of 3-(N-
morpholino)-4-(3-t-butylamino-2-hydroxypropyl)-1,
2,5-thiadiazole was isolated by ion-exchange and
TLC (7) and subjected to DP-MS. The presence of
a large amount of extraneous mass resulted in
complex spectra and no structural information
concerning the metabolite. An aliquot (2000 dpm)
of the sample was analyzed by GL radiochromatography
using a system in which the column effluent is
also split, but in which the CO_2 arising from
combustion passes through a proportional counter
for continous monitoring of radioactivity.
Although a number of peaks were observed by flame
ionization detection, no radioactive components

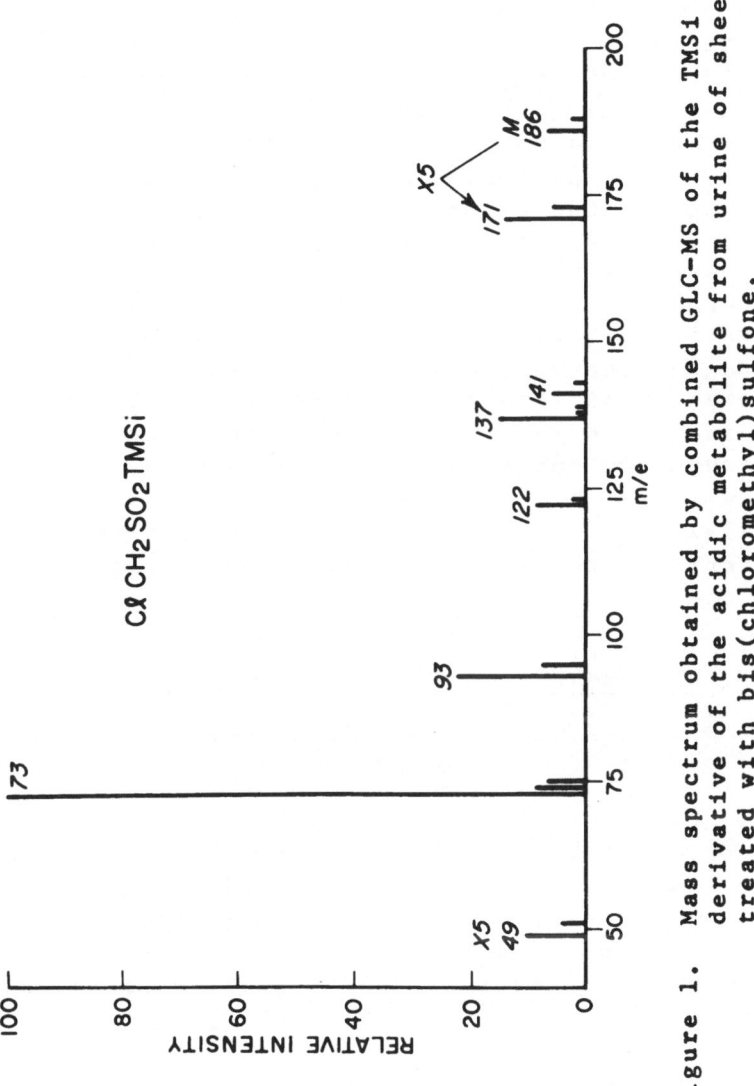

Figure 1. Mass spectrum obtained by combined GLC–MS of the TMSi derivative of the acidic metabolite from urine of sheep treated with <u>bis</u>(chloromethyl)sulfone.

were detected (300 dpm/component is a lower
detection limit). Thus the drug-related compound
was adsorbed by the column. When analysis was
carried out following exposure of the sample to
trimethylsilylation (BSA), a single radioactive
peak was produced, corresponding to a component
with a retention time of 19 min (see Figure 2).
The derivatized sample was then analyzed by
combined GLC-MS; the mass spectrum obtained for
the drug-related compound indicated a molecular
ion of m/e 419. In order to determine the
molecular weight of the underivatized metabolite,
a second portion of the metabolite fraction was
derivatized, this time with BSA-d_{18} to form the
TMSi-d_9 derivative; the resulting radioactive
component exhibited a molecular ion of 437, an
increase of 18 mass units. The metabolite thus
possessed two reactive functional groups, and a
molecular weight of 275 was indicated [419 -
(2x72)=275]. The m/e values of fragment ions of
particular interest are presented below in Table I.

TABLE I

Fragment	TMSi	TMSi-d_9
A	186	186
B	233	251
C	259	268

Fragment A bears neither reactive functional
group whereas B carries both; together they
comprise the entire molecule (186 + 233=419).
Fragment C is fragment A plus a TMSi group.
These data require that the metabolite as shown
below:

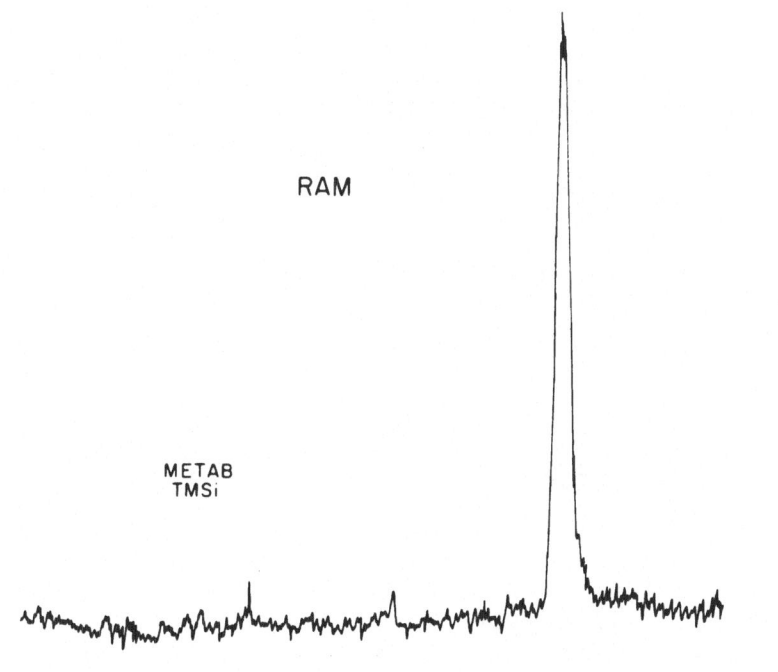

RAM

METAB
TMSi

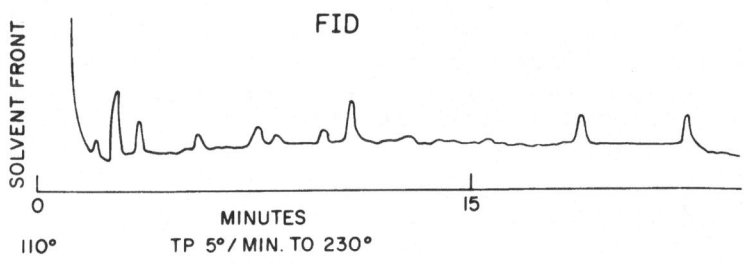

FID

SOLVENT FRONT

0   MINUTES   15

110°   TP 5°/ MIN. TO 230°

Figure 2.   Radioactivity monitor (RAM) and flame
ionization detector (FID) responses to
a trimethylsilylated radioactive urinary
metabolite fraction.

Scission of the side chain ethereal O-C bond
would produce fragments A & B, and transfer of a
TMSi group [occasionally observed (8)] from the
side chain onto the oxygen atom of fragment A
would result in production of fragment C. This
α-hydroxy acid was synthesized (9) and was
found to be identical to the metabolite.

Two metabolites of ^{14}C-rafoxanide have been
found in monkey bile (10). The two compounds

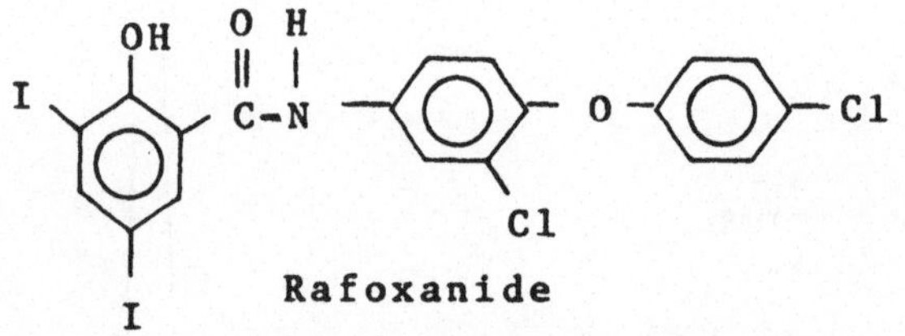

Rafoxanide

(A and B) were located on a TLC plate by radioscanning, eluted from the silica gel with
methanol, partitioned between ethyl acetate and
water, and then subjected to DP-MS. As each was
only $\sim$ 3% pure (on the basis of dpm/μg) the
spectra resulting from scans taken at low to
moderate probe temperatures ($<$ 150°) were highly
complicated. Two factors made possible our
obtaining useful information from these samples.
First, because of the high probe temperatures
($>$ 200°) required to volatilize the metabolites,
it proved possible to fractionate the sample, with
most of the extraneous mass evaporating from the
probe tube before volatilization of the metabolites.
Second, the dichloro nature of the metabolites
provided an easily recognizable isotope pattern to
characterize a spectrum as arising from a drug-
related compound. Rafoxanide itself exhibits
intense, key ions (Table II), at m/e 625 (M), 373

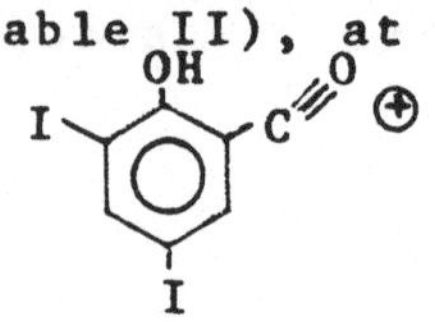

m/e 373

and 253

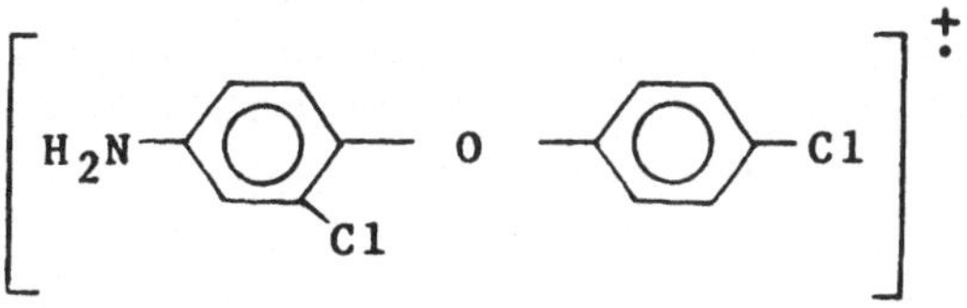

m/e 253

The corresponding ions for the metabolites, which exhibit virtually identical spectra, are also reported in the table.

TABLE II

m/e

Rafoxanide	M,625	373	253
Metabolites	M,515	263	253

The m/e values indicate that the metabolic transformations involve the salicylic acid ring (replacement of an iodine atom by an hydroxyl group), and one is led to hypothesize that the two metabolites possess isomeric structures. As rafoxanide has been successfully analyzed by GLC as its TMSi derivative (11), the metabolite fractions were subjected to trimethylsilylation and combined GLC-MS. Both derivatives were found to chromatograph satisfactorily on a 1 m, 1.8% OV-17 column at 250°; that from A exhibited a slightly longer retention time. The metabolites formed tri-TMSi derivatives (M,731), supporting the proposed structures. The two spectra are very similar, and for the most part are dominated by fragmentations involving the TMSi groups. The presence of a moderately intense dichloro ion of m/e 554 (M-177) in the spectrum of the TMSi derivative of metabolite A, and its virtual absence in the spectrum of B-TMSi, clearly distinguishes between the two compounds. The production of this ion can be rationalized by

elimination of 89 mass units, loss of OTMSi, from
the derivatized amide group (the TMSi and TMSi-d_9
derivatives of benzanilide show losses of 89 and
98, respectively) and 88 mass units, formally
tetramethylsilane, from two ortho TMSiO groups
[trimethylsilylated ortho disubstituted amines,
phenols, and aminophenols show loss of tetramethylsilane (12)]. The proposed structure for
the M-177 ion was confirmed by use of BSA-d_{18};

$$\left[\begin{array}{c} \text{CH}_3 \\ \text{CH}_3-\text{Si} \overset{\text{O}}{\underset{\text{O}}{\big\langle}} \end{array} \hspace{-0.5em} \bigcirc \hspace{-0.3em} -\text{C=N-R} \right]^{+}$$

M-177

the m/e value shifted, as required, from 554 to
560. The two model compounds (replacement of
either the 3- or 5- iodine by hydroxyl) were
prepared (13) and employed in reverse isotope
dilution analysis to demonstrate the presence of
both the metabolites in bile. The DP- and GLC-MS
properties of the ortho dihydroxy isomer matched
perfectly with those of metabolite A, whereas
the properties of the authentic para dihydroxy
isomer coincided with those of metabolite B
(both free and TMSi).

    The two synthetic compounds were also allowed
to react with benzeneboronic and 1-butaneboronic
acids (14), and the products examined by DP-MS.
Both isomers showed shifts in molecular ion from
515 to 601 (benzene) and 581 (butyl) indicating
the formation of cyclic derivatives. Only the
two derivatives of the ortho isomer, however,
produced intense (base peak) fragment ions at
m/e 349 and 329, for which the structures
presented below are likely.

X = C_6H_5, m/e 349   X = C_6H_5, m/e 601 (M)

X = C_4H_9, m/e 329   X = C_4H_9, m/e 581 (M)

The _para_ isomer forms a cyclic boronate, probably involving the enol of the amide, which cannot fragment to the acylium ion. As expected, rafoxanide and salicylanilide form cyclic boronates, but benzanilide and p-hydroxybenzanilide do not. When the two metabolite fractions were exposed to benzeneboronic acid, metabolite A was found to exhibit ions of m/e 601 and 349, whereas metabolite B gave only the ion of m/e 601.

The use of radioactive and stable isotopes proved useful in the identification of N,N-dimethyltryptamine (DMT) as the product of an _in vitro_ enzymic reaction (15). It was desired to prove that indoleamine-N-methyltransferase from human lung (16) catalyses the _in vitro_ methylation of N-methyltryptamine (NMT) by S-adenosylmethionine (SAM) to DMT. SAM-methyl-^{14}C was employed so that the isolation of the reaction product and its purification could be followed by radioactivity detection. As only a few hundred nanograms of DMT were produced per incubation, an excess of deuterium-labeled DMT, $R-CD_2CH_2N(CH_3)_2$, was used as a carrier during the isolation procedure. The radioactive zone with the appropriate R_f from the final TLC separation was trimethylsilylated and analyzed by combined GLC-MS. The presence of the DMT-d_2 in the isolate should not interfere with

the identification of DMT, as the molecular ions
and indole ring-containing fragment ions from the
two compounds differ by two mass units (m/e 262
and 260, and 204 and 202, respectively).

$$CR'_2 \longrightarrow CH_2N(CH_3)_2$$

TMSi

M, 260(R'=H), 262 (R'=D)
M-58 [m/e 202 (R'=H), m/e 204 (R'=D)]

The m/e value for the side chain fragment is 58
with both DMT and DMT-d_2; however, because of the
high specific activity (>40 mC/mM) of the SAM-
$^{14}CH_3$ employed as methyl donor, this fragment
from the biosynthesized DMT should show an
enhanced isotope peak at m/e 60. The trimethylsilylated isolate was subject to combined GLC-MS,
and the MS obtained from a peak with the same
retention time as an authentic mixture of DMT-TMSi
and DMT-d_2-TMSi gave the spectrum shown in Figure
3. The presence of the ions at m/e 202 and 260,
at the appropriate retention time, plus the
accompanying fragment ion of m/e 60, demonstrate
conclusively the enzymic side chain methylation of
NMT to DMT.

     In vitro experiments are also carried out
with drugs, often using liver microsomal enzyme
preparations. These enzyme systems are responsible for much of the metabolism of drugs and
other foreign compounds. Correlations between
in vivo and in vitro metabolism may prove helpful
in describing the pathways of metabolic degradation (17). As microsomal oxidations are
generally carried out with the production of μg,
rather than mg, amounts of metabolite mass
spectrometric techniques are ideally suited for
ascertaining what structural transformations
occur (18).

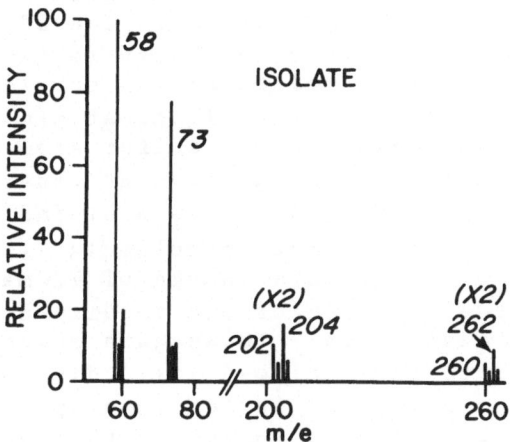

Figure 3. Mass spectrum obtained by combined GLC-MS of a radioactive trimethylsilylated isolate from the <u>in vitro</u> methylation of NMT.

We have recently studied the in vitro
metabolism of ^{14}C-dapsone [4,4'-bis(aminophenyl)-
sulfone] with a rat liver microsome preparation
(19). A radioactive TLC zone with an R_f not
that of parent drug or known metabolites was
observed. This compound was purified by high
pressure liquid chromatography. The DP mass
spectrum of the isolate was characterized by an
intense (base peak) molecular ion of m/e 278;
the intensity (6%) of the signal at m/e 280
suggested the presence of a sulfur atom.
Trimethylsilylation of the product under
conditions where dapsone forms a di-TMSi derivative resulted in a shift in molecular ion from
m/e 278 to m/e 350; this change of 72 mass units
indicates the presence of one readily derivatiz-
able functional group. These data suggested that
the metabolite is 4-amino-4'-nitrodiphenylsulfone,
and the spectrum of an authentic sample of this
compound was virtually identical to that of the
metabolite. TLC, UV and IR data all supported
the validity of this proposal.

The possibility that the nitro compound was
present as an impurity in the ^{14}C-dapsone would
not be expected on the basis of the route of
synthesis, and there was no evidence of any nitro
compound in the parent drug. As the yield of the
product was time dependent, and much reduced when
the enzyme preparation was heated at 80° prior to
incubation, the oxidation of an amino group
definitely appeared to be enzymic in nature. The
in vitro oxidation of dapsone to N-hydroxydapsone
has been reported recently (20). It was possible
that the conversion of dapsone to the nitro
compound was the result of a non-enzymic oxidation
of the enzmically-formed N-hydroxylamine.
Incubation of the latter, with or without microsomal protein, was shown to produce the nitro

compound. Mass spectrometric evidence has also
disclosed that dapsone may be further oxidized to
the corresponding dinitro compound. Although a
crude isolate from an incubation gave a complex
DP mass spectrum which surely resulted from a
mixture of compounds, signals were observed at
the m/e values found in the spectrum of authentic
4,4'-dinitrodiphenylsulfone, including the
characteristic base peak at m/e 170. Thin layer
radiochromatography also indicated formation of
this dinitro compound. Whether examining
isolates from in vivo or in vitro sources, the
mass spectrometer alone can give structural
information on the microgram level. There is no
guarantee of instant identification when using a
mass spectrometer, but the opportunity for success
is sufficient to warrant the attempt with many
samples.

A frequently observed route of urinary
excretion for drugs is hydroxylation followed by
conjugation with either glucuronic or sulfuric
acids to form the water-soluble glucuronides or
sulfates. The common practice is to hydrolyze
(chemically or enzymically) the conjugate to the
corresponding hydroxy compound, and then carry
out structure determination. Although formidable-
looking molecules, intact glucuronides can be
successfully analyzed by combined GLC-MS if they
are converted to the appropriate derivative.
Steroid glucuronides have been chromatographed
both as their methyl ester-TMSi ether (21,22) and
TMSi ester-TMSi ether (23) derivatives. If the
sample is of reasonable purity, the derivatives
also yield useful spectra with the DP technique.
The chromatogram resulting from analysis of the
trimethylsilylated glucuronide of 5-hydroxy-
thiabendazole, a urinary metabolite of thiabendazole in the sheep (24) is presented in Figure 4.

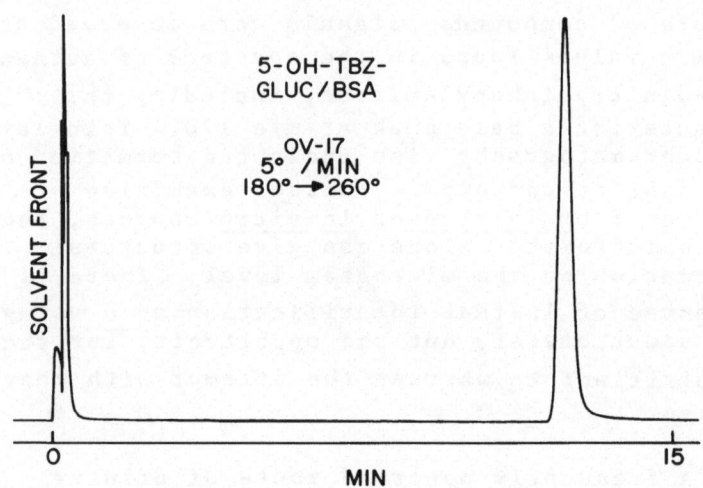

Figure 4.   GLC analysis of the trimethylsilylated
             glucuronide of 5-hydroxy-thiabendazole.
             Column conditions: 3 ft x 3 mm glass
             column; 2% OV-17; temperature programmed
             from 180° to 260° at 5°/min; 30 ml/min
             helium.

The metabolite was isolated in a rather high
state of purity, as was often necessary prior to
the advent of combined GLC-MS techniques.   The
highly polar compound has been converted to a
much less polar substance by virtue of trimethylsilylation of the carboxyl and hydroxyl groups.

5-hydroxy-thiabendazole

Although the molecular weight of the tetra-TMSi
derivative is high (681), the GLC conditions
required are not radically different from those
employed with compounds of lower molecular
weight.   The mass spectrum of the compound is
presented in Figure 5.   The molecular ion of m/e
681 is low in intensity, but readily recognizable;
the M-15 ion is significantly more intense.
Several of the fragment ions (e.g.; m/e 217,375)
arise from the conjugate portion of the molecule.
The base peak, m/e 289 (TMSi-d$_9$, 298) results
from scission of the ether linkage and rearrangement of a TMSi group onto the oxygen atom of the
benzene ring (analogous to the rearrangement ion
of m/e 259 found with the metabolite discussed
on page 10.   This ion (M-392) can be seen to
possess the same m/e value as the TMSi ether of
the parent phenol, and is highly informative with
respect to the structure of the metabolite.

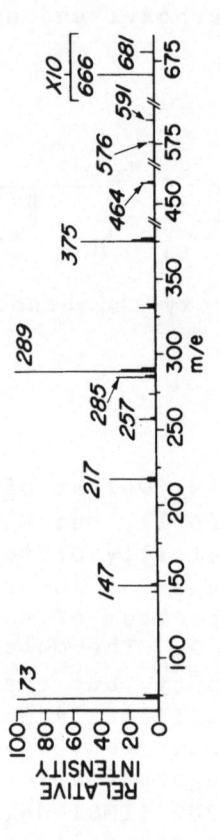

Figure 5.   Mass spectrum obtained by combined GLC-MS of the trimethylsilylated glucuronide of 5-hydroxy-thiabendazole.

Fenselau et al. (25) have reported on the value
of this fortuitous rearrangement. Although the
TMSi derivatives of ethereal sulfates can be
prepared, these are thermally labile species and
lose SO_3 when subjected to GLC at elevated
temperature, forming the TMSi ether of the parent
alcohol (23). [The analogous transformation to
the methyl ether occurs with the methyl derivative
of a sulfate conjugate (26)]. Thus when the O-
sulfate of ⍺-methyldopa (27) is subjected to
trimethylsilylation conditions and analyzed by
GLC it is the TMSi derivative of ⍺-methyldopa
that is eluted from the column (4). Conversion
of an O-sulfate to the stable corresponding
O-acetyl derivative with acetic anhydride prior
to GLC has been studied by several groups of
workers (28,29). GLC of sterol sulfates (and
their ammonium salts) results in elution of
olefinic products (30,31), rather analogous to the
results obtained with the methane- and p-toluene-
sulfonates of sterols (32,33). DP-MS of the
ammonium salt of 3β-hydroxyandrost-5-ene-17-one
sulfate fails to yield a molecular ion; thermal
conversion to the corresponding olefin(s) occurs,
and an intense ion is observed at M-115 (34).

The characterization of amino acids by GLC-MS
requires conversion of these water-soluble
compounds to non-polar, volatile derivatives. A
number of approaches have been employed (35,36),
including trimethylsilylation (37). The separations observed between closely related amino acids
are usually satisfactory and often superior to
those obtained by other techniques. Thin-layer
chromatographic characterization of the amino acid
content of ion-exchange chromatography fractions
from hydrolyzed algae grown on ^{13}C-enriched
(∿ 15%) CO_2 was found to be unsatisfactory for
certain pairs of amino acids -- e.g., serine/
threonine, and leucine/i-leucine. Trimethylsilylation of fractions containing mixtures of these

amino acids followed by GLC clearly demonstrated
the resolving power to this method (see Figure 6).
Identification of amino acids was achieved on the
basis of retention time and also mass spectrum.
The molecular ion is not normally seen with these
compounds (37), but four fragment ions are usually
observed: a, M-15 (loss of a TMSi methyl); b,
M-43 (loss of a TMSi methyl and CO with retention
of the silyloxy group); c, M-117 (loss of COOTMSi);
d, m/e 218, [HC-COOTMSi]$\overline{+}$ These ions are readily
                NHTMSi
observed in the mass spectrum of phenylalanine
di-TMSi (see Figure 7). The mass spectrum of the
di-TMSi derivative of phenylalanine from the
algae grown on ^{13}C-enriched CO_2 is presented in
Figure 8. As these two compounds differ only in
their ^{13}C content, the differences in their
isotope peak intensities are a direct reflection
of the extent of random incorporation of ^{13}C into
the molecule. An expression was derived (JSC) to
calculate the ^{13}C content of a fragment ion from
the relative intensity of its ^{13}C isotope peak
(37). The ^{13}C content of the four types of ions
discussed were calculated; Table III indicates a
trend between the ^{13}C content of an ion and the
number of amino acid carbon atoms in that ion.

TABLE III

Amino Acid Carbon Atoms in Ion	Total ^{13}C Percent
1,2	14.2 ± 0.4
3,4	13.7 ± 0.4
5,6	13.3 ± 0.4
7-9	12.6 ± 0.2

Thus the value for the M-15 ion from glycine, the
simplest amino acid (two C atoms), is 14.2%,
whereas the value is 12.4% for tyrosine (nine C
atoms). The ions of m/e 218 contain only two
amino acid carbon atoms, and from these it is

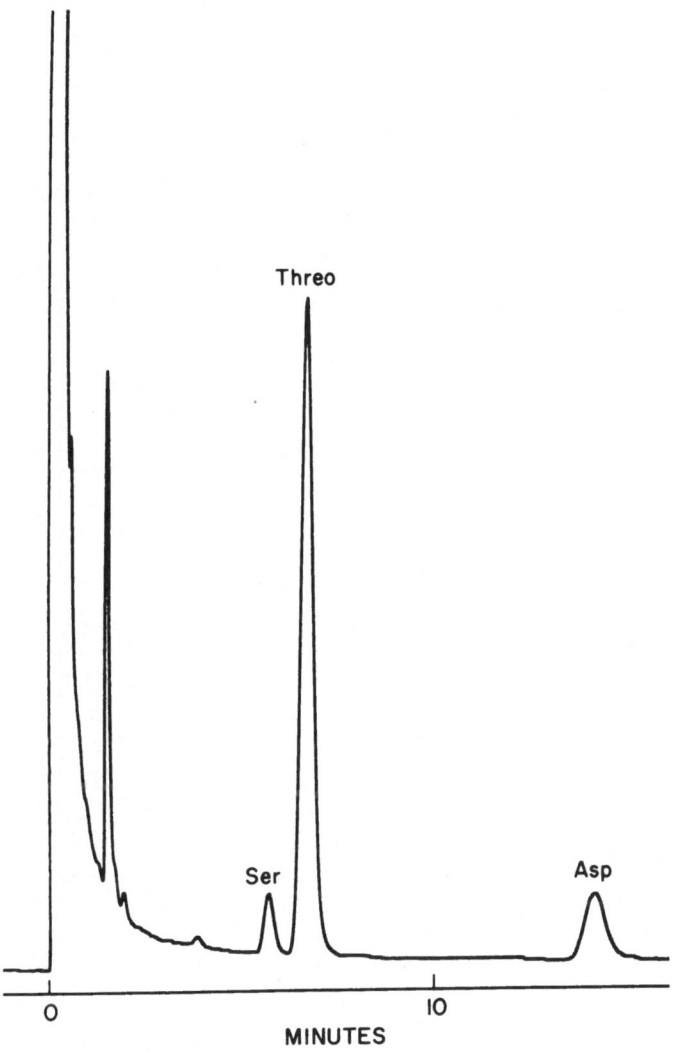

Figure 6.   GLC analysis of a trimethylsilylated
            ^{13}C-enriched amino acid fraction
            obtained by ion-exchange chromatography.

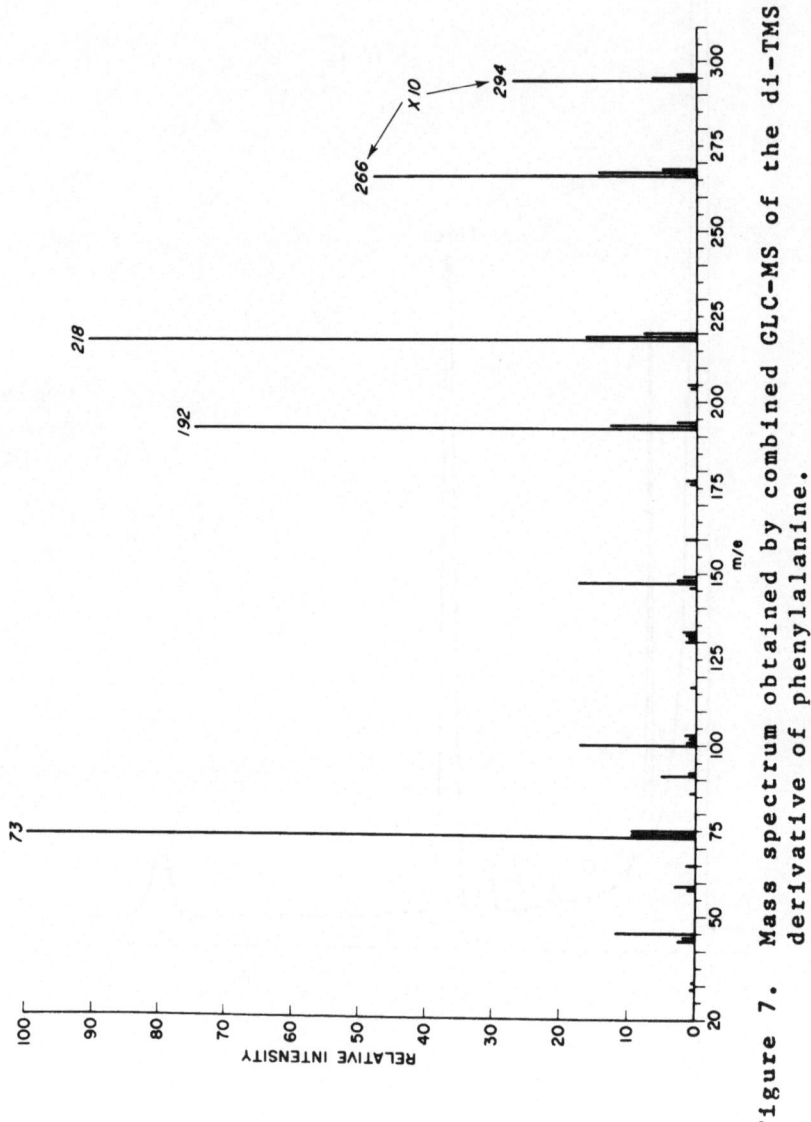

Figure 7.   Mass spectrum obtained by combined GLC-MS of the di-TMSi
derivative of phenylalanine.

possible to calculate ^{13}C content independent of the side chain. These values were found to exhibit little spread and were all approximately 14%, even that for tyrosine (14.2%). The trend observed in Table III cannot simply be explained in terms of the higher molecular weight amino acids containing less ^{13}C throughout the molecule, for the m/e 218 fragments derived from them give the higher ^{13}C content (tyrosine, 14.2%). The difference is related to the number of amino acid carbon atoms in the side chain, and is best explained in terms of a small biosynthetic preference for ^{12}C over ^{13}C. Catch (38) reported in 1954 an isotope effect for some separated ^{14}C-labeled amino acids (obtained from algae grown on ^{14}C-enriched CO_2) by measurement of their specific activities. As with the ^{13}C amino acids, the effect was greatest for phenylalanine and tyrosine. Hallowes et al. (39) later found by use of the ninhydrin reaction that the carboxyl groups of the separated amino acids were uniformly labeled (measurement of liberated $^{14}CO_2$). This finding is consistent with the observation for the m/e 218 fragments from the ^{13}C amino acids. Based on the results of Hallowes et al. (39) Catch is reported in reference 39 to have retracted his findings on the ^{14}C amino acid specific activities, but in retrospect it is likely that his results were correct.

Abelson and Hoering (40) carried out a study on ^{13}C incorporation into amino acids, and using classical mass spectrometric techniques (e.g., combustion to carbon dioxide followed by analysis with a double collector instrument) also found evidence for a ^{13}C isotope effect. Of course their approach is capable of yielding isotope ratios with great accuracy, but the values only reflect the overall ^{13}C content. Another disadvantage of this approach is that it does

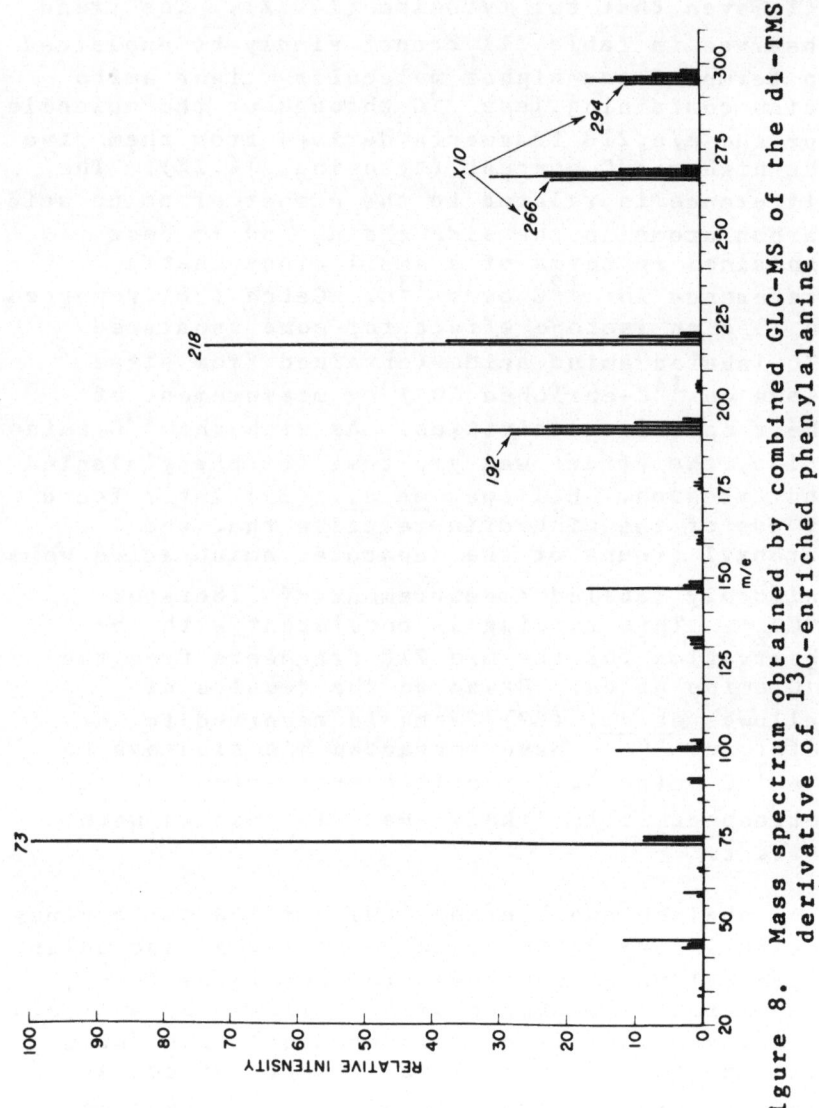

Figure 8. Mass spectrum obtained by combined GLC-MS of the di-TMS1 derivative of 13C-enriched phenylalanine.

not allow differentiation between carbon dioxide
from the compound of interest and from sample
contaminants. The possibility of the introduction
of impurities during combustion and of incomplete
combustion also exists (40). Direct mass
spectrometric analysis of the intact compound of
interest does not suffer from these disadvantages,
although less accurate data are inevitable. The
direct approach in addition allows isotope
content determination on the molecular ion and
also appropriate fragment ions (41), providing a
contaminant does not produce interfering ions.
A GLC column operating under the appropriate
conditions combined with a mass spectrometer
offers a superior technique for yielding isotope
information on μg quantities of components in a
mixture. The level of isotope ratio accuracy is
admittedly much less than what can be obtained
with the double collector approach, but in many
chemical and biological studies employing limited
amounts of partially purified sample values of
$\pm$ 5-10% or even a "yes or no" answer is perfectly
adequate.

Chemical ionization (CI) MS has been applied
by Fales, Milne and co-workers (42,43) to the
fields of drugs and natural products. These
authors found that the CI spectra often display
an intense quasimolecular ion when the electron
impact spectrum shows little or no molecular ion.
On the other hand, sterol TMSi derivatives do not
produce such desirable CI spectra (44). The ion
of m/e 369 is base peak for cholesterol TMSi
ether and signals in the molecular ion region
are of low ($\sim$ 5%) intensity; it is the electron
impact spectrum which exhibits an intense (45%)
molecular ion. The quasimolecular ion (m/e 310)
is base peak in the CI spectrum of the di-TMSi
derivative of phenylalanine (Figure 9) and this
should be compared to Figure 7, the corresponding
electron impact spectrum (no molecular ion).

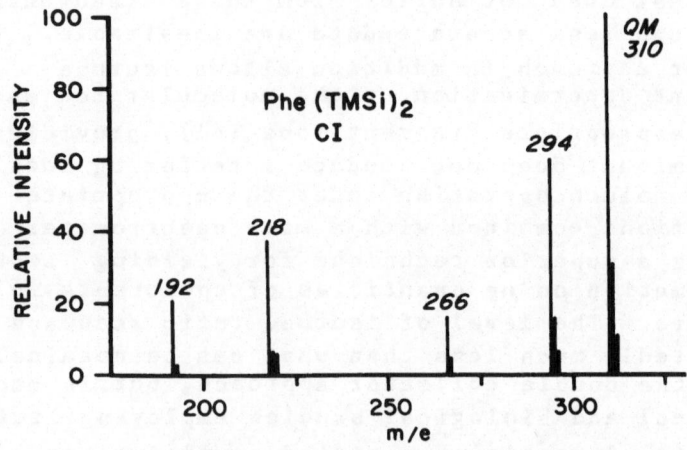

Figure 9. Chemical ionization (isobutane) mass spectrum of the di-TMSi derivative of phenylalanine.

An especially valuable application of
combined GLC-MS is validation of a GLC assay for
a drug or metabolite from a biological source.
Retention time and peak shape with a flame or
electron capture detector are useful indications
of, but do not guarantee, identity and homogeneity.
Combined GLC-MS, however, is a method possessing
a much higher level of confidence for the
identification of GLC components and indication
of peak homogeneity. We have used this approach
with a recently develop electron capture GLC
method for the coccidiostat pyrimethamine
(2,4-diamino-5-p-chlorophenyl-6-ethylpyrimidine)
in chicken tissue (45). The mass spectrum of
this drug is presented in Figure 10. There are
few fragment ions, and the intense ions are
found in the molecular ion region. Although
pyrimethamine possesses a molecular weight of 248,
it is the M-1 ion which is base peak (note the
metastable ion at m/e 246). Combined GLC-MS
demonstrated conclusively with selected extracts
that the component exhibiting the retention time
of pyrimethamine was indeed the drug.

A mass spectrometer can also serve as a
selective GLC detector. One of several intense,
characteristic ions can be monitored continously
to yield quantitative information at the nanogram
level. Repetitive scanning over a narrow mass
range across a GLC peak is also an attractive
approach for the characterization and quantification of trace amounts of compounds. The
characteristic pattern and high intensity of the
signals in the molecular ion region of pyrimethamine make this drug particularly well suited to
this type of analysis. A comparison of the
response of the GLC-mass spectrometer (repetitively scanned from m/e 247-250) to 20 ng of
pyrimethamine and to an aliquot (20 ng by electron
capture GLC) of an isolate is presented in
Figure 11. The results are practically

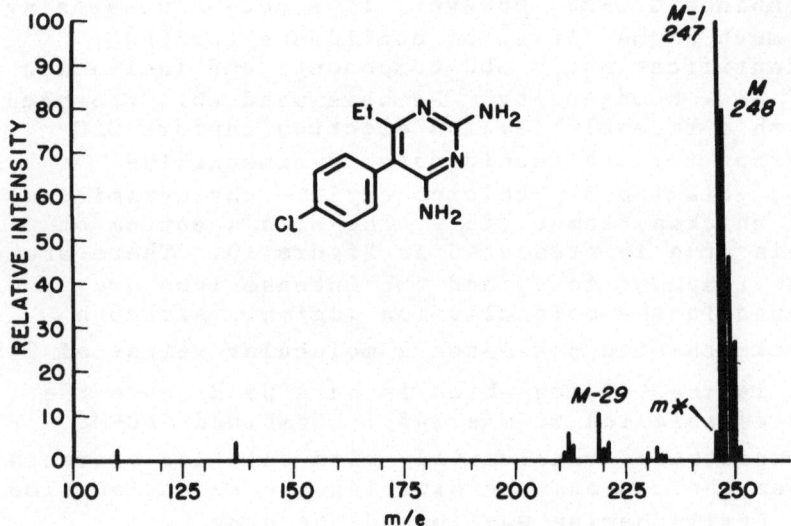

Figure 10.   Mass spectrum obtained by combined
GLC-MS of pyrimethamine.

REPETITIVE PARTIAL MASS SCAN
m/e 247–250

Liver Extract

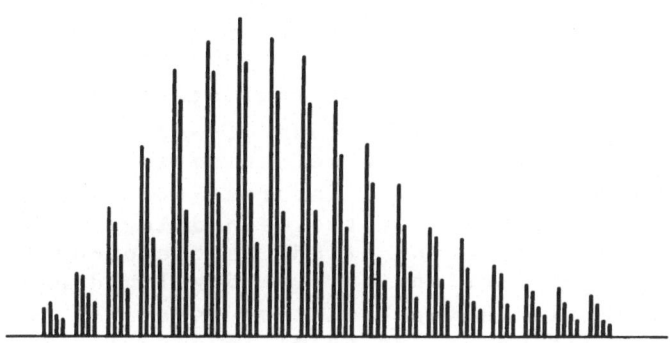

20 ng Pyrimethamine

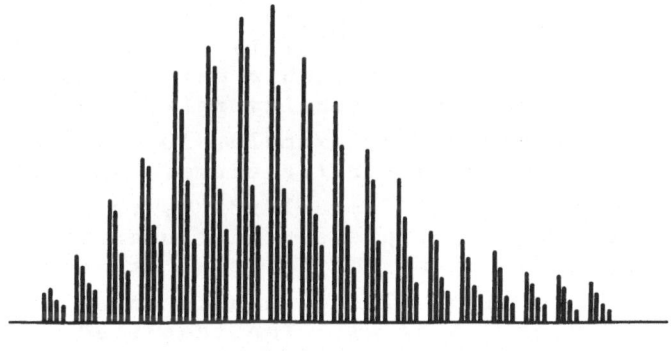

TIME ⟶

Figure 11.   Mass spectrometric response at m/e
247–250 (scanned repetitively) during
elution from a GLC column of 20 ng of
pyrimethamine (lower) and an aliquot
of a 1-day off-drug liver extract
previously shown by election capture
GLC to contain 20 ng of drug (upper).

superimposable, and confirm the specificity and
accuracy of the electron capture GLC assay.

Internal standards are frequently employed
in quantitative GLC, and as indicated by
Gaffney et al. (46)   the internal standard
of choice for a GLC-MS assay is very likely to
be the isotopically-labeled compound of interest.
The use of deuterated internal standards has
been found by several groups of workers to lead
to highly sensitive and selective assays for
compounds of biological interest (47-49).   As
a logical extension of the methodology employed
in our identification of DMT as the product of
an in vitro reaction (15), we developed a mass
spectrometric isotope dilution assay for DMT
in human plasma at the ng/ml level, with DMT-d_2
serving as the internal standard (50).   The ion
pairs 202/204 and 260/262 are monitored for intensity ratios; when both change to the same extent
with an isolate, high confidence can be placed
in the resulting data.   A 15/85 (^{1}H/^{2}H) isotope
mixture is employed as the internal standard.
The amount of endogenous DMT is calculated by
use of the linear relationship which exists
between the isotopic intensity ratios and amount
of DMT present (in, for example, 10 ml of plasma);
the plot serves as a working curve.   A dilution
of the ratio from 15/85 to 17/83 is the minimum
change detectable by present instrumentation;
addition of 200 ng of the internal standard
mixture to 10 ml of plasma results in an overall
sensitivity of 0.5 ng/ml.   The idealized
accelerating voltage alternator response for the
ions of m/e 260 and 262 from analysis of the
internal standard is shown in Figure 12.   The
response at the same m/e values from a normal
plasma (previously found not to contain any DMT;
limit of detection, 0.5 ng/ml) to which was
added 2 ng/ml of DMT is shown in Figure 13.
Note the difference in intensity ratios between

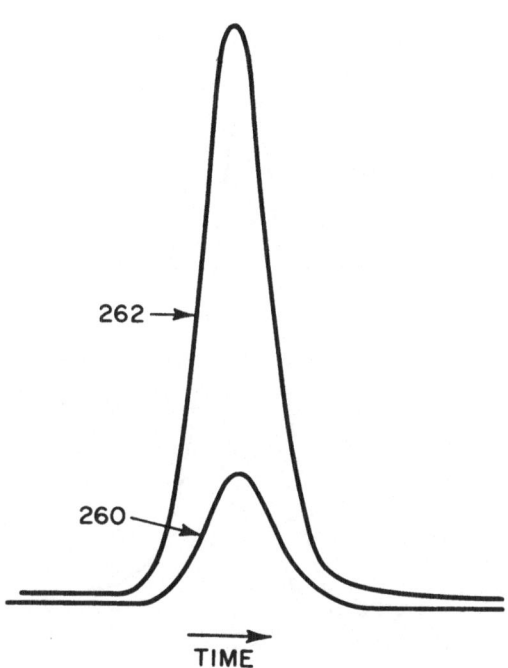

Figure 12.   Accelerating voltage alternator (multiple ion detector) recording of ions of m/e 260 and 262 from combined GLC–MS of a 15/85 mixture of the TMSi derivatives of DMT and DMT-d$_2$.

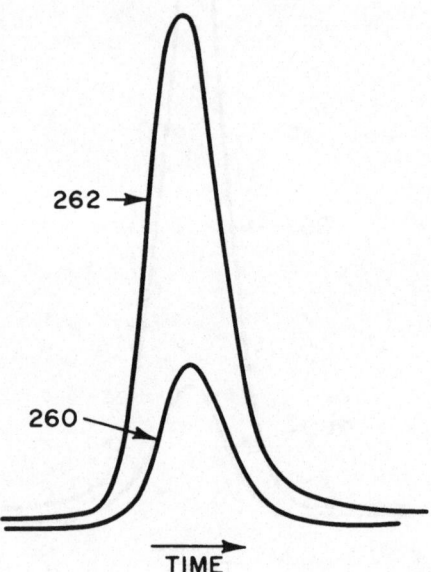

Figure 13.   Same as Figure 12, except analysis of
              an isolate from normal human plasma
              spiked with 2 ng/ml of DMT.

the two figures. When applied to the working
curve the ratio of the two intensities (Figure 13)
indicated a plasma DMT content of 1.8 ng/ml; a
value of 1.9 ng/ml was found by using the m/e 202/
204 ratio. Only six specimans of human plasma
out of 45 from normal individuals have been
found to be positive (levels of 1-2 ng/ml) for
this hallucinogen.

The value of MS has become apparent to the
biologically oriented scientist. He may not
understand the inner workings of this instrument,
but he is aware that mass spectrometric techniques
may answer his questions, and allow him to ask
new questions. The role of the mass spectrometrist is to serve as an interface between a
sophisticated instrument and an enthusiastic
neophyte who is anxious for information -- information the spectrometrist did not realize was
needed, or perhaps even realize that he could
supply.

<u>ACKNOWLEDGEMENT</u>

It is a pleasure to thank the following
Merck scientists for their invaluable contribu-
tions to the work reported in this article:
H.S. Ahn, N. Allen, P. Belanger, R.P. Buhs,
J.R. Carlin, J.S. Cohen, R.L. Ellsworth,
V.B. Gruber, T.A. Jacob, F.R. Koniuszy,
L.R. Mandel, H.E. Mertel, B. Plazonnet,
A. Rosegay, E. Sestokas, J.L. Smith O.C. Speth,
D.J. Tocco, N.R. Trenner, T.R. Tyler, R.W. Walker,
D.E. Wolf and F.J. Wolf.

REFERENCES

1. D.R. Knapp and T.E. Gaffney, Clin. Pharmacol.
    Therap. 13, 307 (1972).

2. W.J.A. VandenHeuvel, R.P. Buhs, J.R. Carlin,
    T.A. Jacob, F.R. Koniuszy, J.L. Smith,
    N.R. Trenner, R.W. Walker, D.E. Wolf and
    F.J. Wolf, Anal. Chem. 44, 14 (1972).

3. J.R. Carlin and T.A. Jacob, unpublished
    results.

4. R.W. Walker and W.J.A. VandenHeuvel,
    unpublished results.

5. D.E. Wolf, W.J.A. VandenHeuvel, F.R. Koniuszy,
    T.R. Tyler, T.A. Jacob and F.J. Wolf, J. Agr.
    Food Chem. 20, 1252 (1972).

6. N.R. Trenner, O.C. Speth, V.B. Gruber and
    W.J.A. VandenHeuvel, J. Chromatog. 71, 415
    (1972).

7. D.J. Tocco, unpublished results.

8. D.J. Harvey, M.G. Horning and P. Vouros,
    Tetrahedron 27, 4231 (1971).

9. P. Belanger, unpublished results.

10. D.E. Wolf, F.R. Koniuszy, M. Hite,
    W.J.A. VandenHeuvel, A. Rosegay, F.J. Wolf
    and T.A. Jacob, unpublished results.

11. C.P. Talley, N.R. Trenner, G.V. Downing and
    W.J.A. VandenHeuvel, Anal. Chem. 43, 1379
    (1971).

12. J.L. Smith, J.L. Beck and W.J.A. VandenHeuvel,
    Org. Mass Spectrom. 5, 473 (1971).

13.   A. Rosegay, unpublished results.

14.   W.J.A. VandenHeuvel, Anal. Letters 6, 51
      (1973).

15.   R.W. Walker, H.S. Ahn, L.R. Mandel and
      W.J.A. VandenHeuvel, Anal. Biochem. 47, 228
      (1972).

16.   L.R. Mandel, H.S. Ahn, W.J.A. VandenHeuvel
      and R.W. Walker, Biochem. Pharmacol. 21,
      1197 (1972).

17.   G.L. Tindell, T. Walle and T.E. Gaffney, Life
      Sciences 11, Part II, 1029 (1972).

18.   H.B. Hucker, B.M. Michniewicz and R.E. Rhodes,
      Biochem. Pharmacol. 20, 2123 (1970).

19.   T.R. Tyler, R.P. Buhs and W.J.A. VandenHeuvel,
      Biochem. Pharmacol., in press.

20.   S. Tabarelli and H. Uehleke, Xenobiotica 1,
      501 (1971).

21.   P.I. Jaakonmaki, K.A. Yarger and E.C. Horning,
      Biochim. Biophys. Acta, 137, 216 (1967).

22.   W.J.A. VandenHeuvel, J. Chromatog. 28, 406
      (1967).

23.   W.J.A. VandenHeuvel, J.L. Smith, G. Albers-
      Schonberg, B. Plazonnet and P. Belanger, in
      Modern Methods of Steroid Analysis,
      E. Heftmann, Ed., Academic Press, New York,
      in press.

24.   D.J. Tocco, R.P. Buhs, H.D. Brown, A.R. Matzuk,
      H.E. Mertel, R.E. Harman and N.R. Trenner,
      J. Med. Chem. 7, 399 (1964).

25. S. Billets, P.S. Lietman and C. Fenselau,
     J. Med. Chem. 16, 30 (1973).

26. M.W. Anders, personal communication.

27. R.P. Buhs, J.L. Beck, O.C. Speth, J.L. Smith,
     N.R. Trenner, P.J. Cannon and J.H. Laragh,
     J. Pharmacol. Expt. Therap. 143, 205 (1964).

28. G.D. Paulson and C.E. Portnoy, J. Agr. Food
     Chem. 18, 180 (1970).

29. B. Plazonnet, unpublished results.

30. G. Bleau and W.J.A. VandenHeuvel, unpublished
     results.

31. P.I. Jaakonmaki, Acta Endocrinol., Suppl.
     119, 118 (1967).

32. W.J.A. VandenHeuvel, R.N. Stillwell,
     W.L. Gardiner, S. Wikstrom and E.C. Horning,
     J. Chromatog. 19, 22 (1965).

33. W.J.A. VandenHeuvel, J. Chromatog. 43, 215
     (1969).

34. J.P. Joseph, J.P. Dusza and S. Bernstein,
     Steroids 7, 577 (1966).

35. A. Islam and A. Darbre, J. Chromatog. 43,
     11 (1969).

36. J.P. Thenot and E.C. Horning, Anal. Letters
     5, 519 (1972).

37. W.J.A. VandenHeuvel, J.L. Smith and
     J.S. Cohen, J. Chromatog. Sci. 8, 567 (1970).

38. J.R. Catch, Proc. 2nd Radioisotope Conf.,
     Oxford, I, Med. and Physiol. Applications,
     258 (1954).

39. K.H. Hallowes, F.P.W. Winteringham and
    W.J. LeQuesne, Nature 181, 336 (1958).

40. P.H. Abelson and T.C. Hoering, Proc. N.A.S.
    47, 623 (1961).

41. R.G. Cooks and S.L. Bernasek, J. Am. Chem.
    Soc. 92, 2129 (1970).

42. H.M. Fales, G.W.A. Milne and M.L. Vestal,
    J. Am. Chem. Soc, 91, 3682 (1969).

43. G.W.A. Milne, H.M. Fales and T. Axenrod,
    Anal. Chem. 43, 1815 (1971).

44. J.L. Smith and W.J.A. VandenHeuvel, Anal.
    Letters 5, 51 (1972).

45. P.C. Cala, N.R. Trenner, R.P. Buhs,
    G.V. Downing, J.L. Smith, and
    W.J.A. VandenHeuvel, J. Agr. Food Chem. 20,
    337 (1972).

46. T.E. Gaffney, C.G. Hammar, B. Holmstedt and
    R.E. McMahon, Anal. Chem. 43, 307 (1971).

47. B. Samuelsson, M. Hamberg and C.C. Sweeley,
    Anal. Biochem. 38, 301 (1970).

48. U. Axen, K. Green, D. Horlin and
    B. Samuelsson, Biochem. Biophys. Res. Commun.
    45, 519 (1971).

49. L. Bertilsson, A.J. Atkinson, J.R. Althaus,
    A. Harfast, J.E. Lindgren and B. Holmstedt,
    Anal. Chem. 44, 1434 (1972).

50. R.W. Walker, H.I. Ahn, G. Albers-Schonberg,
    L.R. Mandel and W.J.A. VandenHeuvel, Biochem.
    Med., in press.

CHEMICAL IONIZATION MASS SPECTROMETRY
Recent Developments in Techniques and Instrumentation

David P. Beggs

Scientific Research Instruments
Subsidiary of G. D. Searle & Co.
6707 Whitestone Road
Baltimore, Maryland 21207

The chemical ionization technique relies on the presence of a relatively high pressure in the ion source region. This high pressure promotes various ion molecule reactions which result in a highly sensitive method for ionization of sample molecules via ion molecule reactions. To obtain the high pressures desired in the ion source, one has to design and construct the source such that it will allow the internal pressures to reach several Torr without interfering with the electron beam or the ion beam exiting from the ion source. Pressures inside the source may reach several Torr while pressures outside the ion source are required to be below 10^{-3} Torr. Outside the source reduced pressure is required to provide maximum ion transfer from the source to the analyzer. Once the ion beam enters the analyzer region, pressure should again drop below 10^{-4} Torr. Hence, there should be three different pressure regions. These regions are satisfied by the presence of a dual pumping system and a differential pressure barrier. This barrier consists of a small hole between the source and analyzer region which will allow the ion beam to enter the analyzer region and also will allow the analyzer pump to provide a high vacuum in the analyzer region. Figure 1 demonstrates the proper pressure requirements for a chemical ionization system.

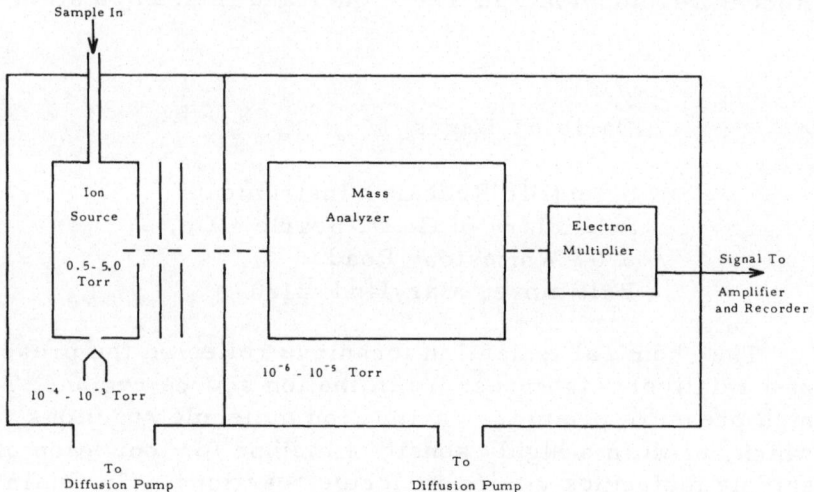

Figure 1   Pressure Regions in CI System

The design of a chemical ionization source is based on the premise that it will provide the conditions which will allow the best compromise of high source pressure with good resolution and sensitivity. In general a CI source should allow the source pressure to reach 1 Torr or greater and at the same time define a residence time between 10^{-5} and 10^{-3} seconds. Under such conditions between 100 - 10,000 ion-molecule collisions will occur. It can be shown that close to one hundred percent of the sample can be ionized.

Calculations show that 0.1% of the sample ions leave the ion source and approximately 0.01% of the total sample reach the detector after passing through the analyzer. Comparing this with data for electron impact it can be shown that about 10% of the sample is ionized by electron impact methods, 0.001% leave the source, and 0.0001% reach the detector. This relative comparison of sensitivities between CI and EI has been empirically affirmed and values for sensitivity ratios of CI to EI differ between one and two orders of magnitude.

Generally, the conditions which are present in a CI source are the following: the pressure is between 0.5 and 5 Torr, the temperature is between 0 and $350^{0}C$, the electron energy is between 70 and 2000 electron volts, and the repeller voltage is between 0 and 10 volts. This range of parameters covers most CI sources presently in use.

Figure 2 shows the schematic diagram of the CHEMSPECT chemical ionization source. The dimensions are given and one can see that the main difference between a CI source and an EI source is the size of the ion exit slit and electron entrance slit. Generally these slits are between 0.1 and 0.6 square millimeters in area for CI. The size and shape differ from one unit to another; however, these must be matched to the speed of the pumping system. Usually a gas flow between 10 and 20 cc/min can be accommodated by such a source. It has been found that the ion lenses have a greater focusing effect upon the ion beam exiting from CI source as opposed to one exiting from the electron impact source.

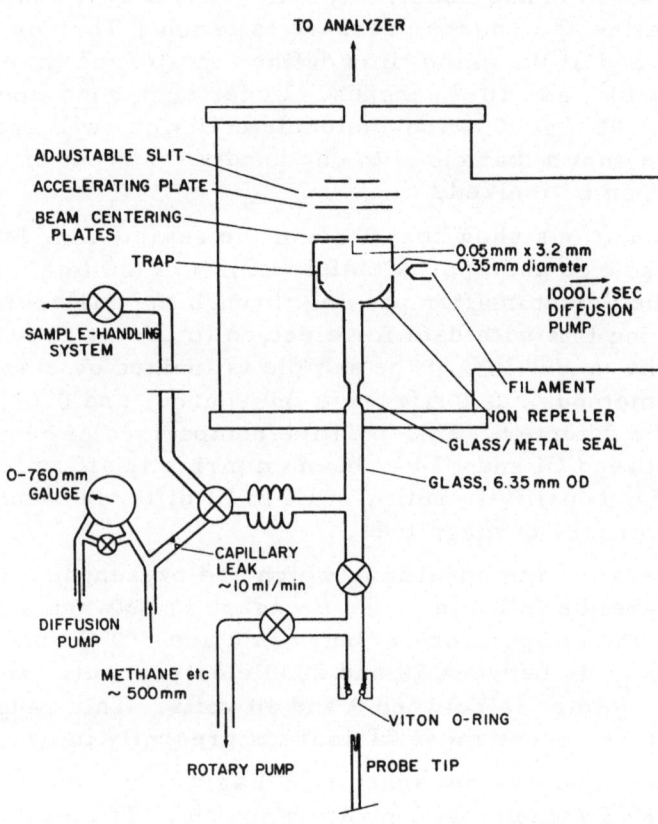

Figure 2   Schematic Diagram - Chemical Ionization Source

The optimum pressure used in the CI source is usually found by empirical methods. By increasing the pressure one will observe a curve similar to that shown in Figure 3 where the ion intensity increases, reaches a peak, and then decreases. The increase in intensity is due to the increase in ionization efficiency with pressure. The decrease in sensitivity at higher pressures is due to the combination of the detrimental effect of ion-molecule reactions outside the source and a decrease in electron beam penetration into the source region due to the higher pressures inside the source. If one operates at too low a pressure, a mixed EI-CI spectra will be observed. At pressures below 0.3 Torr there is insufficient amount of reagent gas to completely ionize the sample via chemical ionization. Some of the sample is ionized by electron impact and a mixed spectra results. An example of such a spectrum is shown in Figure 4. The normal CI spectrum of seconal would consist of an M + 1 ion only. As is evident, there is the M + 1 ion; however, there are many fragments due to the electron impact nature of the ionization process.

The repeller is a very important part of electron impact source design mainly because it is the main method of moving the ions out of the source and into the accelerating field or the analyzer. The usual fields produced by the repeller are of the order of 20-50 volts/centimeter. However, in CI the potential on the repeller should be minimized since the field that is produced acts contrary to the chemical ionization process. A large repeller field in a chemical ionization source moves the ions out of the source quickly thereby reducing the residence time of the ions in the source region. This limits the amount of reagent ions produced as well as reduces the time for ion-molecule reactions to produce sample ions. Michnowicz and Munson[1] have demonstrated this effect of the repeller upon the chemical ionization of p-hydroxy benzophenol. As the repeller voltage is increased the amount of molecular ion M + 1 is reduced drastically and the amount of fragment ions produced by the increase in energy of the ions from the repeller field increased.

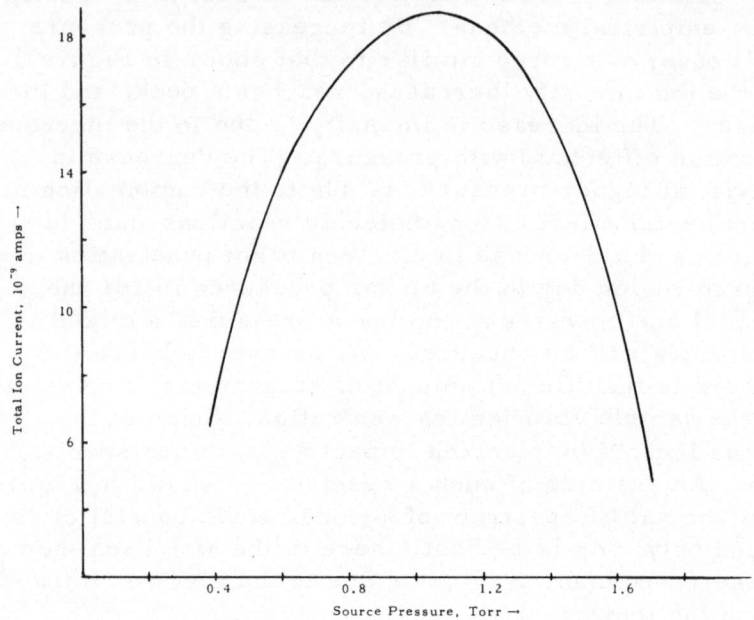

Figure 3   Effect of Source Pressure on Total Ion Current

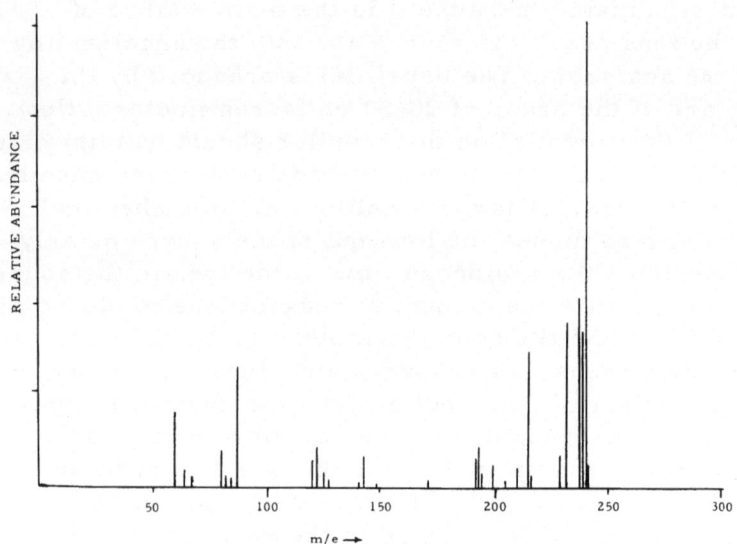

Figure 4   Mixed EI - CI Spectrum of Seconal
Source Pressure = 0.3 Torr

It can be seen then, that it is desirable to keep the repeller voltage as small as possible and in most cases the optimum conditions are obtained by zero repeller voltage

The most important parameter of the chemical ionization source is the reagent gas. One of the main advantages of chemical ionization is that the user may change the reagent gases to suit his particular needs. The type of reagent gas which one uses will determine the amount of fragmentation that one achieves upon ionization. Also, one can vary the type of ionization, utilizing proton transfer, charge exchange or electron impact type of ionization. Some reagent gases or gas mixtures will specifically ionize a particular group site of the sample.[2] Table I lists many of the reagent gases which have been used to date. Those gases which promote proton exchange reactions are listed in order of increasing proton infinity.

The amount of fragmentation is dependent upon the proton affinity of the reagent gas. The lower the proton affinity of the reagent gas the greater the amount of energy which is transferred upon ionization. This greater amount of excess energy results in a greater amount of fragmentation. In the case of isobutane there is almost no fragmentation for many of the samples which have been run to date. This small degree of fragmentation can be used to advantage when one is analyzing a mixture. If a mixture of compounds is injected directly into the ionization source via a direct probe inlet, one can observe one peak from each of the compounds in the mixture, that peak usually corresponding to the M + 1 ion when using isobutane as the reagent gas. Thus, one can predict the components of a mixture without having to separate the mixture prior to mass spectrometric analysis. An example of this is shown in Figure 5. The lack of fragmentation eliminates the interference problems which would be present in a corresponding electron impact spectrum of the same sample. The charge exchange process contributes a greater amount of energy to the sample molecule thus promoting fragmentation. And in effect, one can obtain electron impact type spectra using helium as the reagent gas. In this case, ionization is initiated by the interaction

TABLE I

REAGENT GASES USED TO DATE

Proton Exchange	Charge Exchange
H_2	N_2
CH_4	O_2
C_2H_6	Rare Gases
H_2O	CO
C_3H_8	CO_2
$i-C_4H_{10}$	NO
CH_3OH	CF_4
CH_3CHO	CCl_4
$(CH_3)_2CO$	
NH_3	
CH_3NH_2	
$(CH_3)_2NH$	

Increasing Proton Affinity (downward along the Proton Exchange column)

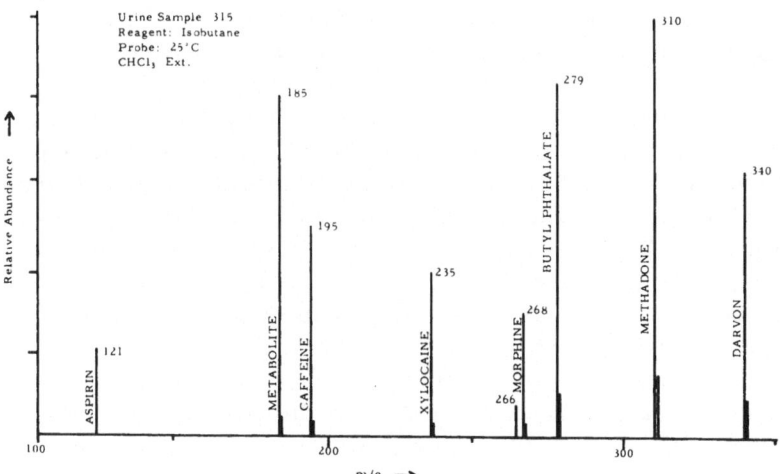

Figure 5   CI Spectrum of Drug Mixture

of sub-electronic state electrons of helium with the sample
molecules. Comparison of EI and helium CI spectra show
that the same fragments are observed and the intensity
ratios are very close such that the difference is no more
than the difference which would be observed in the variance
from one EI mass spectrometer to another.

Generally, the direct inlet probes used by most chemical
ionization practitioners are very similar to the probes used
with   electron impact sources. They consist of a heated
cup at the end of a rod which includes a thermocouple
temperature monitor. In most CI applications the source
region is connected directly to the inlet system so that the
tightness of the source remains intact. The reagent gas
flows over the probe tip or comes in contact with the probe
tip inside the ion source. Dr. Sanford Markey[3] has developed
a unique probe inlet system wherein there is a teflon leak
seal in the glass inlet tube of the probe. The reagent gas
flows through the leak, over the probe tip and into the mass
spectrometer source.

One must be careful to insulate the probe when operating
at high voltage potentials in the source since the reagent
gas will transfer the source potential to the probe even if it
is not in direct contact with the source. The problem of
potential transfer is a great hazard when designing a gas
inlet system. The gas must be at 1 Torr when it reaches
the source, but must reach a higher pressure somewhere
outside the source region so that the high voltage associated
with magnetic deflection instrument sources is not carried
to the sample inlet system. Several schemes have been
developed to solve this problem. Professor Jean Futrell[4]
at the University of Utah has developed a gas system which
uses a resistor string to provide a voltage gradient between
the inlet system and the ion source. Glass enclosed resistors
connected by stainless-steel turnings provide an effective
resistance to drop the potential from the high voltage of
this source to ground in the inlet system. Some problems
have been observed with the presence of the large effective
surface area. One must be careful to provide uniform heat
to the whole inlet system while at the same time insulating

against discharge from the stainless-steel turnings to the
heating system outside the glass inlet lining. Another
approach used by Scientific Research Instruments
Corporation utilizes an all-glass inlet design. An 1/8"
glass tube connects the inlet system to the ion source
itself. This glass tube is wrapped with a heater and connected to a thermocouple so that a uniform temperature
may be achieved. This unit is encased in a stainless steel
quarter-inch tube which is connected via an o-ring gland to
a bellows arrangement so that it may be easily inserted
and extracted from the ion source. The bellows arrangement allows for proper placement of the inlet in relationship to the ion source. This allows the placement of the
inlet system to be orthogonal to the direction of the ion
beam so that direct line-of-sight entrance from the gas
chromotograph can be achieved. The sample and reagent
gas mixture passes through a restricting orifice before
entrance to the 1/8" glass tube. This orifice provides a
pressure differential which effects a potential barrier to
the high voltage of the source region.

One of the problems involved with chemical ionization
source design is due to the restrictions of the orifice sizes
for electron entrance and ion exit. These restrictions
limit the electron impact capability of CI sources. This
can be overcome in several ways: one is to use a suitable
gas, such as helium, which will provide an electron impact
spectrum via chemical ionization. Another way is to design
a source which can be switched rapidly from a CI design to
a EI design. Such a source was developed by Dr. G.
Arsenault[5] at MIT and is shown in Figure 6. In this design
there are two separate sources each with a separate filament.
The reagent gas-sample mixture passes into the CI source
and from thence flows into the EI source. The "closed" CI
source provides pressures high enough for CI while the
"open" source allows the total pressure to drop low enough
for EI. CI and EI can be obtained alternately by switching
the power to either of the filaments.

Another method is to provide a movable slit such that
the electron entrance beam can be changed from a small

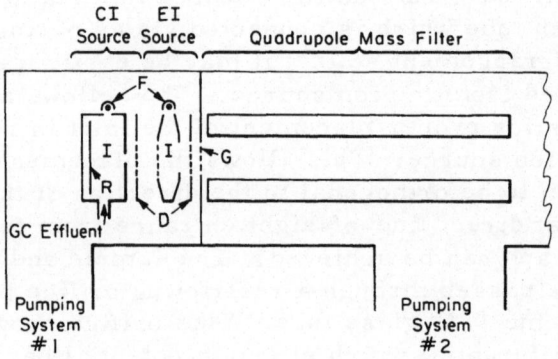

Figure 6   Combination EI - CI Source

orifice or CI to a larger orifice or electron impact. Such a source has been designed and built by Scientific Research Instruments. A movable slide mechanism operates a plate which moves from an open to closed position thus allowing the electron entrance aperture to change from an 1/8" slit to four 0.013 holes. In the former configuration electron impact takes place but in the latter configuration CI takes place. These designs give good EI and CI performance with one source thus providing optimal capability for the mass spectrometric system.

REFERENCES

1.   J. Michnowicz and B. Munson, Org. Mass Spectrom. 4, 481 (1970).
2.   D. F. Hunt et al., Tetrahedron Letters 47, 4539 (1971).
3.   S. P. Markey et al., Proceedings 20th Annual Conf. Mass Spect. Allied Topics, 318 (1972).
4.   J. H. Futrell and L. H. Wojcik, Rev. Sci. Instr. 42, 244 (1971).
5.   G. P. Arsenault et al., Anal. Chem. 43, 1720 (1971).

SELECTIVE REAGENTS FOR CHEMICAL IONIZATION MASS SPECTROMETRY

Donald F. Hunt

Department of Chemistry
University of Virginia
Charlottesville, Virginia   22901

In chemical ionization mass spectrometry[1] (CIMS) a set of reagent ions is generated by bombarding a suitable gas at a pressure of ca. 1 torr with high energy electrons (300-1000eV). Sample molecules are introduced in the usual manner and are ionized by ion-molecule reactions with the reagent ions. Because the sample concentration in the ion source is always maintained at less than 1% that of the reagent gas, ionization of sample molecules by electron impact is not observed.

To date most CI spectra have been recorded using methane as the reagent gas. As shown in eq 1-3 electron bombardment of methane at 1 torr produces CH_5^+ and $C_2H_5^+$ in high abundance and these two ions function as either Brönsted acids or hydride abstractors toward neutral organic samples.

$$CH_4 \xrightarrow{e^-} CH_4^{+\cdot} + CH_3^+ \qquad (1)$$

$$CH_4^{+\cdot} + CH_4 \longrightarrow CH_5^+ + CH_3^{\cdot} \qquad (2)$$

$$CH_3^+ + CH_4 \longrightarrow C_2H_5^+ + H_2 \qquad (3)$$

In comparison with the conventional electron impact (EI) technique, chemical ionization with methane as the reagent affords spectra which contain much more abundant

ions at high mass and fewer low molecular-weight fragments.
Ionization by electron impact affords odd-electron radical
cations initially and delocalization of the unpaired electron
occurs readily in most of these molecular ions. Much of the
fragmentation observed in EI spectra can be rationalized as
having been triggered by the unpaired electron. In contrast
CI with methane affords even electron cations. Since the
exothermicity of the proton transfer and hydride abstraction
reactions is considerably less than that involved in production of radical cations by the electron impact method, these
even-electron cations tend to be more stable toward fragmentation. Furthermore, those that fall apart generally do so
by pathways different from those available to radical cations
so the structural information obtained from CI fragmentation
patterns frequently complements that produced by the conventional electron impact method.

Reactions between the methane reagent ions, CH_5^+ and
$C_2H_5^+$, and the organic sample occur preferentially at sites
possessing high dipole moments. Since delocalization of the
positive charge in the product ions often involves movement
of nuclei rather than electrons, and, therefore, requires an
activation energy relatively few rearrangements occur. Most
fragmentation takes place in the vicinity of the ionized
functional group. As a consequence of this ionization
mechanism, many polyfunctional organic molecules which fail
to show molecular ions under electron impact do exhibit
abundant M+1 ions in $CI(CH_4)$ spectra. In the case of a
sample containing tertiary hydroxyl and ketone groups, proton
transfer to the carbonyl moiety gives rise to a stable M+1
ion whereas reaction at the hydroxyl group promotes dehydration to give an M-17 ion. Under electron impact, where the
ionization process is governed by Frank-Condon considerations,
electron delocalization and distribution of the excess heat
of reaction among the various vibrational modes promotes
dehydration of the tertiary alcohol to produce M-18 ions at
the expense of the molecular ions.

Perhaps the most exciting feature of the chemical ionization technique is the finding that the nature of the mass
spectrum produced is dependent on both the nature of the

reagent gas and the type of ion-molecule reaction used to ionize the sample. Different structural information can be obtained with different reagent gases. As a consequence, it is possible to control the type and quantity of structural information obtained from a mass spectrum by varying the nature of the reagent gas used in the CI mode of operation.

As indicated in Figure 1, the extent of fragmentation produced from a particular sample can be controlled by varying the exothermicity of the ionizing reaction.[1c] Electron bombardment of H_2, CH_4, and $(CH_3)_3CH$ produces H_3^+, CH_5^+, and $(CH_3)_3C^+$ and each of these ions functions as a Brönsted acid toward the neutral organic molecule.

Since the proton affinity (PA) of H_2 and, therefore, the Brönsted acidity of H_3^+ is considerably greater than that of CH_5^+, $CI(H_2)$ spectra exhibit much more fragmentation than

$$H^+ + H_2 \longrightarrow H_3^+ \qquad \Delta H = -101 \text{ kcal} \qquad PA = 101$$

$$CH_4 + H^+ \longrightarrow CH_5^+ \qquad \Delta H = -127 \text{ kcal} \qquad PA = 127$$

$$(CH_3)_2C=CH_2 + H^+ \longrightarrow (CH_3)_3C^+ \qquad \Delta H = -195 \text{ kcal} \quad PA = 195$$

those obtained with methane. In the case of dihydrotestosterone (Figure 1) ionization with H_3^+ causes fragmentation in the vicinity of both functional groups as well as on the carbocyclic ring system. With CH_5^+ fragmentation is largely restricted to loss of water from both the ketone and alcohol functional groups. Ionization with the still weaker Brönsted acid, $(CH_3)_3C^+$ affords a spectrum where most of the ion current is carried by the M+1 ion. Loss of water, presumably from the alcohol moiety, generates an M-17 ion which only accounts for 10% of the total sample ion current. Because isobutane CI spectra contain a paucity of fragment ions, this reagent gas is ideally suited for high sensitivity, quantitative analysis by the isotopic dilution method.

As part of a continuing effort to develop CIMS into a powerful method for the identification and structure elucidation of organic compounds, we have recently explored the utility of a number of reagent gases including argon-water

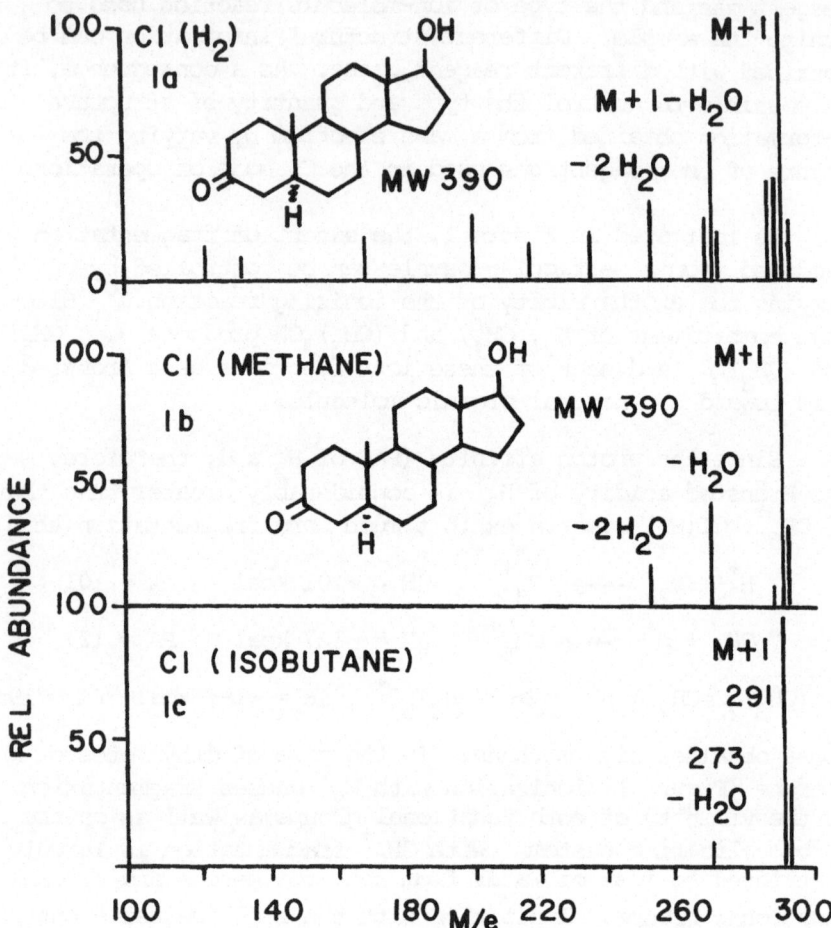

Figure 1.   Dihydrotestosterone CI mass spectra recorded with
(a) hydrogen, (b) methane, and (c) isobutane as the reagent
gases.

deuterium oxide, ammonia, and nitric oxide. Argon-water is
a particularly useful CI reagent since the spectra generated
with this mixture exhibit features characteristic of both
CI (CH$_4$) and conventional electron impact spectra.[2]

Electron bombardment of argon-water at 1 torr produces
ions at m/e 40 (Ar$^+$), 80 (Ar$_2^+$) and 19 (H$_3$O$^+$). Excited argon

neutrals (Ar*) and large numbers of low energy electrons are also generated under the above conditions. As expected H_3O^+ functions as a Brönsted acid in the gas phase and protonates most organic compounds containing a heteroatom. Since the proton affinity of water is quite high (PA = 164±4 kcal/mole), the energy transferred to the sample in the proton transfer reaction is relatively small and the abundant M+1 ions that result rarely undergo extensive fragmentation. In contrast to the situation with H_3O^+, electron transfer (oxidation) occurs when sample molecules encounter Ar$^+$ in the ion source. Since the recombination energy of Ar$^+$ is 15.8eV and the ionization potentials of most organic compounds are in the range 9-12eV, most of the radical cations produced in the oxidation reaction undergo fragmentation along pathways very similar to those observed in conventional electron impact mass spectrometry. Collisions between organic sample molecules and excited argon neutrals or low energy electrons may also result in sample ionization but here the exothermicity of the reactions is low and the molecular ions generated are usually stable toward further fragmentation.

For the purpose of comparison, CI(H_2O), conventional EI, and CI(Ar-H_2O) spectra of dihydrocorticosterone are shown in Figure 2. Ionization of this steroid with H_3O^+ produces an abundant M+1 ion and three fragment ions corresponding to loss of 1, 2, and 3 molecules of water from the M+1 species. In contrast, the EI spectrum displays a relatively weak molecular ion and a series of fragments at m/e 330(M-H_2O), 317(M-CH_2OH), 299(m/e 317-H_2O), 271(317-H_2O,CO), 253(271-H_2O) and 161(271-$C_7H_{10}O$). CI with Ar-H_2O as the reagent affords a spectrum which is essentially the sum of those produced by EI and CI with water. All of the fragment ion produced by EI as well as the abundant M+1 and three fragment ions found in the CI(H_2O) spectra are displayed in Figure 2c.

In an effort to probe the utility of other inert gases in the above procedure, and to compare the sensitivity of the CI and EI methods, we have recently measured the total sample ion current produced from a fixed amount of sample such as 4-decanone using the conventional EI and CI techniques with argon-water, helium-water, and nitrogen-water as the

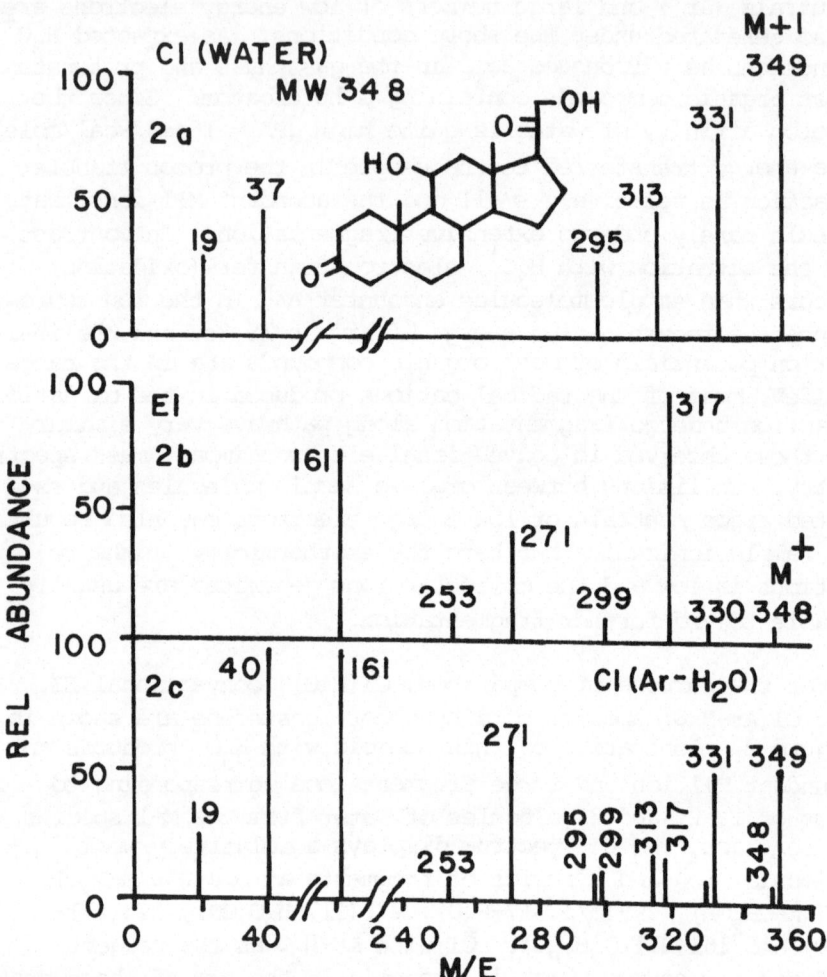

Figure 2.   (a) CI water, (b) EI, and (c) CI argon–water mass spectra of dihydrocorticosterone.

reagent gases. The sensitivity of the CI method with nitrogen–water and argon–water was found to be the same and about 30 times greater than that with EI or CI helium–water.[3] The nitrogen–water system is preferred slightly over the argon–water combination only because nitrogen is cheaper than argon. Both reagent mixtures afford essentially identical spectra.

It should also be mentioned that the CI argon-water (nitrogen-water) system is ideally suited for obtaining accurate mass measurement data by either the peak matching technique or with the aid of an on-line automatic data acquisition and processing system. Perfluorokerosene (PFK) is employed as the internal standard and affords a spectrum identical to that produced in the EI mode of operation. Proton transfer from H_3O^+ to the fluorocarbon does not occur and the only ions generated result from electron transfer between Ar^+ and PFK. Operation of the spectrometer in the high resolution mode (greater than 10,000) with narrow source and collector slits is greatly simplified since the intensity of the ion beam is ca. 30 times greater than that formed by electron impact at low pressures. In addition the abundance of the M+1 peak can be dramatically enhanced relative to that of the fragment ions by simply increasing the percentage of water in the argon-water reagent gas mixture. Argon-water (nitrogen-water) mixtures are also excellent reagent gases for GC-MS work. Since the ion-source is operated at a pressure of 1 torr, the entire flow from a capillary column can be handled without the need for a sample enrichment device between the GC and the mass spectrometer. Argon is used as the GC carrier gas and water is metered into the gaseous effluent at a point beyond the GC but before the ion source in order to generate the CI($Ar-H_2O$) reagent.

In addition to our work with the argon-water system, we have also explored the utility of water and deuterium oxide as CI reagent gases. Initial findings indicate that the determination of active hydrogen in organic compounds can be conveniently carried out by CIMS using deuterium oxide as the reagent gas.[4]

At a pressure of 0.4 torr electron bombardment of D_2O affords abundant ions at m/e 22(D_3O^+), 42($D_2O)_2D^+$, 62($D_2O)_3D^+$, 82($D_2O)_4D^+$ and 102($D_2O)_5D^+$. These ions in turn function as Brönsted acids and deuterate most organic compounds. In addition as a consequence of the relatively high source pressure (0.4 torr), sufficient collisions occur

between sample and neutral deuterium oxide to exchange all
of the hydrogens bonded to oxygen, sulfur and nitrogen atoms
in the organic molecule for deuterium.

As indicated in Figure 3b, the most abundant ion in the
CI(D_2O) spectrum of the nucleoside, adenosine, occurs at m/e
274 and corresponds to d_3-adenosine + D^+. Since this same
ion appears at m/e 268 in the CI(H_2O) spectrum (Figure 3a),
the above result clearly indicates that all five acidic hydrogens in the nucleoside suffer exchange in the ion source
when D_2O is employed as the reagent gas. In addition, two
fragment ions resulting from cleavage of the glycosidic
linkage are also observed. These appear at m/e 136 and 140
and correspond to a d_3-sugar moiety and d_3-adenine + D^+.

In addition to the above example, we have also obtained
CI(D_2O) spectra of a number of compounds containing one or
more common organic functional groups. Our findings indicate
that hydrogens bonded to heteroatoms in alcohols, phenols,
carboxylic acids, amines, amides, and mercaptans undergo
essentially complete exchange in the ion source when D_2O is
employed as the reagent gas at a pressure of 0.4 torr. Small
amounts of deuterium incorporation (less than 15%) occur in
ketones, aldehydes, and esters but, in general, this does not
complicate the analysis.

Besides our work on the determination of active hydrogen,
we are also exploring the possibility of using CIMS to identify organic functional groups in molecules of unknown
structure. Two reagents which show considerable promise in
this respect are ammonia and nitric oxide.

Electron bombardment of ammonia at 1 torr produces the
set of ions, $(NH_3)_nH^+$ (n = 1, 2, and 3) which occur at m/e
18, 35, and 52. These ions, in turn, function as weak
Brönsted acids and weak electrophiles toward other organic
compounds in the gas phase. Proton transfer from NH_4^+ to the
organic sample is observed if the proton affinity of the compound is greater than that of ammonia (PA = 207 kcal/mole).
Of the compounds studied to date only amides[5], amines[5,6], and
some α,β-unsaturated ketones[7] fit into this category and,

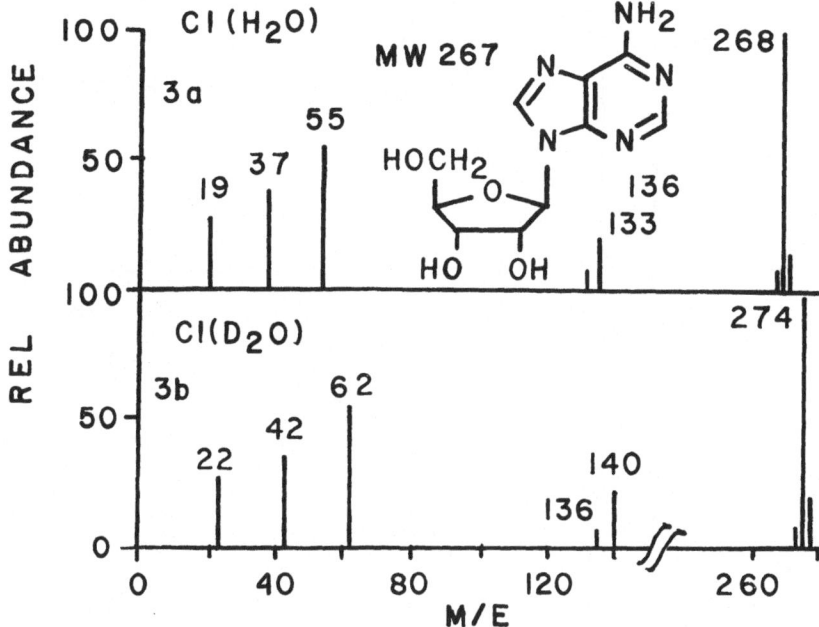

Figure 3.   (a) CI water and (b) CI deuterium oxide mass spectra of adenosine.

therefore, exhibit M+1 peaks in their ammonia CI spectra. As indicated in Figure 4a almost no fragmentation accompanies the proton transfer reaction since the process is only mildly exothermic.

In addition to M+1 ions, spectra of compounds derived from the above categories also display peaks corresponding to the electrophilic attachment of NH_4^+ to the organic sample. Ketones, aldehydes, esters and acids also add NH_4^+ (Figure 4) but are not sufficiently basic to remove a proton from the ammonium ion. Spectra of these compounds, therefore, exhibit a single ion corresponding to $M+NH_4^+$. Neither proton transfer or electrophilic attachment is observed in the $CI(NH_3)$ spectra of ethers, alcohols, phenols, nitro compounds, hydrocarbons, or aromatics. No ions are produced from these compounds. The above results suggest that ammonia will prove to be a useful reagent for the selective ionization of certain compounds in complex organic mixtures. Furthermore because

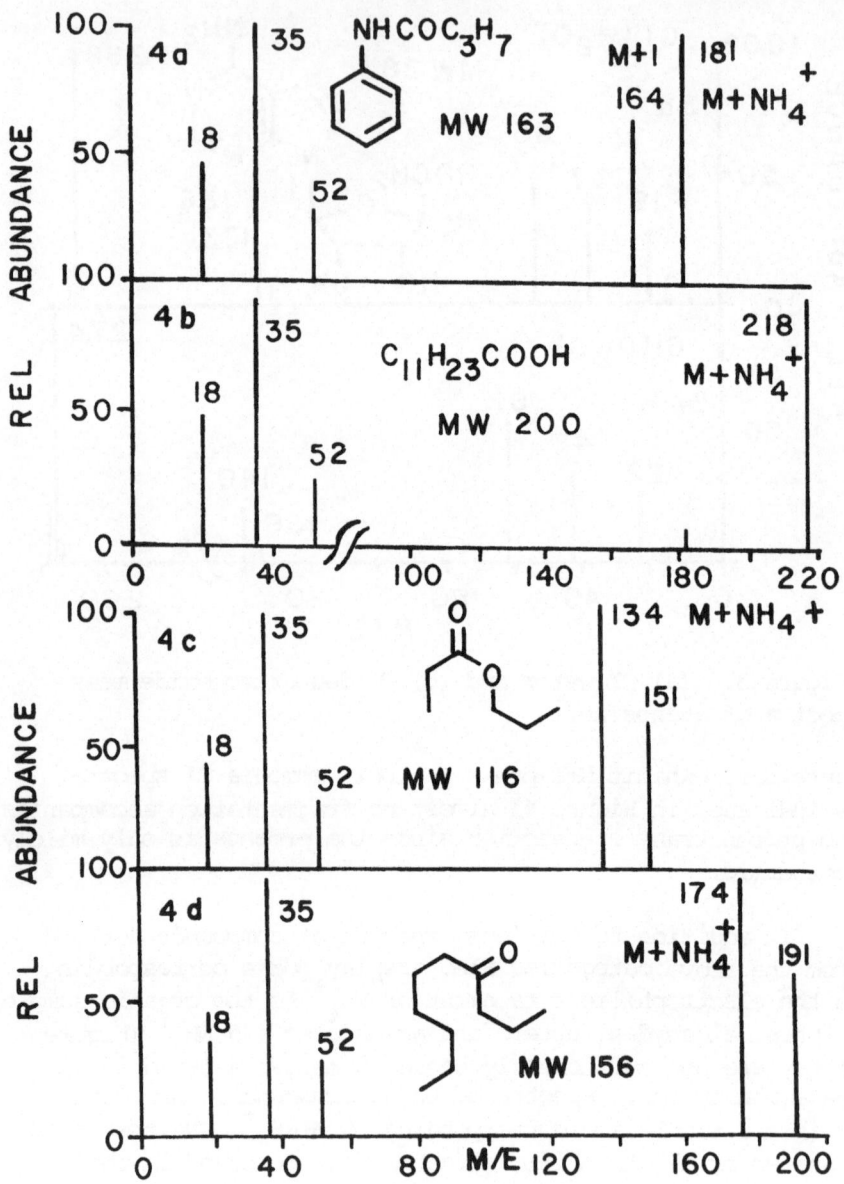

Figure 4. CI ammonia mass spectra of (a) n-butyranilide, (b) lauric acid (c) n-propyl propionate, and (d) 4-decanone.

water is not ionized by NH_4^+, it should be possible to use $CI(NH_3)$ mass spectrometry for the direct analysis of organics in aqueous mixtures.

With respect to the above discussion, it should be mentioned that the $CI(NH_3)$ spectra of aldehydes change as the ammonia pressure is raised above 1 torr. If the aldehyde is premixed with the reagent gas at a pressure of ca. 50 torr and the mixture is then bled into the CI source, a prominent ion appears having a m/e equal to the molecular weight of the sample.[8] This ion corresponds to the protonated Schiff base of the aldehyde. The Schiff base is formed in a neutral reaction between ammonia and the aldehyde[9] and is then protonated by NH_4^+ in the CI source. In contrast to ammonia

$$RCHO + NH_3 \longrightarrow RCH = NH + H_2O \xrightarrow{NH_4^+} RCH = \overset{+}{N}H_2 \qquad (4)$$

which only reacts with aldehydes, methylamine forms Schiff bases with both aldehydes and ketones. Furthermore the reaction takes place readily under the normal operating conditions of the CI source (1 torr). As shown in Figure 5, the reaction is quite sensitive to the structural environment of the carbonyl group. No Schiff base formation is observed in the case of the sterically hindered 5α-pregnan-11-one[9].

Another example of the utility of ammonia as a CI reagent is shown in Figure 6. Recorded here in the $CI(NH_3)$ spectrum of glucose which shows abundant ions at m/e 198, 180, and 162[10]. Even though this molecule contains a number of labile hydroxyl groups, very little fragmentation occurs and the molecular weight can be easily determined from the $M+NH_4^+$ ion.[9] In contrast to this result, the EI and $CI(CH_4)$ spectra are completely devoid of ions in the molecular weight region of the spectrum. It should be pointed out that we have yet to encounter a thermally-stable polyfunctional organic molecule which fails to give an ion characteristic of the molecular weight in CI spectra recorded with ammonia as the reagent gas.

Like ammonia, nitric oxide is also a useful reagent for functional group identification. In addition, nitric oxide

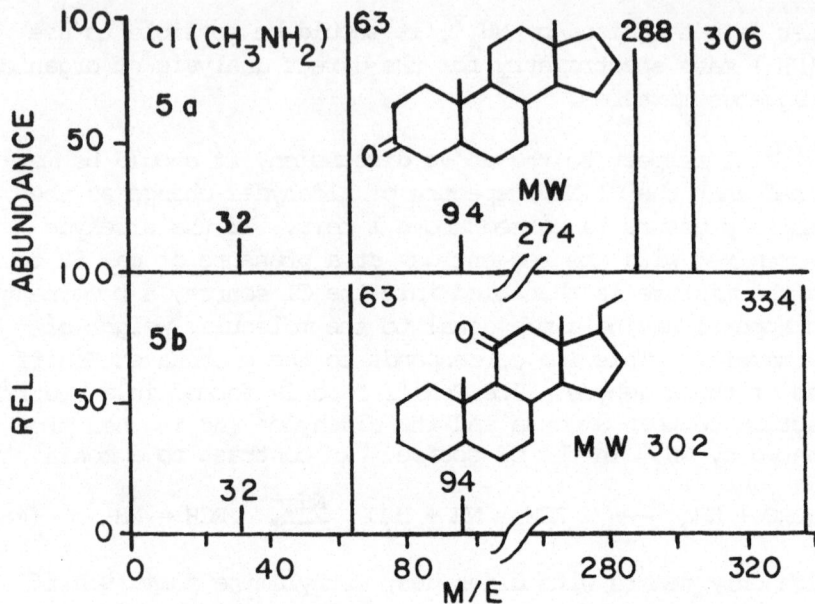

Figure 5. CI methylamine mass spectra of (a) 5α-androstan-
3-one and (b) 5α-pregnan-11-one.

can also be employed for qualitative analysis of hydrocarbons.

        Under electron impact at 1 torr, nitric oxide affords a
high abundance of NO$^+$ ions. Preliminary studies on the ion-
molecule reactions of this reagent species indicate that NO$^+$
functions as an electrophile, hydride abstractor, and one-
electron acceptor toward organic samples.[11] As shown in
Figure 7, CI(NO) spectra of simple ketones and esters exhibit
a single peak at M+30 corresponding to the electrophilic
addition of NO$^+$ to the sample molecule. Aldehyde spectra
(Figure 7b) show two ions, M+30 and M-1. The latter species
is produced by NO$^+$ abstraction of the hydrogen attached to
the carbonyl group. Acids suffer electrophilic addition and
also lose a hydroxyl group to give an M-17 ion (Figure 7d).

        As indicated in Figure 8a, n-amyl alcohol and primary
alcohols in general afford nitric oxide spectra containing
three ions, M-1, M-2+30, and M-3. The latter two ions

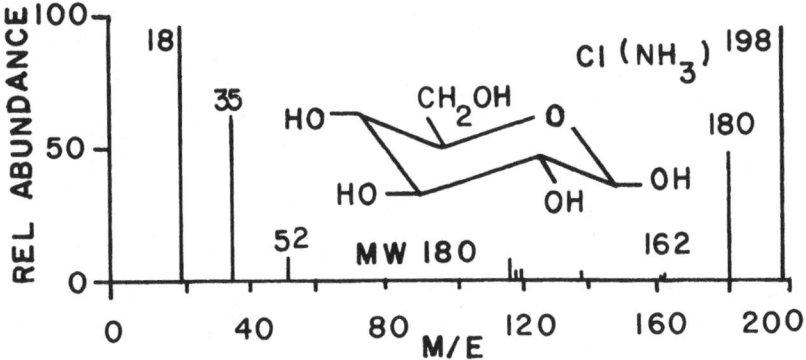

Figure 6. CI ammonia mass spectrum of glucose.

presumably result from oxidation of the alcohol to the corresponding aldehyde which then undergoes electrophilic attachment and hydride abstraction on collision with NO^+. Support
for these reaction pathways was obtained from the CI(NO)

$$\underline{n}\text{-}C_5H_{11}OH \xrightarrow{\overset{+}{NO}} CH_3CH_2CH_2CH_2\overset{+}{CHOH} \longrightarrow CH_3CH_2CH_2CH_2\overset{+}{CHO}$$

$$M\text{-}1$$

$$\xrightarrow{\overset{+}{NO}} CH_3CH_2CH_2CH_2\overset{+}{CO} + CH_3CH_2CH_2CH_2\overset{+}{CHONO}$$

$$M\text{-}3 \qquad\qquad M\text{-}2\text{+}30$$

spectrum of $1,1,10,10\text{-}d_4\text{-}$decane$\text{-}1,10\text{-}$diol which shows three
ions corresponding to M-2, M-5, and M-3+30. Although the
alcohol contains two primary hydroxyl groups, oxidation only
occurs at one end of the molecule. This result suggests
that neutral nitric oxide is not the oxidizing agent in the
above reaction. By a process of elimination, NO^+ is implicated in the reaction sequence although the exact mechanism
of its action remains obscure.

As shown in Figure 8b secondary alcohols afford spectra
containing ions corresponding to M-1, M-17, and M-2+30. This
last ion is produced by oxidation of the sample to a ketone
followed by electrophilic addition of NO^+ to the carbonyl
group. Tertiary alcohols give spectra containing a single

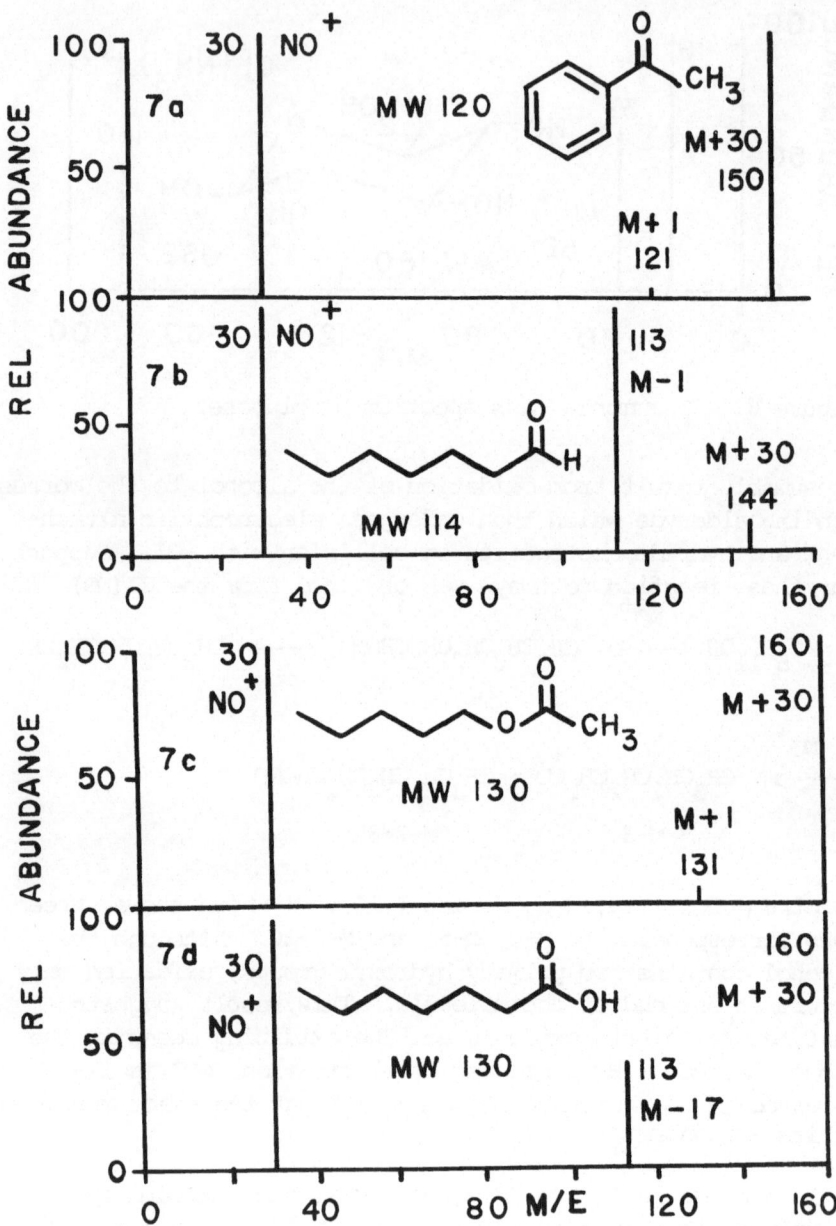

Figure 7. CI nitric oxide spectra of (a) acetophenone, (b) n-heptaldehyde, (c) n-amyl acetate, and (d) heptanoic acid.

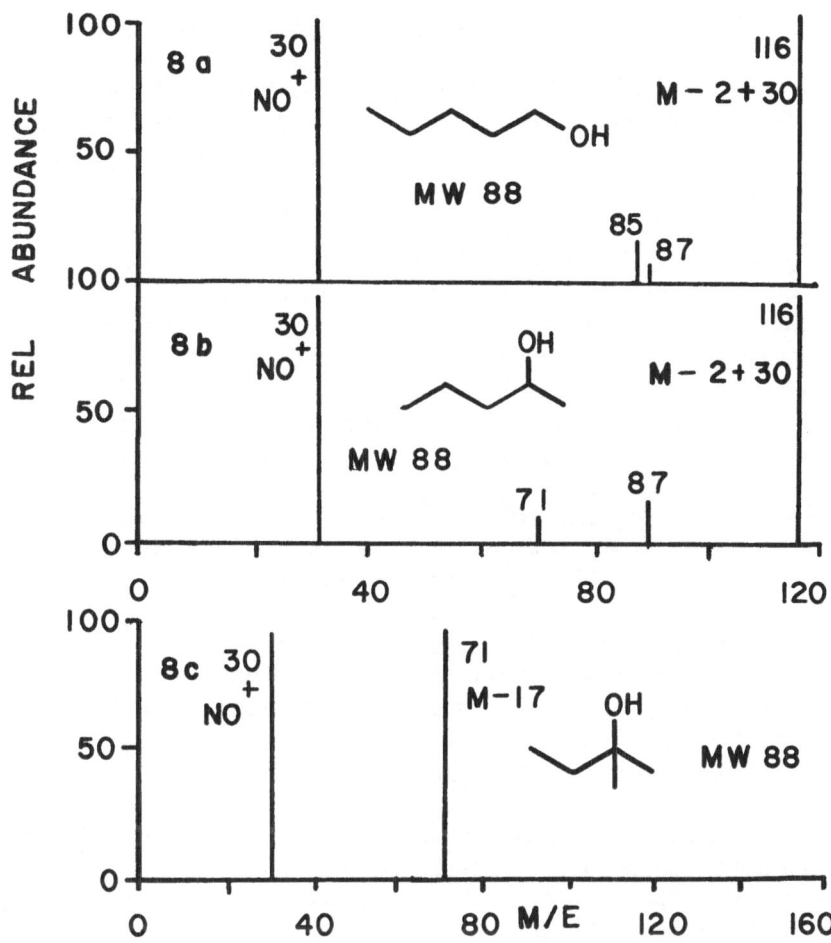

Figure 8. CI nitric oxide spectra of (a) n-amyl alcohol,
(b) sec-amyl alcohol and (c) t-amyl alcohol.

ion which corresponds to M-OH (Figure 8c). Thus it is
possible to use CI(NO) spectra to differentiate primary,
secondary and tertiary alcohols. Although preliminary in
nature, the above results clearly suggest a very promising
future for NO$^+$ as a reagent for identifying organic functional
groups on a nanogram scale.

Another exciting feature of the CI(NO) method is the

finding that many hydrocarbons afford spectra which are devoid of charged fragments and often contain only one or two
ions. The CI(NO) spectrum of decane (Figure 9), for example,
shows an ion which carries over 96% of the total ion current.
In contrast CI(CH$_4$) and EI spectra of decane exhibit large
numbers of fragment ions.

p-Di-t-butylbenzene is also an interesting example
(Figure 9b). EI and CI(CH$_4$) spectra of this molecule display
the base peak at m/e 57, (CH$_3$)$_3$C$^+$, and fail to show an ion in
the molecular weight region. With nitric oxide as the reagent
gas more than 90% of the sample ion current is carried by
the M+30 ion which results from electrophilic addition of
NO$^+$ to the benzene ring.

Nitric oxide CIMS is also useful for differentiating
cyclic alkanes from olefins. M-1 is the only ion produced
from cyclohexane. Olefins such as 3-decene, on the other
hand, afford spectra containing two ions, M-1 from hydride
abstraction and M+30 from electrophilic addition of NO$^+$ to
the double bond (Figure 9c). Spectra of dienes show the
above two ions plus a third species corresponding to M$^+$.
Generation of this molecular ion presumably occurs by transfer of an electron from the olefin to the NO$^+$ ion.

In summary, the above results suggest that chemical ionization mass spectrometry has the potential to become an extremely powerful analytical tool for the structure elucidation
and identification of organic compounds. Not only is the
technique 10-100 times more sensitive than conventional EI
mass spectrometry, it is also considerably more versatile.
By generating several CI spectra with different reagent gases,
it is possible to greatly increase the structure information
available from a particular sample. Operation of the ion
source at high pressures obviates the need for a sample enrichment device in GC-MS work, and the degree of fragmentation
observed in the mass spectrum can be conveniently controlled
by varying the exothermicity of the ionizing reaction. In
addtion to further research on selective reagent gases for
CIMS, we are also exploring the analytical utility of negative ion CIMS and developing a gaseous discharge CI source.

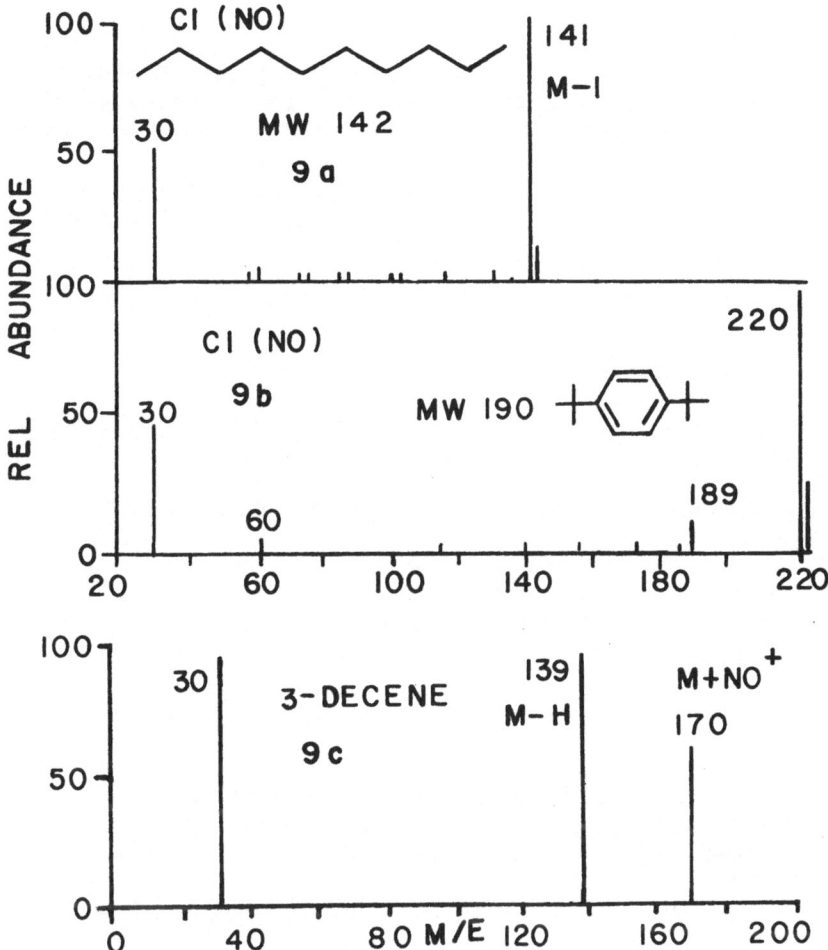

Figure 9. CI nitric oxide spectra of (a) n‑decane, (b) p‑di‑t‑butylbenzene, and (c) 3‑decene.

Acknowledgement is made to the donors of the Petroleum Research Fund, administered by the American Chemical Society, and the National Science Foundation, GP35128, for support of this research.

REFERENCES

1. For recent reviews see: (a) F. H. Field, "Ion-Molecule
    Reactions", J. L. Franklin, Ed., Plenum Press, New York
    (1972) pp. 261-312, (b) F. H. Field, "MTP International
    Review of Science, Physical Chemistry, Series One"
    Vol. 5, A. Maccoll, Ed., Butterworth (1972) pp. 133-
    182, (c) M. S. B. Munson, Anal. Chem., 43, 28A (1971)

2. D. F. Hunt and J. F. Ryan III, Anal. Chem., 44, 1306
    (1972)

3. G. P. Arsenault, J. Amer. Chem. Soc. 94, 8241(1972)

4. D. F. Hunt, C. N. McEwen, and R. A. Upham, Anal. Chem.,
    44, 1292(1972)

5. I. Dzidic, J. Amer. Chem. Soc., 94, 8333(1972)

6. (a) D. F. Hunt, C. N. McEwen, and R. A. Upham, Tetra-
    hedron Lett, 4539(1971); (b) M. S. Wilson, I. Dzidic,
    and J. A. McCloskey, Biochim. Biophys. Acta. 240,
    623(1971)

7. I. Dzidic and J. A. McCloskey, Org. Mass Spectrom.,
    6, 939(1972)

8. D. Beggs, Abstracts of Papers, Twentieth Annual Confer-
    ence on Mass Spectrometry and Allied Topics, Dallas,
    Texas, June 1972, Paper No. N4

9. D. F. Hunt and C. N. McEwen, unpublished results

10. A. M. Hogg and T. L. Nagabhushan, Abstracts of Papers,
    Twentieth Annual Conference on Mass Spectrometry and
    Allied Topics, Dallas, Texas, June 1972, Paper No. P2

11. D. F. Hunt and J. F. Ryan, J. C. S. Chem. Comm. 620
    (1972)